OPERATIONS RESEARCH

AN INTRODUCTION

P. Mariappan
Department of Mathematics
Bishop Heber College
Tiruchirappalli, Tamil Nadu

Pearson

ISBN978-81-317-9934-5

FirstImpression

PublishedbyPearsonIndiaEducationServicesPvt.Ltd,CIN:U72200TN2005PTC057128.

HeadOffice:15thFloor,Tower-B,WorldTradeTower,PlotNo.1,Block-C, Sector16, Noida201 301, UttarPradesh, India.RegisteredOffice:7thFloor,SDB2,ODC 7,8&9,SurveyNo. 01ELCOTIT/ITES-SEZ,Sholinganallur, Chennai–600119,Tamilnadu,India.
Phone: 044-66540100
Website:in.pearson.com,Email:companysecretary.india@pearson.com

Printer Manipal Technologies Limited, Manipal

I am dedicating this book to my beloved
Sister Mrs M. Krishnakumari,
Brother-in-law Mr O. Ravichandran,
and Mr O. R. Sivabalan

Contents

Foreword

Operations Research - An Introduction, written by Dr P. Mariappan, takes into account the whole gamut of undergraduate, postgraduate and professional courses that will require a good knowledge of operations research. It has a singular merit of catering to the requirements of students of introductory as well as advanced level courses.

The contents have been arranged on the principle of gradation. The ten chapters of the book are arranged in a logical sequence of Introduction, Linear Programming Problems and so on, before the concept of Dynamic Programming is discussed at the end.

There are two distinctive features that make the book stand apart—first, it is based on the self-taught learning method and second, the book has an exclusive section on examples and exercises related to the standard problems. I am sure the book will be very useful to the teaching and the student community.

I congratulate Dr Mariappan for the service he has rendered to the student community.

Dr E. C. Henry Amirtharaj
Dean for Training and Placement Division
and
Associate Professor of Mathematics
Bishop Heber College
Tiruchirappalli
Tamil Nadu

Preface

In recent years, operations research (OR) is being used widely for decision making in management. As with most tools, however, OR is useless unless the user understands its application and purpose. The user has to ensure that the mathematical input accurately reflects the real-life problems to be solved and that the numerical results are correctly applied to solve them. With this in mind, this text emphasizes model-formulations, model-building skills and manual solution methods.

Targeted Audience

This book is intended for the use of beginners as well as advanced learners, as a text in operations research, management science, or mathematical programming. The following groups of students stand to benefit from using this book:

- Undergraduate students of B.E., B.Sc. (Mathematics), B.Sc. (Computer Science), B.B.A. and B.Com.
- MBA, MBM students enrolled in applications oriented operations research, management science or mathematical programming courses.
- MCA, M.Com. and CA students who need an overview of the major topics in operations research and management science.
- Operations research users who need a comprehensive reference.

Organisation

The topics in this book are fashioned to allow for ample flexibility when teachers adapt the text to suit their individual course plans. For instance, courses in which certain units would be applicable include:

- **Deterministic Models:** Linear programming, Transportation problem, Assignment problem, CPM, Non-linear programming and Sequencing.
- **Probabilistic Models:** PERT and Queuing theory.
- **Dynamic Programming Problems**.

Highlights

- The problems denoted by asterisks in the examples and exercises have appeared in the operations research paper of Indian university examinations.
- This book is designed to encourage the self-taught learning method.
- Algorithms are clearly explained using the flowcharting methodology.
- All units of the book have a rich pedagogy comprising of short questions, objective-type questions and fill in the blanks, along with the answers.

I hope that this book will meet the needs of those for whom it has been designed.

Online Resources

Solutions for the exercise problems provided in the textbook are available at www.pearsoned.co.in/pmariappan

Acknowledgments

Many people have played significant roles in the development and release of this text. My views on teaching Operations Research were greatly influenced by the many illustrious teachers who had guided me, including Prof. R. Antonysamy, Prof. J. D. Samuel, Prof. A. Hirudaya Raj, Prof. S. K. Jeyakar, Dr Rajeswari Thiyagarajan, Prof. V. Natarajan and Dr Rajavelu Srinivasan.

I thank Dr E. C. Henry Amirtharaj, Associate Professor in Mathematics, Bishop Heber College, Tiruchirappalli, for consenting to write the foreword for this book.

I am obliged to the management and staff members of Pearson Education for publishing this book.

I am indebted to the members of my family and friends who had helped me in this achievement, either directly or indirectly.

While every care has been taken to ensure that this book does not have any mistakes in it, I take responsibility for any errors that may have inadvertently crept in. I would love to hear the suggestions and comments from the readers of this text.

P. Mariappan
mathmari@yahoo.com

About the Author

Dr P. Mariappan, is Associate Professor of Mathematics at Bishop Heber College, Tiruchirappalli. He pursued his postgraduate degree in Applied Mathematics with specialization in Decision Sciences from the same college and holds a Ph.D. degree in Applied Mathematics, for his research on Manpower Flow Models, from Bharathidasan University. He is also a management graduate from National Institute of Technology Tiruchirappalli (formerly Regional Engineering College Tiruchirappalli), who obtained his Master of Philosophy degree in Management Studies from Madurai Kamaraj University and proceeded to earn a doctorate for his work on Mathematical Models for Financial Decision-Making from Bharathidasan University. Dr Mariappan has a rich experience spanning over twenty years, teaching Applied Mathematics, Statistics and Business Administration in India and the UAE.

Dr Mariappan's teaching and research interests extend over a wide range of topics that include Business Mathematics, Business Statistics, Decision Sciences, Operations Management, Management Information System, Quantitative Methods in Operations Management, Numerical Methods, Optimization Techniques, Differential Equations, and Computer Programming. He is a recipient of the prestigious "Shiksha Rattan Puraskar" award for the year 2012 proffered by the India International Friendship Society [IIFS], New Delhi. He was also conferred with the "Best Teacher" award for the year 2004 by The Association of Centre for Indian Intellectuals [CII], India.

A distinguished academician, Dr Mariappan has published thirty-four articles in leading academic journals such as *International Journal of Management and System*, *OPSEARCH* and *Indian Journal of Pure and Applied Mathematics*. He has also authored eight textbooks. Besides, he has presented numerous research articles at international conferences held in the Philippines, Singapore, Costa Rica, Thailand and the USA. He was part of the jury for the IFI International Panel and Conference held at the Groupe ESC Rouen, France. Currently, he is guiding doctoral research students in the field of Management Science and reviews leading journals of national and international stature.

About the Author

1

Introduction

1.1 THE HISTORY OF OPERATIONS RESEARCH

In order to understand what is Operations Research (OR), one must know its history and evolution. It is generally agreed that the discipline began during World War II. Many strategic and technical problems associated with military effort were highly complicated. In response to these complex problems, groups of scientists with varied educational backgrounds assembled as special units within the armed forces.

Since the scientists are talented men, pressure due to war time necessity and the synergism generated from the interactions of different disciplines, they were remarkably successful in improving the effectiveness of complex military operations. By 1941, each of the three wings of British Armed Forces were utilizing such scientific teams. Due to immediate success of the idea, other allied nations adopted the same approach and organised their own teams. Because the problems assigned to these groups were in the nature of military operations, their work was called Operational Research in U.K., and in other nations called Operations Research. Even though the Americans put their effort at later date, produced many mathematical techniques for analysing military problems.

After the World War II, many soldiers belong to the military OR groups turned their attention to the possibilities of applying similar approach to day to day problems in the society.

Initially in a large scale profit making organisations like petroleum companies, etc. have introduced this OR techniques. Then slowly the researchers designed the model even to suit the small industries. Now, we can apply this OR techniques even for the day to day problems of any size and any category. Today OR is having more applications in Engineering, Business Management, Agriculture, Transporting system, etc.

It is very interesting to note that the modern perception of Operations Research as a body of well established models and techniques. A change of this kind is to be expected in any emerging field of scientific inquiry.

1.2 THE MEANING OF OPERATIONS RESEARCH

The term "Operations Research" has variations of meaning. In general OR is certain body of problems, techniques and solutions that has been accumulated under the name of OR for the past 30 years. We can see few accepted definitions of OR.

H.M. Wagner

Operations Research is a scientific approach to problem solving for Executive Management.

Hamdy A. Taha

Operations Research is a scientific knowledge through interdisciplinary team effort for the purpose of determining the best utilisation of limited resources.

OR Society of America

Operations Research is concerned with scientifically deciding how to design and operate man-machine systems in the best manner, usually under conditions requiring the allocation of scarce resources.

Based on all the definitions of various experts in this OR field, one may conclude that "Operations Research is nothing but a tool for decision making to solve any kind of problem".

1.3 MODELS OF OPERATIONS RESEARCH

We have many Models of OR like

1. Linear Programming Model
2. Transportation Model
3. Assignment Model
4. Network Model
   (i) CPM (ii) PERT
5. Game Theory and Decision Theory
6. Queuing Theory
7. Simulation Model
8. Dynamic Programming Model
9. Replacement Model
10. Sequencing Model
11. Integer Programming Model
12. Non-Linear Programming Model
13. Inventory Control Model
14. Goal Programming Model
15. Simulation
16. Markov Chain Model

1.4 SCOPE OF OPERATIONS RESEARCH

It has got wide scope, more specifically in developing nations, Mathematicians, Statisticians, Economists, Management experts, etc. use the O.R. techniques to take the better decision. Now we shall discuss the scope based on its applications.

1.4.1 Agriculture

Due to high increase in population size there is a need to increase the output and agriculture. Optimum allocation of land to different crops based on the climatic conditions and optimum distribution of water from various resources used scientific evaluation. This kind of problem can be easily solved using operation research technique.

1.4.2 In Organisation/Industry

Based on the latest concept "systems approach" in the Industries, Management functions are divided based on the department wise and given individual responsibilities. For example the production manager minimizes the cost of production and first to maximize the profit or tries to minimize the wastages of the raw material used for production and maximizes the output. O.R. techniques help the executives to take the best decision regarding the allocation of various resources like men, machines, material, money, time, etc.

1.4.3 In Military Operations

In modern warfare the military operations are carried out by a number of independent groups. The application of modern warfare requires expertiseness In this O.R helps to coordinate all groups in order to attain the optimum benefit.

1.4.4 Planning

Planning is a must for any type of organisation either government or non-government. To take proper decision regarding the planning aspect O.R helps a lot. For example O.R techniques can be applied to maximise the per capita income with minimum sacrifice of time.

1.4.5 In Transport

Monte Carlo Method O.R technique can be applied to regulate aircraft arrivals and service times. Likewise to arrive at a suitable transportation policy Linear Programming Method can be applied.

1.4.6 In Hospitals

Queuing theory can be applied to reduce the service time for the patients. It can take care of administrative problem of hospitals as well.

1.4.7 In Production Management

O.R Models can be used in the production management in many ways. For example,

1. to decide the optimum product mix.
2. to decide the assignment of persons to various jobs to be done.
3. to evaluate the project completion time for the analysis purpose.

1.4.8 In Marketing

O.R Methods can be very well applied in the Advertising Media selection problem, in deciding the level of finished goods inventory, etc.

1.4.9 In Finance

In the case of project selection based on Investment and its reform O.R Models are very well applied.

1.4.10 L.I.C.

O.R. approach can be used to find the premium rates for various policies and in what way the profits can be distributed among the policy holders. On the whole we find that Operations Research has a wide scope and can be applied in all areas.

1.5 PHASES OF OR

Regarding the application of Operations Research as a tool for any situation, we have to consider the following phases/steps very carefully.

O.R. Methodology

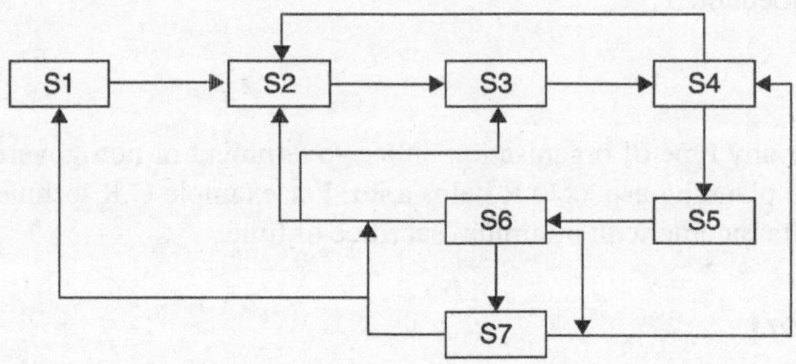

S1: Formulate the problem
S2: Observe the system carefully
S3: Construct the Mathematical model
S4: Verify the model and use the model for prediction
S5: Choose the best alternative
S6: Submit the results to organisation
S7: Implement and evaluate recommendations

1.6 LIMITATIONS OF OPERATIONS RESEARCH

Operations Research has a few limitations. Most of them are based on money and time as factors, which are listed below.

High Degree of Calculation

All the models require high degree of computations. Sometimes manual things are not possible. We seek the help of systems which require heavy investments. Hence the application of this kind is used in large concerns only.

Quantitative Nature

All kinds of problems require variables of quantitative nature. In reality it is not possible to quantify certain things. Those factors which cannot be converted are not taken into account.

Wide Gap Between the Manager and the OR User

Operations Research user means he/she must be well versed in Mathematics/Statistics. The difficulty is that they may not fully aware of the existing business problems. On the other hand there is a possibility of a Manager not understanding the complex procedure of all OR Models. Hence there is a big gap between the person who is going to make use of the solution and one who actually arrives at such a solution. So, here each party must have a full knowledge of it.

EXERCISE PROBLEMS

1. Explain the use of OR techniques in Management decision making.
2. Discuss the Limitations of OR.

REVIEW QUESTIONS

1. OR discipline began during _________. (Worldwar II)
2. OR Group Comprises _________, _________ and _________. (Scientists, Mathematicians, Management experts)
3. Today OR is having more application in _________ and _________ etc. (Business Management, Engineering)
4. State the definition of OR given by (i) H.M. Wagner (ii) Hamdy A. Taha (iii) O.R. Society of America. (Section 1.2)
5. State the scope of OR. (Section 1.4)
6. State the various phases of OR. (Section 1.5)
7. What are the limitations of OR. (Section 1.6)
8. Define OR. (Section 1.2)

2

Linear Programming Problem (LPP)

2.1 INTRODUCTION

Linear Programming constitutes a set of Mathematical Methods specially designed for the Modelling and solution of certain kinds of constrained optimization problems.

The Mathematical presentation of a Linear Programming Problem in the form of a linear objective function and one or more linear constraints with equations or inequations constitutes a Linear Programming Problem. The process leading to the construction of this model is referred to as the Model Building or Mathematical formulation of Business problem given. In this model, a linear objective function of the decision variables are maximized/minimized subject to a set of linear constraints with equations/inequations. This technique has met outstanding success in the solution of wide range problems like production planning and scheduling, personal assignment, finance planning, agricultural applications, competitive decision making and so on.

2.2 GENERAL MODEL OF THE LINEAR PROGRAMMING PROBLEM

Maximize/Minimize $Z = c_1x_1 + c_2x_2 + \ldots + c_nx_n$
subject to the constraints

$$
\begin{aligned}
&a_{11}x_1 + a_{12}x_2 + \ldots + a_{1r}x_r + \ldots + a_{1n}x_n \ \{\leq, = \geq\}\ b_1 \\
&a_{21}x_1 + a_{22}x_2 + \ldots + a_{2r}x_r + \ldots + a_{2n}x_n \ \{\leq, = \geq\}\ b_2 \\
&\ldots\ldots\ldots\ldots\ldots\ldots\ldots\ldots\ldots\ldots \\
&\ldots\ldots\ldots\ldots\ldots\ldots\ldots\ldots\ldots\ldots \\
&\ldots\ldots\ldots\ldots\ldots\ldots\ldots\ldots\ldots\ldots \\
&a_{m1}x_1 + a_{m2}x_2 + \ldots + a_{mr}x_r + \ldots + a_{mn}x_n \ \{\leq, = \geq\}\ b_m \\
&\qquad x_1, x_1, \ldots x_n \geq 0
\end{aligned}
$$

The above problem can be expressed using mathematical symbol as follows:

Maximize/Minimize $Z = \sum_{j=1}^{n} c_j x_j$

Subject to

$$\sum_{j=1}^{n} a_{ij}x_j \{\leq, =, \geq\} b_i$$

(for i = 1, 2 .... m); $x_j \geq 0$ (for j = 1, 2, ... n)

2.3 CHARACTERISTICS OF AN LPP

a) Regarding the symbol used in the General Model each constraint can take either ≤ or = or ≥.
b) The decision variables x_j's should take non-negative values only. (j = 1, 2, ..., m)
c) The values c_j; b_j; a_{ij}.

(for i = 1, 2, ... m; j = 1, 2 .... n) can be got from the given information. These values are called parameters and assumed to be fixed constants.

2.4 ASSUMPTIONS OF LINEAR PROGRAMMING

The Model of Linear Programming is based on Linearity (which contains proportionality and additivity), certainty, and devisability assumptions.

Linearity: It requires that the objective function and the constraints of linear programming problems lead themselves to a mathematical formulation in linear form. The term proportionality means that the contribution of each decision variable to the value of the objective function is proportional to the value assumed by the decision variable. Also additivity implies that the overall contributions of the objective are obtained by the algebraic sum of the individual contributions.

Certainty: It implies that the parameters c_j, b_j and a_{ij} (for i = 1, ... m; j = 1, ... n) of a Linear Programming model are fixed i.e., known with certainty.

Devisability: It requires that the decision variables be continuous, such that they can assume any value (fractional or Integer), based on the limitations imposed on the decision variables.

- Common errors one can think of while constructing the constraints are
  i. Insertion of inconsistent constraints
  ii. Omission of some constraints
  iii. Usage of different dimensions in right and left hand side of a constraint
  iv. Clerical errors
- Areas in which we can apply Linear Programming Techniques.
  i. Product-Mix
  ii. Use of Resources
  iii. Blending Problems
  iv. Staffing Problem
  v. Production Scheduling
  vi. Transportation Problem
  vii. Assignment problem
  viii. Networks etc.

2.4.1 Limitations of Linear Programming

1. The applications of LPP is possible iff we can express the given situation in the form of Linear objective and linear constraints.
2. The coefficients of the variables preserved in the objective and the constraints must be known well in advance.

3. Regarding the solution of an LPP, there is a chance for us to have fractional values, which can not be admissible in the decision making process.
4. Whenever if the Number of variables and constraints are more, we may require the help of computers to solve it.
5. Using this technique management make decision for only one objective at a time. That is multiple goals can not be achieved through this.

2.5 FORMULATION OF AN LPP

***Example:* 1**

Product-Mix Problem

The Krishna Company wishes to schedule the production of two items namely chairs and tables. The Management of the company has to decide as to how many chairs and Tables to be produced per day to maximize the profit. The following information is available to the Management. The profit per item is ₹35 and ₹55 respectively. In order to produce the two items that requires resources-wood, manpower and machine hour. The supply of wood is restricted to 1500 kgs per day. The manpower available per day is 75 carpenters. Also the maximum of 120 machine hours available per day. The production formula is as given in the following Table.

	Requirement per Unit		
	Wood (kg)	Manpower (persons)	Machine (hr)
Chair	6	2	4
Table	12	3	6

Formulate an LP model to determine the daily production rate of various Models in order to maximize the profit.

Formulation:

Step 1: Identify the decision variables. Let the number of units of chairs and tables to be produced be x_1 and x_2 respectively.

Step 2: Identify the constraints. In this problem the constraints are the limited availability of the three resources wood, manpower and machine hour. In addition to that we restrict the variables x_1, x_2 to have only non-negative values.

Step 3: Identify the objective. Here it is maximization of profit.

Then tabulate the given items:

Items	No. of Units to be Produced	Profit/Unit in ₹	Requirement of Raw Materials/Unit		
			Wood (kg)	Men	Machine (hr)
Chair	x_1	35	6	2	4
Table	x_2	55	12	3	6
	Maximum availability		1500	75	120

Construction of the objective function:

Profit from chairs	:	$35x_1$
Profit from tables	:	$55x_2$
Total Profit	:	$35x_1 + 55x_2$

(i.e.) $$\text{Max } Z = 35x_1 + 55x_2$$

Where Z is the name of the objective function.

Construction of constraint for wood:

Wood requirement for chairs	:	$6x_1$
Wood requirement for tables	:	$12x_2$
Total wood requirement	:	$6x_1 + 12x_2$

Given maximum availability is 1500 kgs

$$\Longrightarrow 6x_1 + 12x_2 \leq 1500$$

In order to achieve the maximization of the objective function, we can make use of all 1500 kgs or less than 1500 kgs.

Similarly, the other two constraints are

$$2x_1 + 3x_2 \leq 75 \quad \text{(men)}$$
$$4x_1 + 6x_2 \leq 120 \quad \text{(machine)}$$

In addition, we restrict the variables x_1, x_2, x_3 to have only non-negative values.

Thus, the Linear Programming Problem for our product mix problem becomes:

$$\begin{aligned} \text{Max } Z &= 35x_1 + 55x_2 \\ \text{s.t.} \quad 6x_1 + 12x_2 &\leq 1500 \\ 2x_1 + 3x_2 &\leq 75 \\ 4x_1 + 6x_2 &\leq 120 \\ x_1, x_2 &\geq 0 \end{aligned}$$

*Example: 2

A firm manufactures three products A,B and C. The profits are ₹3, ₹2 and ₹4 respectively. The firm has two machines. The processing time in minutes for each machine on each product is given below.

		Products		
		A	B	C
Machines	D	4	3	5
	E	2	2	4

Machines D and E have 2,000 and 2,500 machine-minutes respectively. The firm must manufacture at least 100A's; 200B's and 50C's but not more than 150 C's. Formulate a linear programming model to maximize the total profit.

Let x_1, x_2 & x_3 be the number of units to be produced in product A, B and C respectively.

---

*These questions have appeared in Indian university examination papers.

Tabulate the given information as follows.

Products	No. of Units	Profit (₹)	Machine/mts		Production Condition	
			D	E	Min.	Max.
A	x_1	3	4	2	100	–
B	x_2	2	3	2	200	–
C	x_3	4	5	4	50	150
Max. availability			2,000	2,500		

$$\text{Max } Z = 3x_1 + 2x_2 + 4x_3$$

Subject to the constraints

$$4x_1 + 3x_2 + 5x_3 \leq 2000 \quad \text{(Machine D)}$$
$$2x_1 + 2x_2 + 4x_3 \leq 2500 \quad \text{(Machine E)}$$
$$x_1 \geq 100; \quad x_2 \geq 200; \quad x_3 \geq 50; \quad x_3 \leq 150; \quad x_1, x_2, x_3 \geq 0$$

*Example: 3

Formulate the following problem as an LP model:

A co-operative farm owns 100 acres of land and has ₹25,000 in funds available for investment. The farm members can produce a total of 3,500 man-hours worth of labour during the months September-May and 4,000 man-hours during June-August. If any of these man-hours are not needed, some members of the farm will use them to work on a neighbouring farm for ₹2 per hour during September-May and ₹3 per hour during June-August. Cash income can be obtained from the three main crops and two types of livestock: dairy cows and laying hens. No investment funds are needed for the crops. However, each cow will require an investment outlay of ₹3,200 and each hen will require ₹15.

Moreover, each cow will require 1.5 acres of land, 100 man-hours of work during September-May and another 50 man-hours during June-August. Each cow will produce a net annual cash income of ₹3,500 for the farm. The corresponding figures for each hen are: no acreages, 0.6 man-hours during September-May 0.4 man-hours during June-August, and an annual netcash income of ₹200. The chicken house can accommodate a maximum of 4,000 hens and the size of the cattle-shed limits the members to a maximum of 32 cows.

Estimated man-hours and income per acre planted in each of the three crops are:

Man-Hours:	Paddy	Bajra	Jowar
September-May	40	20	25
June-August	50	35	40
Net Annual Cash Income (₹)	1,200	800	850

The co-operative farm wishes to determine how much acreage should be planted in each of the crops and how many cows and hens should be kept to maximise its net cash income.

Let x_1, x_2, x_3, x_4 and x_5 stand for the number of cows, hens, acre of lands used for paddy, Bajra and jowar respectively. Also x_6 & x_7 stands for the excess man hours available during the months SEP-MAY and JUNE-AUG respectively.

Tabulate the given information

Items	No. Purchased/ Acres of Land Used	Max. Nos.	Cost/ Unit (₹)	Land Required (Acre)	Labour SEP-MAY	Labour JUNE-AUG	Net Income (₹)
Cow	x_1	32	3,200	1.5	100	50	3,500
Hen	x_2	4,000	15	–	0.6	0.4	200
Paddy	x_3	–	–	1	40	50	1,200
Bajra	x_4	–	–	1	20	35	800
Jowar	x_5	–	–	1	25	40	850
	Max.	25,000		Max 100	Max 3,500	Max 4,000	

$$\text{Max } Z = 3500x_1 + 200x_2 + 1200x_3 + 800x_4 + 850x_5 + 2x_6 + 3x_7$$

$$\text{s.t. } 1.5x_1 + x_3 + x_4 + x_5 \le 100$$

$$100x_1 + 0.6x_2 + 40x_3 + 20x_4 + 25x_5 + x_6 = 3500$$

$$50x_1 + 0.4x_2 + 50x_3 + 35x_4 + 40x_5 + x_7 = 4000$$

$$x_1 \le 32,\ x_2 \le 4000,\ 3200x_1 + 15x_2 \le 25000,\ x_j \ge 0 \text{ (for all } j = 1 \ldots 7)$$

*Example: 4

Relax and Enjoy Lake Development Corporation is developing a lake side community at a privately owned lake and in the business of selling property for vacation and rent cottages. The primary market for these lakeside lots are middle and upper income families within 100 miles of the development. Relax and Enjoy has employed the advertising firm Philipines Jackson to design the promotional campaign for the project.

After considering possible advertising media and the market to be covered, the agency has made the preliminary recommendations to restrict the first month's advertising to five sources. At the end of this month, the agency will re-evaluate the strategy based upon the month's results. The agency has collected data on the number of potential purchase families reached, the cost per advertisement, the maximum number of times each medium is available and the expected exposure for each of the five media. The expected exposure is measured in terms of an exposure unit, a management judgement measure of the relative value of one advertisement in each of the media. The measures based on agency's experience in the advertising business take into account such factors as audience profile (age, income and education of the audience reached), image presented and quality of the advertisement. The information collected is presented in table 2.1.

Table 2.1 Advertising Media Alternative for Relax and Enjoy Lake Development Corporation

No.	Advertising Media	Number of Potential Purchase Families Reached	Cost per Advertisement	Maximum Time Available per Month	Expected Exposure Units
1.	Day time T.V. (1 min)	1000	5,000	15	65
2.	Prime time T.V. (seconds)	2000	5,000	10	90
3.	Daily newspapers	1500	4,000	25	40
4.	Sunday magazines	2500	3,000	4	60
5.	Radio (30 seconds)	800	1,500	30	20

Relax and Enjoy has provided with an advertising budget of ₹3,00,000 for the first month's campaign. In addition the company has a wish to have at least 10 television commercials, and atleast 50,000 potential purchasers must be reached during the month. Formulate the problem of media selection.

Let x_1, x_2, x_3, x_4, and x_6 stands for the number of units of advertisement to be given in day time TV, prime time TV, Newspapers, Sunday magazine & Radio respectively. Tabulate the given information as follows.

Medias	No. of Units	Cost per Add (₹)	No. of Potential Purchase Families Reached	Max. Time Available per Month	Expected Exposure Units
Daytime TV	x_1	5,000	1,000	15	65
Prime TV	x_2	5,000	2,000	10	90
Newspapers	x_3	4,000	1,500	25	40
Sunday	x_4				
Magazine	x_5	3,000	2,500	4	60
Radio	x_1	1,500	800	30	20
	Max.	3,00,000	Min. 50,000		

* Atleast 10 TV commercials must be given

$$\text{Max } Z = \frac{65}{1000}x_1 + \frac{90}{2000}x_2 + \frac{40}{1500}x_3 + \frac{60}{2500}x_4 + \frac{20}{800}x_5$$

$$\text{s.t.} \quad 5000x_1 + 5000x_2 + 4000x_3 + 3000x_4 + 1500x_5 \le 300000$$

$$1000x_1 + 2000x_2 + 1500x_3 + 2500x_4 + 800x_5 \leq 50000$$

$$x_1 \leq 15,\ x_2 \leq 10,\ x_3 \leq 25,\ x_4 \leq 4,\ x_5 \leq 30 \quad x_j \geq 0 \text{ (for all } j = 1 \ldots 5)$$

Optimum Solution

$$\text{Max } Z = 2.9210,\ x_1 = 15,\ x_2 = 10,\ x_3 = 25,\ x_4 = 4 \text{ and } x_5 = 30$$

***Example: 5**

Fixed Charge Problem

The River Falls Textile Company can use any, or all of three different processes for weaving its standard white polyester fabric. Each of these production processes has a weaving machine set-up cost and per-square-meter processing cost. These costs and the capacities of each of the three production processes are as follows:

Processing Number	Weaving Machine Set-up Cost (₹)	Processing Cost ₹/m²	Maximum Daily Capacity (m²)
1	50	0.06	20,000
2	80	0.04	30,000
3	100	0.03	35,000

The company forecasts daily demand for its white polyester fabric of 40,000 square meters. The company's production manager wants to make a decision concerning which production processes to utilize to meet the daily demand forecast, and at what capacity, to minimize total production costs. Formulate the suitable mathematical model for the above situation.

Let x_1, x_2 and x_3 stand for the number of square meters of white polyester produced using process-1, process-2 and process-3 respectively.

$$\text{Let } d_j = \begin{cases} 1; & \text{if jth process is used} \\ 0; & \text{if jth process is not used} \end{cases}$$

Tabulate all the information as follows:

Processing Number	No. of Square Meters Produced	Processing Cost ₹/m²	Maximum daily Capacity	Set-up Cost (₹)
1	x_1	0.06	20,000	50
2	x_2	0.04	30,000	80
3	x_3	0.03	35,000	100
		Max.	40,000	

Here the objective is to minimize the overall production cost.

$$\text{Min } Z = 50d_1, + 80d_2 + 100d_3 + 0.06x_1 + 0.04x_2 + 0.03x_3$$
$$\text{s.t. } x_1 + x_2 + x_3 \le 40000$$
$$x_1 \le 20000\, d_1$$
$$x_2 \le 30000\, d_2$$
$$x_3 \le 35000\, d_3$$
$$x_1, x_2, x_3 \ge 0 \text{ and integers } d_1, d_2, d_3 \text{ are either 0 or 1}$$

Example: 6

Mathura Bank has to formulate the loan policy for the financial year 1997-1998. The total loanable amount is ₹ 1,000 crores. The bank grants loan to different types of client to maximise the return subject to the constraints of RBI and the Market. The following table provides the interests rate charged by the bank for various type of Loans and the probability of the bad debts as estimated from the past records. Bad debts are assumed to be unrecoverable and also do not produce any revenue.

Type of Loan	Interest Rate	Likely Proportion of Bad Debts
Jewel Loan	15%	0.001
Car Loan	22%	0.10
Housing Loan	16%	0.04
Agricultural Loan	12%	0.06
Industry Loan	17%	0.09

As per RBI guidelines atleast 15 percentage of funds must be loaned to Agricultural sector. Industry and agricultural Loan should be more than 50 percentage. Jewel Loan should not exceed 10 percentage of funds. Housing Loan should be greater than Jewel Loan and Car Loan put together. The overall bad debt on all loans should not exceed .05.

Formulate the above bank loan policy into a LP problem to maximise the total interest earned.

Let x_1, x_2, x_3, x_4 and x_5 be the amounts in crores given in the form of jewel Loan (JL), Car Loan (CL), Housing Loan (HL), Agricultural Loan (AL) and Industry Loan (IL) respectively. Tabulate the given information as follows:

Type of Loan	Amounts Given (₹ in Crores)	Interest Rate	Proportion Bad Debts	Expected Debts	Min. Amount of Loan %	Max. Amount of Loan %
JL	x_1	15%	.001	0.999		10%
CL	x_2	22%	0.10	0.99		

Type of Loan	Amounts Given (₹ in Crores)	Interest Rate	Proportion Bad Debts	Expected Debts	Min. Amount of Loan %	Max. Amount of Loan %
HL	x_3	16%	0.04	0.96		
AL	x_4	12%	0.06	0.94	15% }	
IL	x_5	17%	0.09	0.91	} 50%	
		Max.	0.05			

* HL ≥ JL & CL

$$\text{Max } Z = (0.15)\,[0.999x_1] + 0.22\,[0.99x_2] + 0.16\,[0.96x_3] + 0.12\,[0.94x_4] + 0.17\,[0.91x_5]$$

Total funds : $x_1 + x_2 + x_3 + x_4 + x_5 \leq 1000$ crores

Agri. Loan : $x_4 \geq 0.15 * 1000$ crores

$x_4 + x_5 \geq 0.5 * 1000$ crores

Jewel Loan : $x_1 \leq 0.1 * 1000$ crores

$x_3 \geq x_2 + x_2$

Limit on bad debts : $\dfrac{0.001x_1 + 0.1x_2 + 0.04x_3 + 0.06x_4 + 0.09x_5}{x_1 + x_2 + x_3 + x_4 + x_5} \leq 0.05$

$x_1, x_2, x_3, x_4\ x_5 \geq 0$

Example: 7

A factory has decided to diversify their activities and data collected by sales and production is summarised below:

Potential demand exists for 3 products A, B and C. Market can take any amount of A and C whereas the share of B for this organisation is expected to be not more than 400 units a month.

For every three units of C produced, there will be one unit of a by-product which sells at a contribution of ₹3 a unit and only 100 units of this by-product can be sold per month. Contribution per unit of products A, B and C is expected to be ₹6, ₹8 and ₹4 respectively.

These products require 3 different processes and time required per unit production is given in the following table:

Process	Product (Hrs/Unit)			Available Hours
	A	B	C	
I	2	3	1	90
II	–	1	2	600
III	3	2	2	1200

Determine the optimum product mix for maximizing the contribution.

Let x_1 x_2 and x_3 be the number of units produced in Product A, B and C respectively. Given that the by-product based on C is 1 unit for every 3 units of C. Also, given the factory can sell only 100 units of the by-product.

We can discuss the formulation in the following two aspects.

1. Production of $x_3 \leq 300$ (by-product ≤ 100)
2. Production of $x_3 \geq 300$ (by-product ≥ 100)

Tabulate all the information as follows:

Product	No. of Units Produced	Demand Max.	Contribution (₹)	Time (Hr.) Product		
				I	II	III
A	x_1	–	6	2	–	3
B	x_2	400	8	3	1	2
C	x_3	–	4	1	2	2
			Max.	900	600	1200

Case (i)

Producing $x_3 \leq 300$

$$\text{Max } Z = 6x_1 + 8x_2 + 4x_3 + 3\,[x_3/3]$$

$$\begin{aligned} \text{s.t.}\quad 2x_1 + 3x_2 + x_3 &\leq 900 \\ x_2 + 2x_3 &\leq 600 \\ 3x_1 + 2x_2 + 2x_3 &\leq 1200 \\ x_3 &\leq 300 \\ x_1, x_2, x_3, &\geq 0 \end{aligned}$$

Case (ii)

$$\text{Max } Z = 6x_1 + 8x_2 + 4x_3 + 300$$

$$\begin{aligned} \text{s.t.}\quad 2x_1 + 3x_2 + x_3 &\leq 900 \\ x_2 + 2x_3 &\leq 600 \\ 3x_1 + 2x_2 + 2x_3 &\leq 1200 \\ x_3 &\geq 300 \\ x_1, x_2, x_3 &\geq 0 \end{aligned}$$

Example: 8

A certain farming organisation operates three farms of comparable productivity. The output of each form is limited both by the usable acreage and by the amount of water available for irrigation. The data for the upcoming season are the following.

Farm	Usable Acreage	Water Available in Acre Feet
1	400	1500
2	600	2000
3	300	900

The organisation is considering three crops for planting, which differ primarily in their expected profit/acre and in their consumption of water. Furthermore, the total acreage that can be devoted to each of the crops is limited by the amount appropriate harvesting equipment available.

Crop	Available Acreage	Water Consumption in Acre Feet/Acre	Expected Profit/ Acre
A	700	5	400
B	800	4	300
C	300	3	100

In order to maintain uniform workload among the farms, it is the policy of the organisation that the percentage of the usable acreage planted must be the same at each farm. However, any combination of the crops may be grown at any of the farms. The organisation wishes to know how much of each crop should be planted at the respective farms in order to maximize the expected profit. Formulate this as a linear programming problem.

Let x_{A1}, x_{A2} & x_{A3} stand for the number of acreages used for the crop A in farms 1, 2 & 3.
Let x_{B1}, x_{B2} & x_{B3} stand for the number of acreages used for the crop B in farms 1, 2 & 3.
Let x_{C1}, x_{C2} & x_{C3} stand for the number of acreages used for the crop A in farms 1, 2 & 3.
Here the objective is to maximize the expected profit.

Crops	Acres Used Farm 1	2	3	Available Acres	Expected Profit/Acre (₹)	Water Consumption in Acre Feet/ Acre	Total Acreages Used for Crop
A	x_{A1}	x_{A2}	x_{A3}	700	400	5	$x_{A1} + x_{A2} + x_{A3}$
B	x_{B1}	x_{B2}	x_{B3}	800	300	4	$x_{B1} + x_{B2} + x_{B3}$
C	x_{C1}	x_{C2}	x_{C3}	300	100	3	$x_{C1} + x_{C2} + x_{C3}$
Available acreage	400	600	300				
Availability of acre in acre/feet	1500	2000	900				
Total acre used	$(x_{A1} + x_{B1} + x_{C1})$	$(x_{A2} + x_{B2} + x_{C2})$	$(x_{A3} + x_{B3} + x_{C3})$				

* The % of the usable acreage planted must be the same at each farm.

$$\text{Max } Z = 400\,[x_{A1} + x_{A2} + x_{A3}] + 300\,[x_{B1} + x_{B2} + x_{B3}] + 100\,[x_{C1} + x_{C2} + x_{C3}]$$

s.t. demand of crops :

$$x_{A1} + x_{A2} + x_{A3} \le 700$$
$$x_{B1} + x_{B2} + x_{B3} \le 800$$
$$x_{C1} + x_{C2} + x_{C3} \le 300$$

Acres of land :

$$x_{A1} + x_{B1} + x_{C1} \le 400$$
$$x_{A2} + x_{B2} + x_{C2} \le 600$$
$$x_{A3} + x_{B3} + x_{C3} \le 300$$

Water consumption :

$$5x_{A1} + 4x_{B1} + 3x_{C1} \le 1500$$
$$5x_{A2} + 4x_{B2} + 3x_{C2} \le 2000$$
$$5x_{A3} + 4x_{B3} + 3x_{C3} \le 900$$

% of usable acreage must be the same

$$\frac{x_{A1} + x_{B1} + x_{C1}}{400} = \frac{x_{A2} + x_{B2} + x_{C2}}{600} = \frac{x_{A3} + x_{B3} + x_{C3}}{300}$$

Multiply by (1200)

$$=> \quad (x_{A1} + x_{B1} + x_{C1})\,3 = (x_{A2} + x_{B2} + x_{C2})\,2 = 4\,(x_{A3} + x_{B3} + x_{C3})$$

Here we have (1) = (2), = (3); This form can be done in the following way

=> (i) (1) = (2); (ii) (2) = (3) and (iii) (3) = (1)

$$3\,[x_{A1} + x_{B1} + x_{C1}] - 2\,[x_{A2} + x_{B2} + x_{C2}] = 0$$
$$2\,[x_{A2} + x_{B2} + x_{C2}] - 4\,[x_{A3} + x_{B3} + x_{C3}] = 0$$
$$4\,[x_{A3} + x_{B3} + x_{C3}] - 3\,[x_{A1} + x_{B1} + x_{C1}] = 0$$
$$x_{A1}, x_{A2}, x_{A3}, x_{B1}, x_{B2}, x_{B3}, x_{C1}, x_{C2}, x_{C3} \ge 0$$

Example: 9

Shift Scheduling Problem

A local hospital administration wishes to determine a work schedule for registered nurses (RNs). The union contract specifies that nurses work a normal day of six hours. The administrator has determined a daily work schedule in which the day is divided into eight three-hour shifts. The following table indicates the estimated minimum requirement for nurses per shift. The Nurses will start work each day at the beginning of one of these shifts and end work at the end of the following shift.

	Period							
Shifts	12-3 a.m.	3-6 a.m.	6-9 a.m.	9-12 a.m.	12-3 p.m.	3-6 p.m.	6-9 p.m.	9-12 p.m.
Minimum number of required RNs	30	20	40	50	60	50	40	40

The administration must specify the number of nurses that is to begin work each shift, so that the required number of nurses is available for each three-hour period.

The objective is to minimize the total number of nurses employed NP. Let x_j stand for the number of nurses report up to work at the beginning of period j.

$$\text{Min } Z = \sum_{j=1}^{8} x_j$$

Consider the staffing requirement of atleast 30 nurses for period 1. Nurses on duty for this period would include those starting work at the beginning of this shift plus those who began work during the previous shift. Thus the constraints for period 1 is

$$x_1 + x_8 \le 30$$

Similarly,

$$x_1 + x_2 \ge 20 \qquad x_4 + x_5 \ge 60$$
$$x_2 + x_3 \ge 40 \qquad x_5 + x_6 \ge 50$$
$$x_3 + x_4 \ge 50 \qquad x_6 + x_7 \ge 40$$
$$x_7 + x_8 \ge 40$$
$$x_j \ge 0 \ (j = 1, 2 \ldots\ldots, 8)$$

Optimum Solution:

$$\text{Min } Z = 170,\ x_1 = 30,\ x_2 = 0,\ x_3 = 40,\ x_4 = 10,\ x_5 = 50,\ x_6 = 0,\ x_7 = 40 \text{ and } x_8 = 0$$

*Example: 10

A toy company manufactures two types of dolls a basic version-doll A and a delux version-doll B. Each type of doll B takes time twice as long as one type A, and the company would have time to make maximum of 2000 in type A per day. The supply of plastic is sufficient to produce 1500 dolls per day (both A and B combined). The delux version requires a fancy dress of which there are only 600 per day available. If the company makes a profit of ₹3 and ₹5 per doll respectively on doll A and doll B, then how many of each doll should be produced per day inorder to maximize the total profit. Formulate it.

Let x_1 and x_2 be the number of products to be produced in doll A and doll B respectively.

Tabulate the given information as follows:

Products	No. of Units	Profit (₹)	Production Time	Fancy Dress Available
Doll A	x_1	3	1	---
Doll B	x_2	5	2	600
	availability		2000	

Plastic is available to produce 1500 dolls per day.

$$\text{Max } Z = 3x_1 + 5x_2$$

$$\text{s.t.} \quad x_1 + 2x_2 \leq 2000; \quad x_2 \leq 600; \quad x_1 + x_2 \leq 1500; \quad x_1, x_2 \geq 0$$

***Example:* 11**

A company has two grades of inspectors, 1 and 2 to undertake quality control inspection. At least 1500 pieces must be inspected in an 8 hour day. Grade 1 inspector can check 20 pieces in an hour with an accuracy of 96%. Grade 2 inspector checks 14 pieces an hour with an accuracy of 92%. The daily wages of Grade 1 inspectors are ₹5.00 per hour, while those of Grade 2 inspectors are ₹4.00 per hour. Any error made by an inspector costs ₹3.00 to the company. If the company is having 10 grade 1 and 15 grade 2 inspectors, find the optimal assignment of inspectors that minimizes the daily inspection Cost. Formulate this problem as an LPP.

Let x_1 and x_2 be the number of Grade 1 and Grade 2 inspectors to be inducted. Tabulate the given information as follows:

Inspector	Inspectors Available	Persons Employed	No. of Pieces Tested/Day	Wage/Day (₹)
Grade 1	10	x_1	$20 \times 8 = 160$	$5 \times 8 = 40$
Grade 2	15	x_2	$14 \times 8 = 112$	$4 \times 8 = 32$

% of Error	No. of Errors (Approximately)	Error Cost (₹)	Total Cost (Error)
4	$160 \times 0.04 = 6.4 = 7$	21	$40 + 21 = 61$
8	$112 \times 0.08 = 8.96 = 9$	27	$32 + 27 = 59$

Here, one day = 8 hours
% of error = 100 – % of accuracy
No. of errors = No. of Tubes tested * error %
error cost = given as ₹3 per error
Total Cost = wage per day + error cost

The objective is to minimize overall cost for the inspection.

$$\text{Min } Z = 61x_1 + 59x_2$$

Subject to, $160x_1 + 112x_2 \geq 1500; \; x_1 \leq 10; \; x_2 \leq 15; \; x_1, x_2 \geq 0$

***Example:* 12**

A firm manufactures two items. It purchases castings which are then machined (M), bored (B) and polished (P). Castings for items A and B costs ₹3 and ₹4 respectively and are sold at ₹6 and ₹7 each respectively. Running cost of the three machines are ₹20 and ₹14 and ₹17.50 per hour respectively. What product mix maximizes the profit. Formulate an L.P. Model, solution is not necessary. Capacities of the machines are

	PART A	PART B
Machining Capacity	25/hr	40/hr
Boring Capacity	28/hr	35/hr
Polishing Capacity	35/hr	25/hr

Let x_1 and x_2 be the number of units to be produced per hour in Part A and Part B respectively. Tabulate the given information as follows;

Products	Units Produced	Processing Time per Unit in mts.			Processing Cost per Unit (₹)		
		M	B	P	M	B	P
Part A	x_1	2.4	2.14	1.71	0.8	0.5	0.5
Part B	x_2	1.5	1.71	2.4	0.5	0.4	0.7
	Max. availability	60	60	60			

Total Processing Cost (₹)	Casting Cost/Unit (₹)	Total Cost/Unit (₹)	Selling Price/ Unit (₹)	Profit/Unit (₹)
1.8	3	4.8	6	1.2
1.6	4	5.6	7	1.4

Processing time in minutes = 60/(no. of units/hr)

Processing cost/unit = (processing time in minutes) * (running cost)/60

Cost/unit = Casting Cost/unit + Processing Cost/unit

Profit/unit = Selling Cost/unit – Cost/unit

$$\text{Max } Z = 1.2x_1 + 1.4x_2$$
$$\text{s.t.} \quad 2.4x_1 + 1.5x_2 \le 60$$
$$2.14x_1 + 1.71x_2 \le 60$$
$$1.71x_1 + 2.4x_2 \le 60$$
$$x_1, x_2 \ge 0$$

Example: 13

Advertising Media Selection

An advertising company wishes to plan an advertising campaign in three different Media – TV, Radio, and Magazines. The aim of the advertising program is to reach as many potential customers as possible. Research of a market study is given below:

Particulars	Television		Radio	Magazines
	Day Time	Prime Time		
Cost of an advertising unit ₹	40,000	80,000	35,000	17,000
Number of Customers reached/unit	5,00,000	9,00,000	5,00,000	2,50,000
No. of women Customers reached/unit	3,50,000	4,25,000	2,00,000	1,00,000

The Company does not want to spend more than ₹7,50,000 on advertising. It also requires that (i) atleast 2.5 million exposures take place among women (ii) advertising on TV be limited to ₹4,50,000 (iii) atleast 4 advertising units be bought on daytime TV and two during prime time and (iv) The number of advertising units on radio and magazine should each be between 4 and 10.

Let x_1, x_2, x_3 and x_4 be the number of advertising units bought in Day TV, Prime TV, Radio and Magazines respectively.

Tabulate the given information as follows:

Advertising Media	No. of Units	Cost/Unit (₹)	No. of Customers	No. of Women
TV day	x_1	40,000	5,00,000	3,50,000
TV prime	x_2	80,000	9,00,000	4,25,000
Radio	x_3	35,000	5,00,000	2,00,000
Magazine	x_4	17,000	2,50,000	1,00,000
	Max.	7,50,000	Min. Women	25,00,000

Advertising Media	Max Amount can be Spent	No. of Advertisement	
		Min.	Max.
TV day }	4,50,000	4	
TV prime }		2	
Radio	--	5	9
Magazine	--	5	9

$$\text{Max } Z = 500000x_1 + 900000x_2 + 500000x_3 + 250000x_4$$

subject to the constraints

$$40000x_1 + 80000x_2 + 35000x_3 + 17000x_4 \leq 750000$$
$$350000x_1 + 425000x_2 + 200000x_3 + 100000x_4 \geq 2500000$$
$$40000x_1 + 80000x_2 \leq 450000$$
$$x_1 \geq 4; \quad x_2 \geq 2; \quad x_3 \geq 5; \quad x_3 \leq 9; \quad x_4 \geq 5; \quad x_4 \leq 9;$$
$$x_1, x_2, x_3, x_4 \geq 0$$

Optimum Solution:

$$\text{Max } Z = 100.0714 \text{ lakhs}, x_1 = 4, x_2 = 2, x_3 = 7.9 \;\&\; x_4 = 9$$

Integer Optimum Solution:

$$\text{Max } Z = 9800000, x_1 = 6, x_2 = 2, x_3 = 6 \;\&\; x_4 = 8$$

***Example:* 14**

The Krishna Paper Company produces paper rolls with standard width of 210 cm, special customers orders with different widths are produced by slitting the standard rolls. The orders are summarized in the following table:

Order	Required Width	Required Number of Rolls
1	100 cm	210
2	50 cm	130
3	30 cm	100

Formulate the above problem as an LPP in order to minimize the wastage.

First we have to decide the various possible ways of cutting the required size.

Here in the given problem, the required sizes can be cut in nine ways, (only using trial and error method)

Let $x_1, x_2, \ldots\ldots\ldots\ldots, \& \; x_9$, be the number of rolls cut in different ways.

While cutting there is a possibility for wastages, hence the objective here is to the minimize the wastages as well as to satisfy the demand.

Various Ways	Number of Rolls Cut	No. of Rolls Got in Different Size			Wastages in cm
		100 cm	50 cm	30 cm	
1	x_1	2	--	--	10
2	x_2	1	2	--	10
3	x_3	1	1	2	0
4	x_4	1	--	3	20

Various Ways	Number of Rolls Cut	No. of Rolls Got in Different Size			Wastages in cm
		100 cm	50 cm	30 cm	
5	x_5	--	4	--	10
6	x_6	--	3	2	0
7	x_7	--	2	3	20
8	x_8	--	1	5	10
9	x_9	--	--	7	0
Required number		210	130	100	

$$\text{Minimize } Z = 10x_1 + 10x_2 + 20x_4 + 10x_5 + 20x_7 + 10x_8$$

subject to the constraints

$$2x_1 + x_2 + x_3 + x_4 = 210; \quad 2x_2 + x_3 + 4x_5 + 3x_6 + 2x_7 + x_8 = 130;$$
$$2x_3 + 3x_4 + 2x_6 + 3x_7 + 5x_8 + 7x_9 = 100;$$
$$x_j \geq 0 \text{ (for } j = 1, 2 \ldots.. 9)$$

Optimum Solution:

$$\text{Min } Z = 1000, x_1 = 60, x_2 = 40 \ \& \ x_3 = 50$$

***Example: 15**

Blending Problem

The Manager of an Oil Company must decide on the optimal mix of two possible blending processes of which the inputs and outputs per production run are as follows:

Process	Input (Units)		Output (Units)	
	Crude A	Crude B	Gasoline X	Gasoline Y
1	5	3	5	8
2	4	5	4	4

The Maximum amounts available for orders A and B are 200 units and 150 units respectively. The Market requirements shows that atleast 100 units of Gasoline X and 80 units of Gasoline Y must be produced. The profit per production run from process 1 and process 2 are ₹300 and ₹400 respectively. Form the LPP.

Let x_1 and x_2 be the number of times the process 1 and process 2 are to be repeated.

Tabulate the given information as follows:

Process	Number of Time	Input		Output		Profit ₹
		Crude A	Crude B	X	Y	
1	x_1	5	3	5	8	300
2	x_2	4	5	4	4	400
	Max. availability	200	150	Min. 100	80	

$$\text{Maximize } Z = 300x_1 + 400x_2$$

Subject to the constraints

$$5x_1 + 4x_2 \leq 200; \quad 3x_1 + 5x_2 \leq 150; \quad 5x_1 + 4x_2 \geq 100$$
$$8x_1 + 4x_2 \geq 80; \quad x_1, x_2 \geq 0$$

*Example: 16

A Manufacturer uses three raw products a, b, c priced at 30, 50, 120 rupees per kg respectively. He can make three different products A, B and C which can be sold at 90,100 and 120 rupees per kg respectively. The raw products can be obtained only in limited quantities, namely 20, 15 and 10 kg per day. Given 2 kg of a plus 1 kg of b plus 1 kg of c will yield 4 kg of A; 3 kg of a plus 2 kg of b plus 2 kg of c will yield 7 kg of B; 2 kg of b plus 1 kg of c will yield 3 kg of C. Make a production plan, assuming that the other costs are not influenced by the choice among the alternatives. Formulate the Linear model of the problem.

Let x_1, x_2 and x_3 be the number of kgs. to be produced in products A, B, and C respectively.

Tabulate the given information as follows:

Product	No. of Units	Raw Materials/ Units in kg			Cost/Unit	Selling Price/Unit	Profit/Unit
		a	b	c			
A	x_1	1/2	1/4	1/4	15 + 12.5 + 30 = 57.5	90	32.5
B	x_2	3/7	2/7	2/7	12.86 + 14.29 + 34.29 = 61.44	100	38.56
C	x_3	–	2/3	1/3	33 – 33 + 40 = 73.33	120	46.67
Max. availability kg		20	15	10			

$$\text{Maximize } Z = 32.5x_1 + 38.56x_2 + 46.67x_3$$

s.t. $1/2x_1 + 3/7x_2 \leq 20$; $1/4x_1 + 2/7x_2 + 2/3x_3 \leq 15$; $1/4x_1 + 2/7x_2 + 1/3x_3 \leq 10$; $x_1, x_2, x_3 \geq 0$

*Example: 17

The Omega Data Processing Company performs three types of activities: Payrolls, account receivables and inventories. The profit and time requirement for keypunch computation and office printing for a "standard job" are shown in the following table:

Job	Profit/Standard Job (₹)	Time Requirement (mts)		
		Keypunch	Computation	Print
Payroll	275	1200	20	100
A/c Receivable	125	1400	15	60
Inventory	225	800	35	80

Omega guarantees overnight completion of the job. Any job scheduled during the day can be completed during the day or night. Any job scheduled during the night, however, must be completed during the night. The capacity for both day and night are shown in the following table:

Capacity (Mts)	Keypunch	Computation	Print
Day	4200	150	400
Night	9200	250	650

Formulate the linear programming problem in order to determine the "mixture" of standard jobs that should be accepted during the day and night.

Let us assume the following variables to refer the number of jobs to be taken in each category.

Job/Time	Day	Night
Payroll	P_D	P_N
A/c Receivable	A_D	A_N
Inventory	I_D	I_N

Tabulate the given information as follows:

Job	No. of Jobs to be Taken	Time Required in Minutes			Profit (₹)	Capacity		
		Keypunch	Computation	Print		Keypunch	Computation	Print
Payroll day	P_D	1200	20	100	275			
A/c receivable day	A_D	1400	15	60	125			
Inventory day	I_D	800	35	80	225	13200	400	1050
Payroll night	P_N	1200	20	100	275			
A/c receivable night	A_N	1400	15	60	125			
Inventory night	I_N	800	35	80	225	9200	250	650

$$\text{Maximize } Z = 275[P_D + P_N] + 125[A_D + A_N] + 225[I_D + I_N]$$

$$\text{s.t.} \quad 1200[P_D + P_N + 1400[A_D + A_N] + 800[I_D + I_N] \le 13200$$

$$1200P_N + 1400A_N + 800I_N \le 9200$$

$$20[P_D + P_J + 15[A_D + A_N] + 35[I_D + I_N] \le 400$$

$$20P_N + 15A_N + 35I_N \le 250$$

$$100[P_D + P_N] + 60[A_D + A_N] + 80[I_D + I_N] \le 1050$$

$$100P_N + 60A_N + 80I_N \le 650$$

$$P_D, P_N, A_D, A_N. I_D, I_N \ge 0 \text{ and integers.}$$

Optimum Solution:

$$P = 4;\ I_D = 1 \text{ and } I_N = 7 \quad \text{Max } Z = 2900$$

2.6 STANDARD FORM OF AN LPP

If an LPP satisfies the following conditions, then it is said to be in a standard form.

1. All the variables are non-negative.
2. All constraints must be equations together with nonnegative right-hand side values.
3. The objective function may be either minimization or maximization.

2.6.1 Conversion of an LPP into Standard Form with Maximization Objective

The various steps to be followed to convert the LPP with a standard form are clearly explained in the Flowchart (2.1).

Notations Used

X_B : Right-hand side value of the constraints
X : Variables used in the LPP
Obj : Objective function
IE : Inequality
Slack : Something extra in our hand.
Surplus : Something extra required which is not in our hand.

Remark: Slack/surplus introduced in the constraints having ≤/≥ symbol to make the inequations into equations.

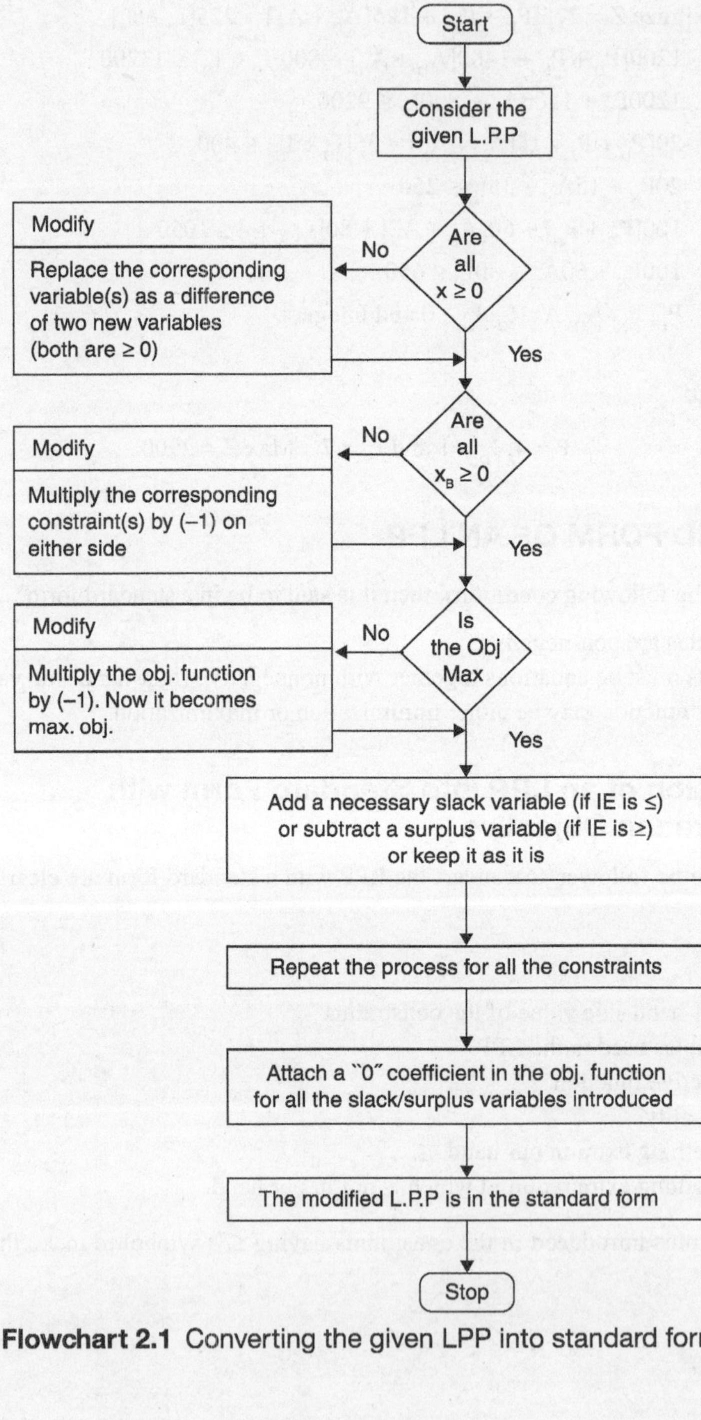

Flowchart 2.1 Converting the given LPP into standard form

Example: 18

Find the standard form of an LPP:

$$\text{Minimise } Z = 3x_1 + 2x_2$$
$$\text{s.t.} \quad 2x_1 + x_2 = 15; \quad 3x_1 - 4x_2 \geq -6; \quad x_1 - 2x_2 \geq 5; \quad x_2 \geq 0 \qquad (1)$$

Non-negative Restrictions

In the given problem, the non-negative restriction for the variable x_1 is missing. Then x_1 is called unrestricted variable, i.e. the variable x_1 can take a (+ve)/(–ve)/0 value. So, x_1 must be replaced as a difference of two new variables. $x_1 = x_3 - x_4$; where $x_3, x_4 \geq 0$. Substitute the value of x_1 in the given problem.

$$\text{Min } Z = 3(x_3 - x_4) + 2x_2$$
$$\text{s.t.} \quad 2(x_3 - x_4) + x_2 = 15; \; 3(x_3 - x_4) - 4x_2 \geq -6$$
$$(x_3 - x_4) - 2x_2 \leq 5; \; x_3, x_4, x_2 \geq 0 \qquad (2)$$

Right-hand Side Value

The second constraint

$3x_1 - 4x_2 \geq -6$; has a negative value in right-hand side.

Multiply the constraint using (–1) on both sides. Note that while multiplying with (–1); the inequality will be reverted.

$$-3(x_3 - x_4) + 4x_2 \leq 6$$

Introduction of Slack/Surplus Variables

Constraint 1: It is already an equation, keep it as it is.

Constraint 2: Since, it has the inequality ≤, add a variable x_5 (slack), $-3(x_3 - x_4) + 4x_2 + x_5 = 6$; $x_5 \geq 0$

Constraint 3: Since, it has the inequality ≥, subtract a variable x_6 (surplus),

$$(x_3 - x_4) - 2x_2 - x_6 = 5; \; x_6 \geq 0$$

The required standard form is,

$$\text{Min } Z = 3x_3 - 3x_4 + 2x_2 + 0x_5 + 0x_6$$
$$\text{s.t.} \quad 2x_3 - 2x_4 + x_2 = 15; \quad 3x_3 - 3x_4 + 4x_2 + x_5 = 6;$$
$$x_3 - x_4 - 2x_2 - x_6 = 5; \quad x_3, x_4, x_2, x_5, x_6 \geq 0$$

2.7 SOLUTION TO AN LPP

After the formulation of an LPP, the next step is to find the solution to the same. We have so many methods to find the solution, that are clearly stated in the Flowchart (2.2).

2.8 TYPES OF POSSIBLE SOLUTIONS TO AN LPP

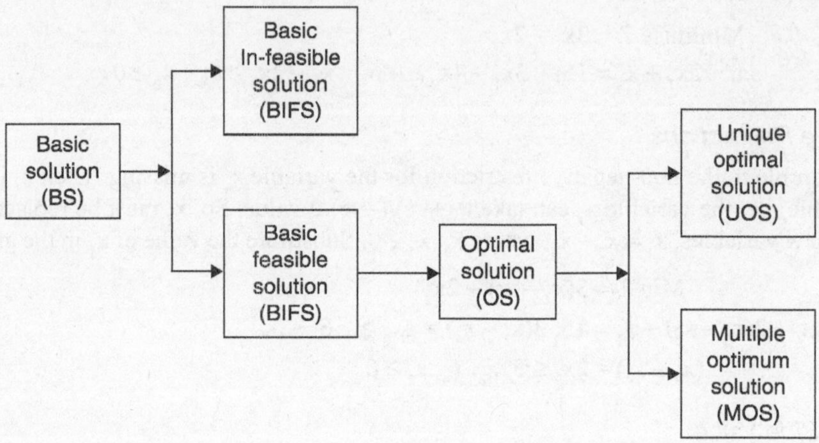

2.8.1 Basic Solution

Assume that the standard form of an LPP has 'm' equations with 'n' variables ($m \leq n$) together with non-negative conditions. The unique solution resulting from setting $(n - m)$ variables to zero such that the remaining m columns **Linearly Independent**, are called Basic Solutions.

2.8.2 Basic Feasible Solution

If a Basic Solution satisfies the non-negative conditions, it is named as Basic Feasible Solution.

2.8.3 Basic Infeasible Solution

If a Basic Solution does not satisfy the non-negative conditions, it is named as Basic Infeasible Solution.

2.8.4 Optimal Solution

If a Basic Feasible Solution maximizes or minimizes the objective function of the problem it is named as an Optimum Solution.

2.8.5 Unique Optimum Solution

It means that the optimum value is attained for only one feasible solution.

2.8.6 Multiple Optimum Solution

It means that the same optimum value is attained for more than one feasible solution.

2.8.7 Basic and Non-Basic Variables

The variables set equal to zero are (refer to Basic Solution Definition) called non-basic variables; the remaining variables are called basic variables.

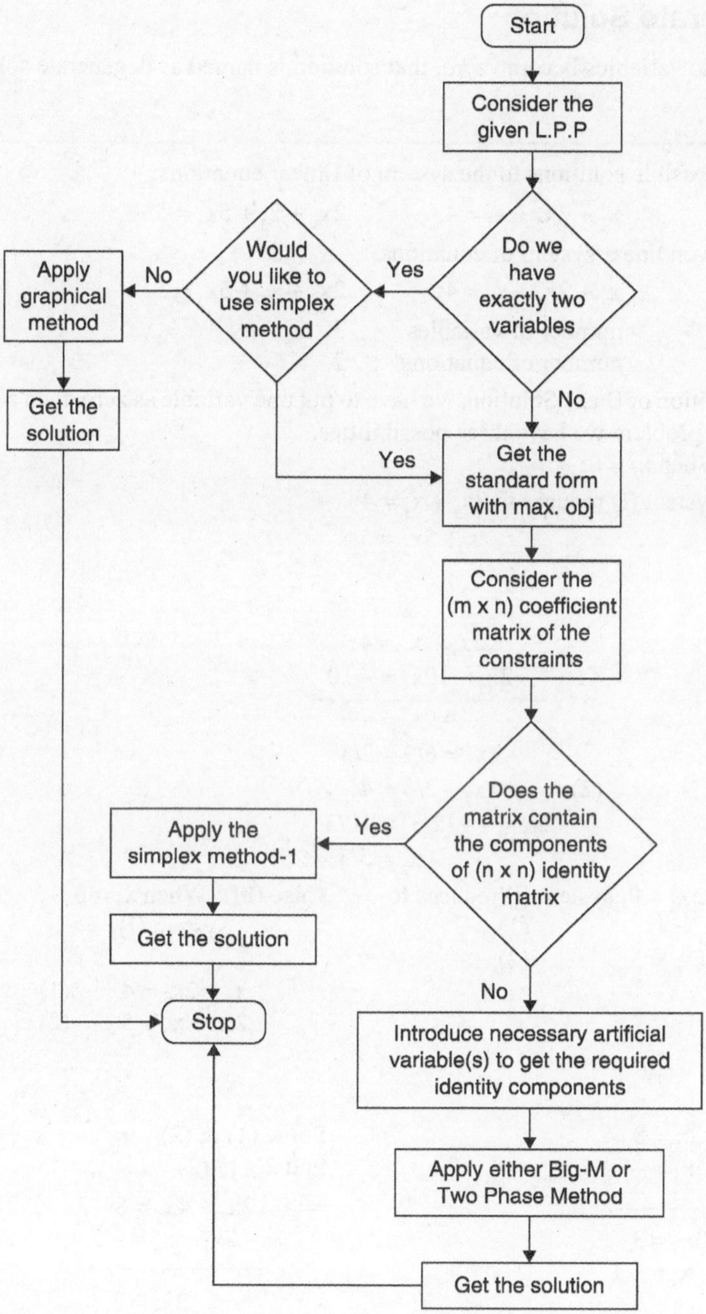

Flowchart 2.2 Methods used to find the solution to an LPP

2.8.8 Degenerate Solution

If one or more basic variables become zero, that solution is named as degenerate solution.

***Example:* 19**

Find all the basic feasible solutions to the system of Linear equations:

$$x_1 + 2x_2 + x_3 = 4; \qquad 2x_1 + x_2 + 5x_3 = 5$$

Consider the given linear system of equations.

$$x_1 + 2x_2 + x_3 = 4; \qquad 2x_1 + x_2 + 5x_3 = 5 \qquad \text{(I)}$$

number of variables : 3
number of equations : 2

As per the definition of Basic Solution, we have to put one variable as zero, then solve for the remaining system. In this problem we have three possibilities.

Case (i): When $x_1 = 0$;
system (I) reduces to $2x_2 + x_3 = 4$ (1)
$x_2 + 5x_3 = 5$ (2)

Solve (1) & (2)
find (1)-2 * (2)

$$\Longrightarrow \quad 2x_2 + x_3 = 4$$
$$-2x_2 - 10x_3 = -10$$
$$-9x_3 = -6$$
$$x_3 = 6/9 = 2/3$$
$$(2) \Longrightarrow 2x_2 + 2/3 = 4$$
$$2x_2 = 4 - (2/3) = 10/3$$
$$x_2 = 5/3$$

Case (ii): When $x_2 = 0$; system (I) reduces to
$x_1 + x_3 = 4$ (1)
$2x_1 + 5x_3 = 5$ (2)

Solve (1) & (2)
find 2 * (1)-(2)

$$\Longrightarrow \quad 2x_1 + 2x_3 = 8$$
$$-2x_1 - 5x_3 = -5$$
$$-3x_3 = 3$$
$$x_3 = -1$$
$$(1) \Longrightarrow \quad x_1 - 1 = 4$$
$$x_1 = 5$$

Case (iii): When $x_3 = 0$
system (I) $\Longrightarrow$

$x_1 + 2x_2 = 4$ (1)
$2x_1 + x_2 = 5$ (2)

Solve (1) & (2)
find 2* (1)-(2)

$$\Longrightarrow \quad 2x_1 + 4x_2 = 8$$
$$-2x_1 - x_2 = -5$$
$$3x_2 = 3$$
$$x_2 = 1$$
$$(1) \Longrightarrow \quad x_1 + 2 = 4$$
$$x_1 = 2$$

Tabulate the results obtain in all the three cases.

Basic Solutions			Remark	Basic Variables	Non-basic Variables
x_1	x_2	x_3			
0	5/3	2/3	Feasible	x_2, x_3	x_1
5	0	–1	Infeasible	x_1, x_3	x_2
2	1	0	Feasible	x_1, x_2	x_3

***Example: 20**

Find all the Basic solutions for the system

$$3x_1 + x_2 + x_3 = 5; \qquad -x_1 + x_2 - x_4 = 1$$

State which among them are,
(i) basic feasible (ii) basic infeasible

Consider the given Linear system of equations.

$$3x_1 + x_2 + x_3 = 5$$
$$-x_1 + x_2 - x_4 = 1 \qquad \text{(I)}$$

number of variables = 4
number of constraints = 2

As per the definition of BASIC solution, we have to put (4-2) variables as zero, then solve for the remaining system.

In this problem we have six possibilities i.e. 4_{C_2}.

Case (i): When $x_1 = 0; x_2 = 0$
(I) ==> $x_3 = 5; x_4 = -1$

Case (ii): When $x_1 = 0; x_3 = 0$
(I) ==> $x_2 = 5$
$x_2 - x_4 = 1$
==> $x_4 = 4$

Case (iii): When $x_1 = 0; x_4 = 0$
(I) ==> $x_2 + x_3 = 5$
$x_2 = 1$
==> $x_3 = 4$

Case (iv): When $x_2 = 0; x_3 = 0$
(I) == $x_1 = 5/3$
$-x_1 - x_4 = 1$
$-x_4 = 1 + 5/3 = 8/3$
$x_4 = -8/3$

Case (v): When $x_2 = 0; x_4 = 0$
(I) ==> $x_1 = -1$
$3x_1 + x_3 = 5$
$x_3 = 8$

Case (vi): When
$x_3 = 0; x_4 = 0$

(I) ==> $3x_1 + x_2 = 5$ (1)
$-x_1 + x_2 = 1$
Solve (1) & (2)
Find 3* (2) + (1)

$$3x_1 + x_2 = 5$$
$$-3x_1 + 3x_2 = 3$$
$$\overline{\qquad 4x_2 = 8 \qquad}$$

$x_2 = 2$
(2) ==> $-x_1 = -3$
$x_1 = 3$

Basic Solutions				Remark	Basic Variables	Non-basic Variables
x_1	x_2	x_3	x_4			
0	0	5	– 1	Infeasible	x_3, x_4	x_1, x_2
0	5	0	4	Feasible	x_2, x_4	x_1, x_3
0	1	4	0	Feasible	x_2, x_3	x_1, x_4
– 1	0	8	0	Infeasible	x_1, x_3	x_2, x_4
5/3	0	0	– 8/3	Infeasible	x_1, x_4	x_2, x_3
3	4	0	0	Feasible	x_1, x_2	x_3, x_4

***Example:* 21**

Show that feasible solution $x_1 = 1$; $x_2 = 0$; $x_3 = 1$; $z = 6$ to the system of equations

$$x_1 + x_2 + x_3 = 2;\ x_1 - x_2 + x_3 = 2 \text{ with}$$
$$\text{Max } Z = 2x_1 + 3x_2 + 4x_3 \text{ is not basic.}$$

Consider the given Linear system of equations.

$$x_1 + x_2 + x_3 = 2; \quad x_1 - x_2 + x_3 = 2 \qquad \text{(I)}$$

Given, $\quad x_2 = 0$

(I)==> $\quad x_1 + x_3 = 2;\quad x_1 + x_3 = 2 \qquad$ (II)

In system (II) columns are not linearly independent. Hence the given solution is not basic.

This contradicts the definition of Basic Solution.
==> $x_1 = 1$; $x_2 = 0$; $x_3 = 1$ is not a Basic Solution.

2.9 CONVEX SET AND EXTREME POINT

Convex set: A set X is convex if for any two points X_1, X_2 in the set, the line segment joining X_1, X_2 is also in the set. (i.e) if X is convex, then $x_1, x_2 \in X \Rightarrow x_1 + (1 - d)\, x_2 \in X;\ 0 \le d \le 1$

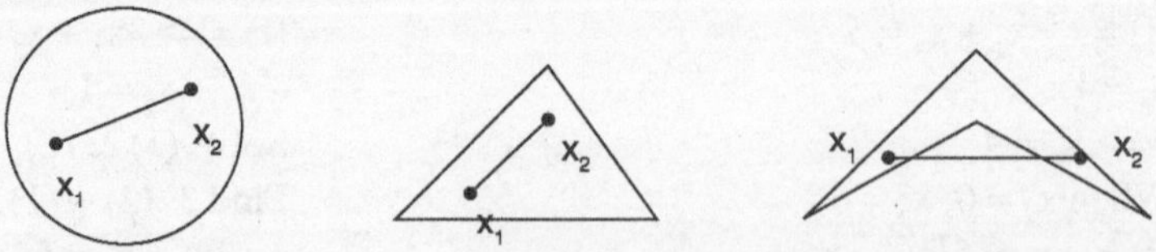

Convex Convex Not convex

Note: By convention, we say that any set containing only one point is a convex set.

Extreme Point (Corner Point)

A point x is an extreme point of a convex set off there do not exist other points x_1, x_2, $x_3 \neq x_2$, in the set such that $$x = \lambda x_2 + (1 - \lambda)x_1, 0 \leq \lambda \leq 1$$

Example: 22

1. In a triangle the three vertices are extreme points.
2. In a circle all boundary points are extreme points.

Remark 1: All boundary points need not be extreme points.

Remark 2: The maximum number of extreme points of a convex set of feasible solution to an LPP is $n_{C_m} = \frac{n!}{m!(n-m)!}$; where 'n' is the number of variables and m be the number of constraints in the problem. Such that $m \leq n$.

2.9.1 Important Points to be Remembered

1. If there is an optimal solution to an LPP then atleast one extreme point of the convex set of feasible solution will be an optimum solution.
2. Every basic feasible solution to an LPP is an extreme point of the convex set of feasible solutions and every extreme point of the convex set of feasible solution is a basic feasible solution.
3. LPP attains its optimum/extremum value at the extreme points.

2.10 GRAPHICAL SOLUTION TO AN LPP

The next step after formulation of an LPP is to solve the problem mathematically to obtain the best possible solution. In this section, a Graphical Method to solve Linear Programming Problem involving only two variables is discussed very clearly using a Flowchart (2.3).

Notations Used

GM - Graphical Method
Pts. - Points
FR - Feasible Region (The set of all feasible solutions)
Max. - Maximum
Min. - Minimum
SM - Simplex Method
LB - Lower Bound
Fun. - Function
NV - Number of Variables
Soln. - Solution
Reqd. - Required
Bounded - Both upper as well as lower bound of the feasible region exist.
Co-pts. - Coordinate Points
Obj. fun. - Objective Function

Example: 23

$$\text{Max } Z = 3x_1 + 9x_2; \quad \text{s.t. } x_1 + x_2 \le 8; \quad x_1 + 2x_2 \le 4; \quad x_1, x_2 \ge 0$$

Here, the given problem consists of exactly two variables and both the variables are positive. So, we can apply the graphical method to get the optimum solution.

Step 1: Consider the constraints with equality symbol

i.e., $x_1 + x_2 = 8$ (1) $\quad$ $x_1 + 2x_2 = 4$ (2)

Step 2: Get the two co-ordinate points for each equation.

Consider the equation (1) $x_1 + x_2 = 8$

put $x_2 = 0 ==> x_1 = 8$; i.e., (8, 0) | put $x_1 = 0 ==> x_2 = 8$ i.e., (0, 8)

Similarly proceeding for the equation (2), we have (4,0) and (0,2).

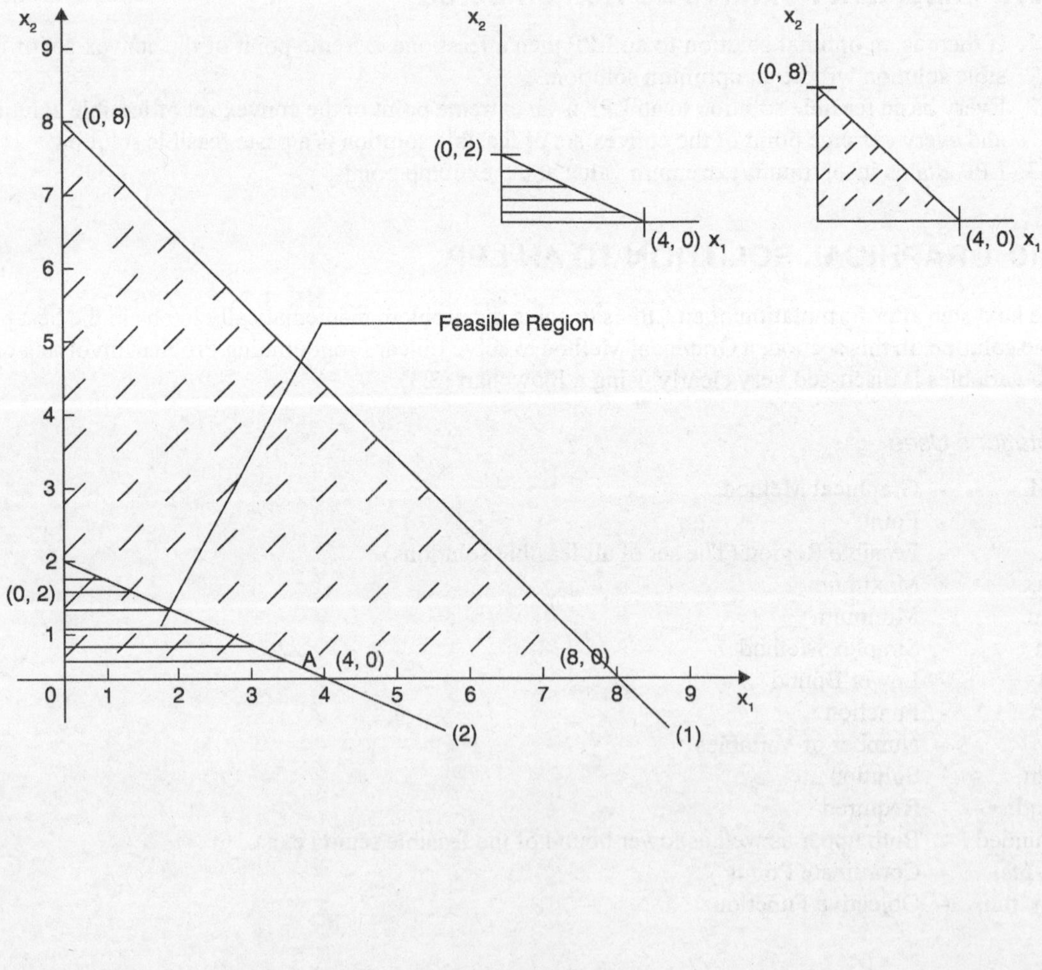

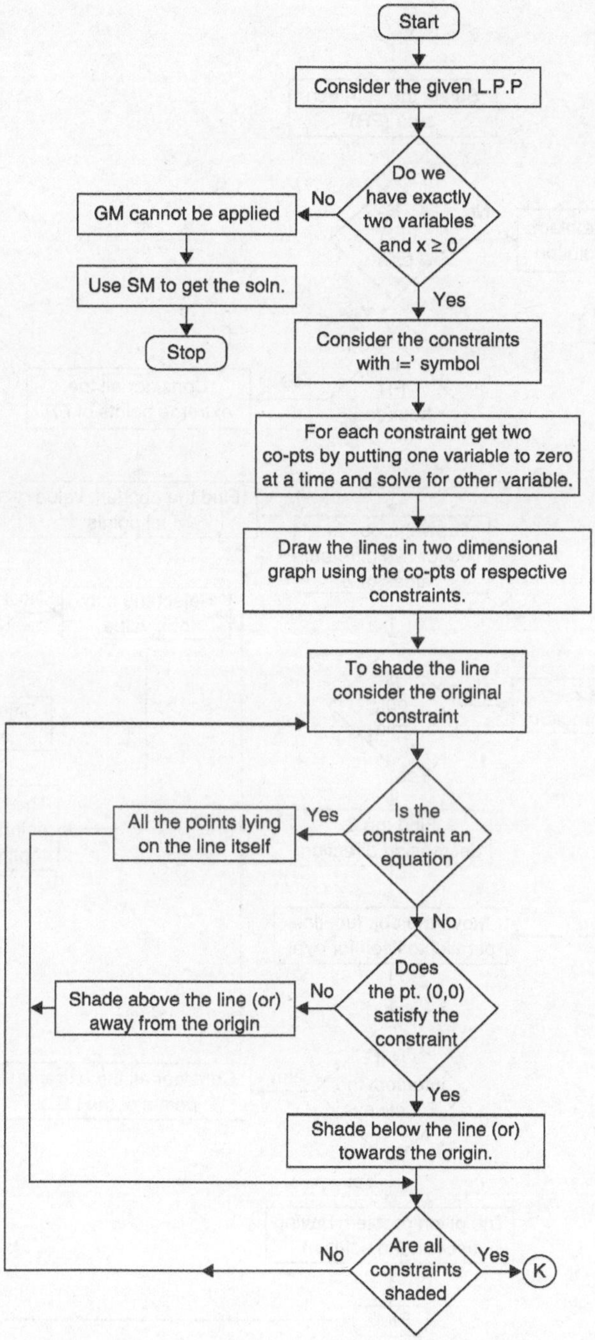

Flowchart 2.3 Graphical method to solve an LPP

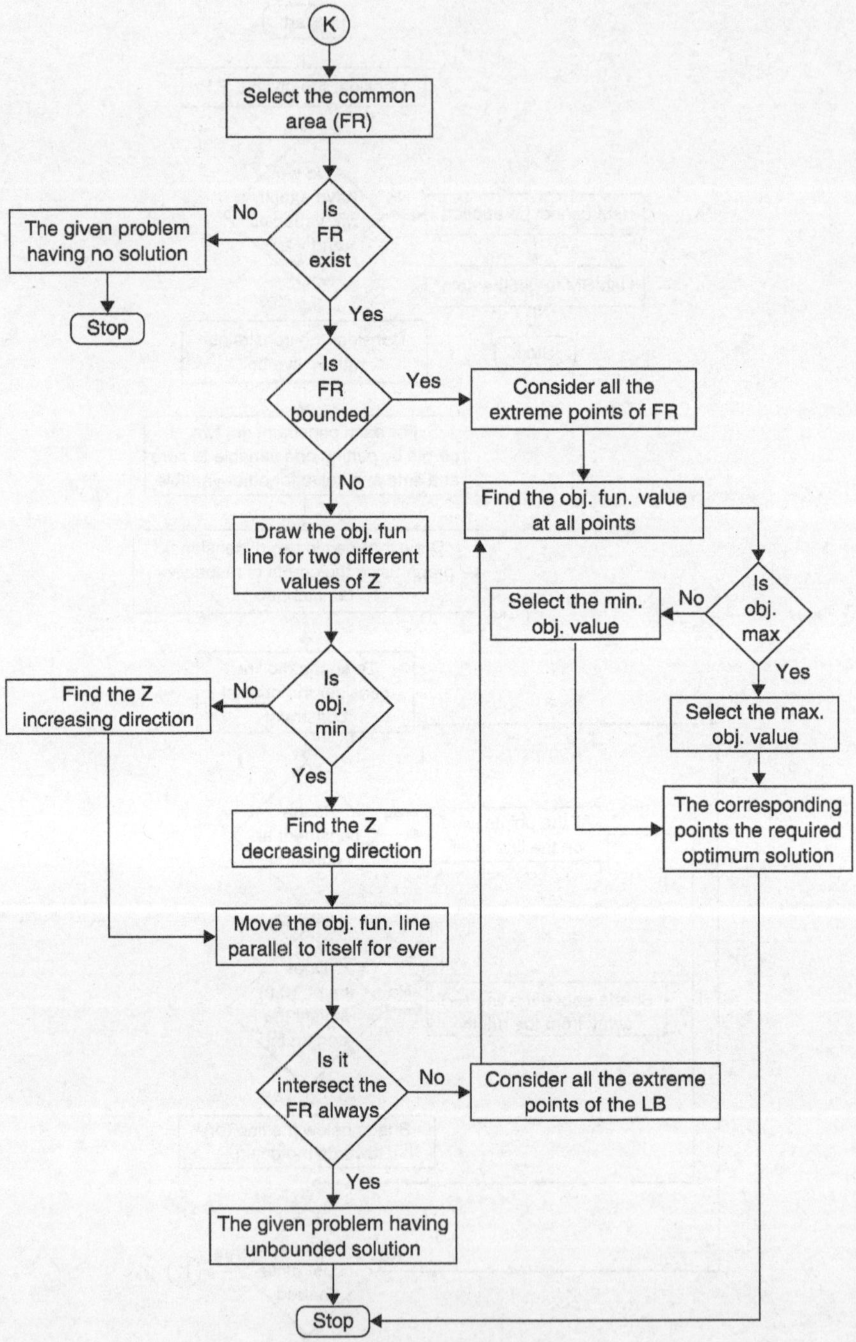

Flowchart 2.3 Continued

Step 3: Using the Co-ordinate points, draw the lines in a two-dimensional graph, and shade the regions.

Step 4: Put $x_1 = 0$ and $x_2 = 0$ in constraint (1)

i.e., $x_1 + x_2 \le 8$ ===> $0 \le 8$;

The constraint is satisfied, shade towards the origin. Also, the 2nd constraint is satisfied when $x_1 = 0$ and $x_2 = 0$ then shade towards the origin.

Step 5: We have a closed intersection area common to both lines, call it OAB.

i.e., Area OAB is the required feasible region.

Step 6: Consider the extreme points of the feasible region. Get the co-ordinates of the extreme points in the feasible region. O (0, 0); A (4, 0) & B (0, 2)

Step 7:

Extreme Points	$Z = 3x_1 + 9x_2$	Max.
0 (0, 0)	$Z_O = 0$	
A (4, 0)	$Z_A = 12$	
B (0, 2)	$Z_B = 18$	18

Optimum $Z = \text{Max}\ \{Z_O, Z_A, Z_B\}$

$= \text{Max}\ \{0, 12, 18\}$

$= 18$

==> at point B, the function Z takes the maximum value.

==> The optimum solution is

$x_1 = 0;\ x_2 = 2$ and Max Z = 18

Example: 24

Min $Z = 2x_1 + 3x_2$; s.t. $x_1 + x_2 \le 4$; $6x_1 + 2x_2 \ge 8$; $x_1+5x_2 \ge 4$; $x_1 \le 3$; $x_2 \le 3$; $x_1, x_2 \ge 0$

Here, the given problem consists of exactly two variables and both the variables are positive. So, we can apply graphical method to get the optimum solution.

Step 1: Consider the constraints with equality symbol.

i.e., $x_1 + x_2 = 4$; (1) $6x_1 + 2x_2 = 8$; (2) $x_1 + 5x_2 = 4$; (3) $x_1 = 3$; (4) $x_2 = 3$ (5)

Step 2: Get the co-ordinate points for each equation.

Consider equation (1)	$x_1 + x_2 = 4$;
put $x_1 = 0$	put $x_2 = 0$
==> $x_2 = 4$	$x_1 = 4$
i.e., (0, 4)	i.e., (4, 0)

Similarly proceeding, we have

for equation (2) : (0, 4) & (4/3, 0)

for equation (3) : (4, 0) & (0, 4/5)
for equation (4) : $x_1 = 3$ implies that whatever be the value of x_2, x_1 is fixed
===> the line is a vertical line at the point $x_1 = 3$
(parallel to x_2-axis at $x_1 = 3$)
for equation (5); $x_2 = 3$
a vertical line at the pt $x_2 = 3$
[parallel to x_1-axis at $x_2 = 3$]

Step 3: Using the co-ordinate points, draw the lines in a two-dimensional graph and shade the regions.

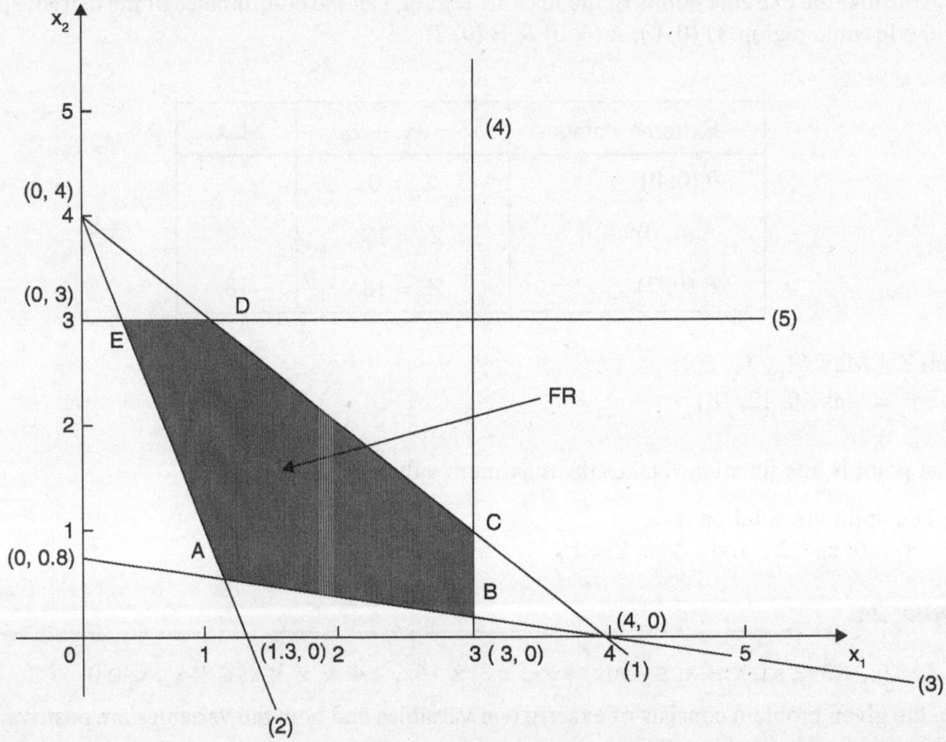

Step 4: Put $x_1 = 0$ & $x_2 = 0$ in all the constraints, constraints (2), (4) and (5) are satisfied then shade the corresponding lines towards origin. And (2) & (3) are not satisfied, then shade the corresponding lines away from the origin.

The common intersection area of all the five lines is ABCDE (a closed feasible region).

Step 5: Consider the extreme points of the feasible region. Get the co-ordinate of the extreme points in the feasible region.

A: Intersection point of the Lines (2) & (3)

Solving equation (2) & (3)

$$6x_1 + 2x_2 = 8; \qquad x_1 + 5x_2 = 4$$
$$(2) \qquad\qquad (3)$$

[Dividing (2) by 2 – multiplying (3) by 3]

$$3x_1 + x_2 = 4$$
$$-3x_1 - 15x_2 = -12$$

$$-14x_2 = -8$$
$$x_2 = 8/14 = 4/7$$

Using in (3)

$$==> x_1 + 5(4/7) = 4$$
$$x_1 + 20/7 = 4$$
$$x_1 = 4 - 20/7 = \frac{28-20}{7} = 8/7$$

(8/7, 4/7)

B: intersection point of the Lines (3) & (4)

$$x_1 + 5x_2 = 4$$
$$x_1 = 3$$
$$==> (3, 0.2)$$

C: intersection point of the Lines (1) & (4)

$$x_1 + x_2 = 4$$
$$x_1 = 3$$
$$==> (3, 1)$$

D: intersection point of the Lines (1) & (5)

$$x_1 + x_2 = 4;\ x_2 = 3 ==> (1, 3)$$

E: intersection point of the Lines (2) & (5)

$$6x_1 + 2x_2 = 8 \quad \text{i.e.,} \quad 3x_1 + x_2 = 4$$
$$x_2 = 3$$
$$==> (1/3, 3)$$

Step 6:

Extreme Points	$Z = 2x_1 + 3x_2$	Min.
A(8/7, 4/7)	$Z_A = 4$	4
B (3, 0.2)	$Z_B = 6.6$	
C (3, 1)	$Z_C = 9$	
D (1, 3)	$Z_D = 11$	
E (1/3, 3)	$Z_E = 9.67$	

Optimum $Z = \text{Min } \{Z_A, Z_B, Z_C, Z_D, Z_E\}$
$= \text{Min } \{4, 6.6, 9, 11, 9.67\}$
$= 4$

==> at point A, the function Z takes the minimum value.
Hence, the optimum solution is $x_1 = 8/7$, $x_2 = 4/7$ and Min Z = 4.

Example: 25

Solve: Max $Z = 4x_1 + 5x_2$ s.t. $x_1 + x_2 \geq 1$; $-2x_1 + x_2 \leq 1$; $4x_1 - 2x_2 \leq 1$; $x_1 x_2 \geq 0$
For the given problem, Graphical method can be applied.

Step 1: Consider the constraints with equality symbol.

$x_1 + x_2 = 1$; (1) $-2x_1 + x_2 = 1$; (2) $4x_1 - 2x_2 = 1$ (3)

Step 2: Get the two co-ordinate points for each equation.
for equation (1); (0, 1) and (1, 0)
for equation (2); (0, 1) and (–1/2, 0)
for equation (3); (0, –1/2) and (1/4, 0)

Step 3: Using the co-ordinate points, draw the lines in a two-dimensional graph and shade the regions.

Step 4: When $x_1 = 0$ and $x_2 = 0$; Constraints (2) & (3) are satisfied then shade towards origin and constraint (1) is not satisfied then shade away from the origin.

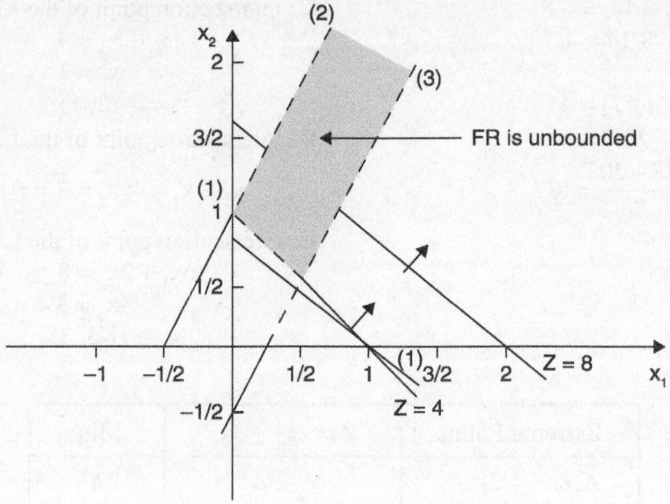

Remark 1: If the line lies in quadrants (2), (3) & (4), then it must be extended into quadrant (1) [dotted line]

Remark 2: Shade only in quadrant (1)

Step 5: The feasible region is not closed, i.e., The feasible region is unbounded.
Select any two arbitrary values for Z; as Z = 8 & Z = 4

i.e.,	$4x_1 + 5x_2 = 8$	|	i.e.,	$4x_1 + 5x_2 = 4$
put	$x_1 = 0; x_2 = 1.6$	|	put	$x_1 = 0; x_2 = 0.8$
put	$x_2 = 0; x_1 = 2$	|	put	$x_2 = 0; x_1 = 1$
	(0, 1.6); (2, 0)	|		(0, 0.8); (1, 0)

represent the Z line when Z = 8 and Z = 4 in the graph using thick line.

Step 6: Clearly the Z increasing direction always intersects the feasible region.
==> The solution to the given problem is unbounded.
i.e., The given LPP is having an unbounded solution.

Example: 26

$$\text{Max } Z = 6x_1 - 2x_2 \quad \text{s.t. } 2x_1 - x_2 \leq 0; \quad x_1 \leq 4; \quad x_1, x_2 \geq 0$$

For the given problem, Graphical method can be applied.

Step 1: Consider the constraints with equality symbol.

$2x_1 - x_2 = 0$ (1) $\qquad x_1 = 4$ (2)

Step 2: Get the two co-ordinate points for each equation.
for equation (1): (0, 0); (1, 2)
for equation (2): vertical line at the point $x_1 = 4$.

Step 3: Using the co-ordinate points, draw the lines in a two dimensional graph and shade the regions.

Step 4: When $x_1 = 0$ and $x_2 = 0$ both the constraints are satisfied. Then shade towards the origin for both lines.

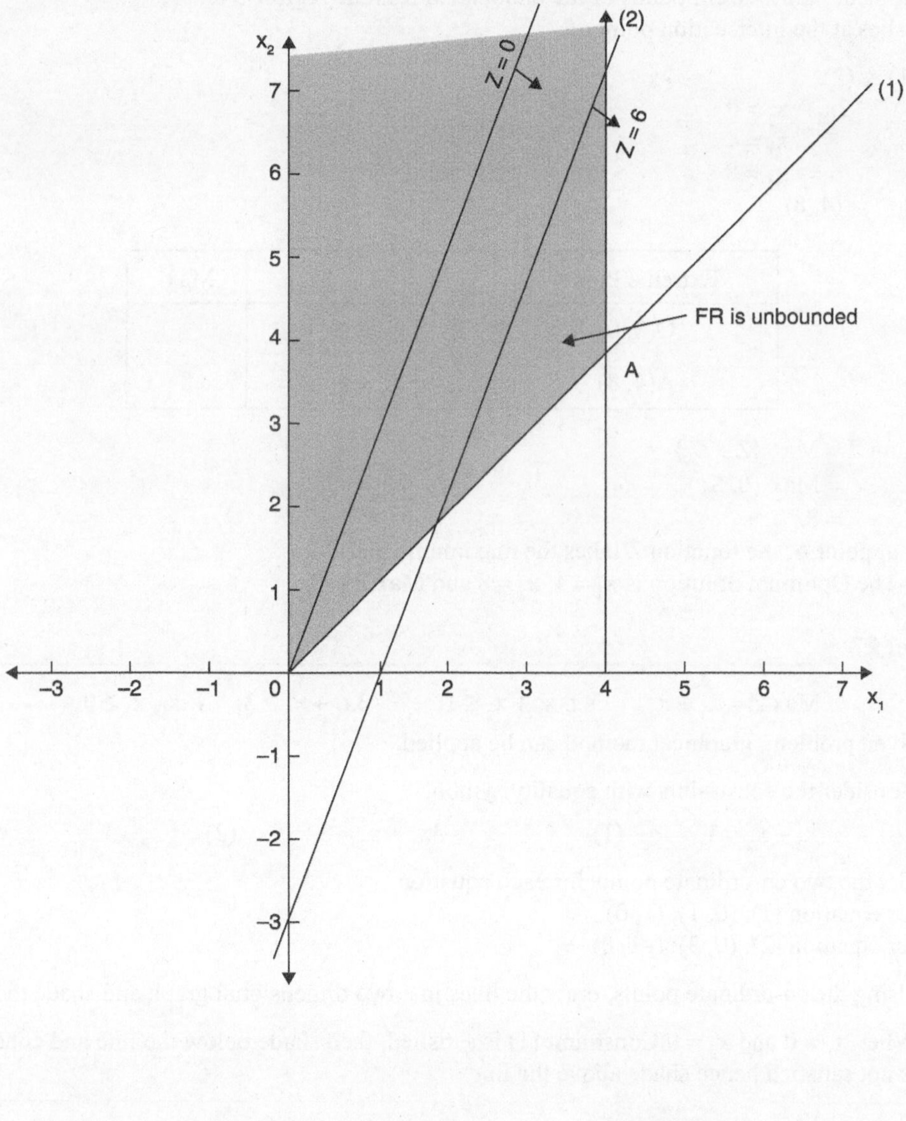

Step 5: Since the feasible region is not closed, select any two arbitrary values for Z, as Z = 0 and Z = 6

Z = 0	Z = 6
	$6x_1 - 2x_2 = 6$
$\Rightarrow 6x_1 - 2x_2 = 0$	$3x_1 - x_2 = 3$
$3x_1 - x_2 = 0$	
(0, 0); (1, 3)	(0, –3); (1, 0)

Represent the 2 lines when Z = 0 and Z = 6 in the graph using thick line.

Step 6: Clearly, the Z increasing direction does not intersect the feasible region, the solution to the LPP is Bounded.

Step 7: Consider the Lower bounds of the unbounded feasible region O and A.
A lies at the intersection point of

(1) & (2)

$2x_1 - x_2 = 0$

$x_1 = 4$

$\Rightarrow x_2 = 8$

(4, 8)

Extreme Points	$Z = 6x_1 - 2x_2$	Max.
O(0, 0)	$Z_O = 0$	
A(4, 8)	$Z_A = 8$	8

Optimum $Z = \text{Max}\ \{Z_0, Z_A\}$

$= \text{Max}\ \{0, 8\}$

$= 8.$

==> at point A, the function Z takes the maximum value.

==> The Optimum Solution is $x_1 = 4$; $x_2 = 8$ and Max Z = 8.

Example: 27

$$\text{Max } Z = x_1 + x_2 \quad \text{s.t. } x_1 + x_2 \le 1; \quad -3x_1 + x_2 \ge 3; \quad x_1, x_2 \ge 0$$

For the given problem, graphical method can be applied.

Step 1: Consider the constraints with equality symbol.

$x_1 + x_2 = 1$ (1) $-3x_1 + x_2 = 3$ (2)

Step 2: Get the two co-ordinate points for each equation.
for equation (1): (0, 1); (1, 0)
for equation (2): (0, 3); (–1, 0)

Step 3: Using the co-ordinate points, draw the lines in a two dimensional graph and shade the regions.

Step 4: When $x_1 = 0$ and $x_2 = 0$ Constraint (1) is satisfied, then shade below the line and constraint (2) is not satisfied hence shade above the line.

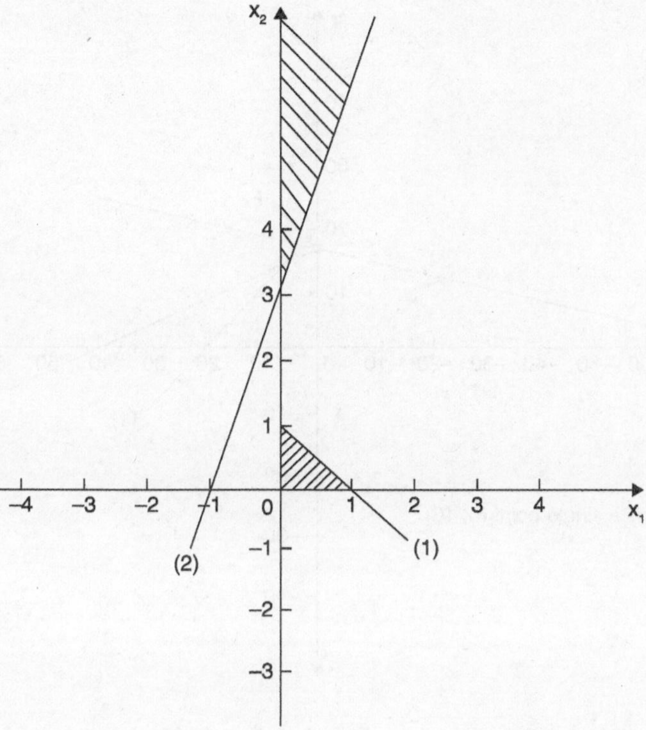

Step 5: There is no common region for both the lines.

==> There is no solution for the given LPP.

Example: 28

$$\text{Max } Z = -0.1x_1 + 0.5x_2$$

$$\text{s.t.} \quad x_1 + x_2 \le 20; \quad 2x_1 + 5x_2 \le 80; \quad -0.1x_1 + 0.5x_2 = 8; \quad x_1, x_2 \ge 0$$

Graphical Method can be applied for the given problem.

Step 1: Consider the constraints with equality symbol

$$x_1 + x_2 \le 20; \quad (1) \qquad 2x_1 + 5x_2 \le 80; \quad (2) \qquad -0.1x_1 + 0.5x_2 = 8; \quad (3)$$

Step 2: Get the two co-ordinate points for each equation.

(1) => (0, 20); (20, 0)
(2) => (0, 16); (40, 0)
(3) => (0, 16); (–80, 0)

Step 3: Using the co-ordinate points, draw the lines in a two dimension! graph and shade the regions.

Step 4: When $x_1 = 0$ & $x_2 = 0$ constraints (1) & (2) are satisfied, so for both the constraints shade below the line, and since the constraint (3) is an equation, all the points lying on the line belong to the feasible region.

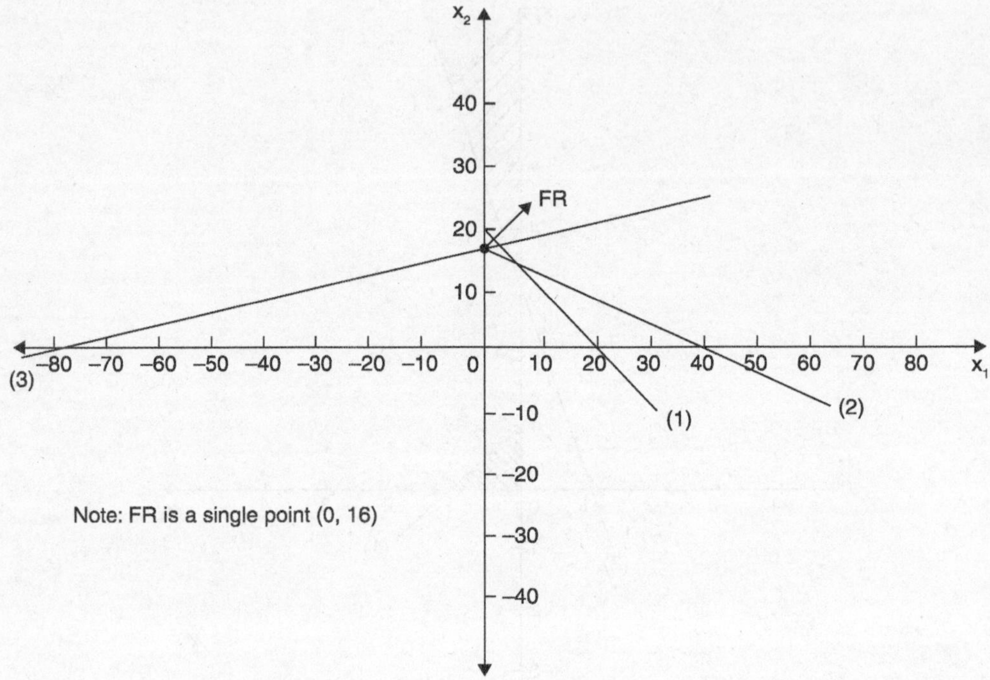

Step 5: Here the feasible region exists, i.e., a single point (0, 16). That point is the required optimum solution. Max Z = 8, $x_1 = 0$; $x_2 = 16$

Example: 29

A sick patient has been advised by his doctor that his diet must contain atleast 4000 units of vitamins, 50 units of minerals and 1400 units of Calories. Two foods, F_1 and F_2 are available at a cost of ₹4 and ₹3 per unit respectively. One unit of food F_1, contains 200 units of vitamins, 1 unit of mineral and 40 units of Calories. One unit of food F_2 contains 100 units of vitamins, 2 units of minerals and 40 units of calories. Formulate the above problem as an LPP and solve using graphical Method.

Let x_1 and x_2 stand for the number of units of foods to be purchased in F_1 and F_2 respectively. The objective is to minimize the cost.

Tabulate all the information as follows:

Food	No. of Units	Cost/U (₹)	Vitamin	Mineral	Calories
F_1	x_1	4	200	1	40
F_2	x_2	3	100	2	40
		Min.	4000	50	1400

$$\text{Minimize } Z = 4x_1 + 3x_2$$

$$\begin{aligned}
\text{s.t.}\quad 200x_1 + 100x_2 &\geq 4000 \quad \text{(vitamin)}\\
x_1 + 2x_2 &\geq 50 \quad \text{(mineral)}\\
40x_1 + 40x_2 &\geq 1400 \quad \text{(calories)}\\
x_1, x_2 &\geq 0
\end{aligned}$$

Step 1: As stated in the previous examples, get the coordinate points based on the constraints.

$$\begin{aligned}
200x_1 + 100x_2 &= 4000 \quad \Rightarrow \quad (0, 40); \quad (20, 0)\\
x_1 + 2x_2 &= 50 \quad \Rightarrow \quad (0, 25); \quad (50, 0)\\
40x_1 + 40x_2 &= 1400 \quad \Rightarrow \quad (0, 35); \quad (35, 0)
\end{aligned}$$

Step 2: Using the co-ordinate points, draw the lines in a two-dimensional graph and shade the regions. Clearly, the feasible region is unbounded. Select two arbitrary values for z and draw the two z lines.

Let $z = 48$; $z = 24$

$z = 48 \Rightarrow 4x_1 + 3x_2 = 48$	$z = 24 \Rightarrow 4x_1 + 3x_2 = 24$
(0, 16); (12, 0)	(0, 8); (6, 0)

Since the z-decreasing direction does not intersect the feasible region always, the given problem is said to have a feasible solution at the Lower Bounds (LB). The LB extreme points are A, B, C & D.

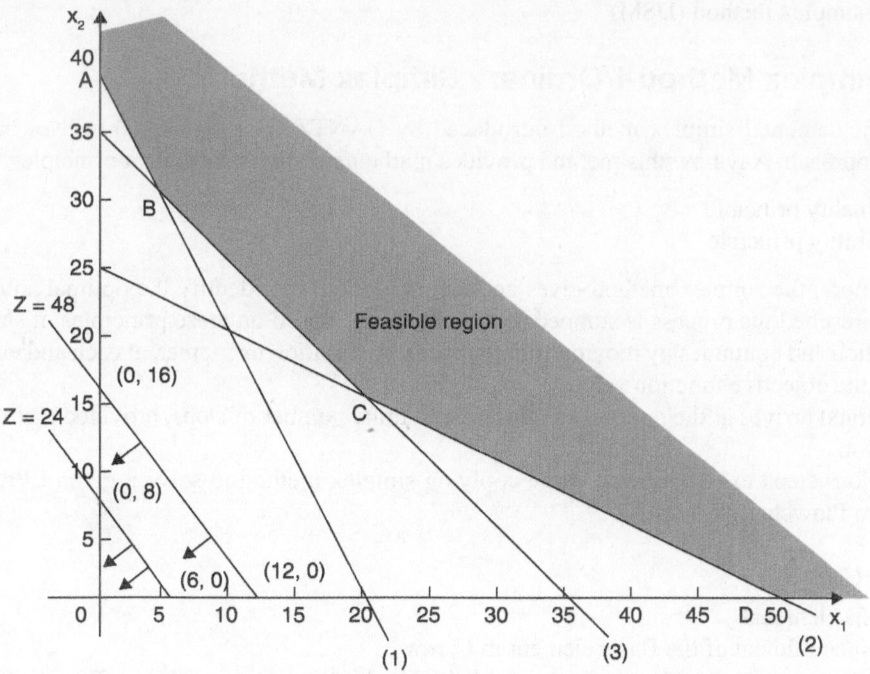

Extreme Point	$Z = 4x_1 + 3x_2$	Min.
A (0, 40)	$Z_A = 120$	
B (5, 30)	$Z_B = 110$	110*
C (20, 15)	$Z_C = 125$	
D (50, 0)	$Z_D = 200$	

Hence the optimum occurs at B. in order to minimize the cost of acquiring the minimum required vitamins, the patient should purchase 5 units of food F_1 and 30 units of food F_2. The minimum cost is ₹110.

2.11 SIMPLEX METHODS

Whenever the graphical method fails to solve an LPP, then we seek the help of simplex method. We have four different methods of this nature. They are

1. Simplex method-I (or) Ordinary simplex method
2. Big-M Method (or) Penalty method (or) Charnes Penalty method
3. Two-phase simplex method
4. Dual simplex method (DSM)

2.11.1 Simplex Method-I/Ordinary Simplex Method

The very fundamental simplex method introduced by DANTZIG (1951, 1963) utilises basically the algebraic approach. Any how this method provides mathematically two efficient principles,

1. Optimality principle
2. Feasibility principle.

Furthermore, the simplex method gives an algebraic criteria to identify the optimal solution. Once it has been reached the process is stopped (optimality test). Based on these principles, it starts with an initial solution and continues by moving from one feasible solution to another, at each and every step by improving the objective function value.

This method arrives at the optimal solution after a finite number of steps, provided that the problem has a solution.

The various steps to be followed while applying simplex method to solve a given LPP are clearly explained in Flowchart (2.4).

Notations Used

B - Basis elements
C_B - Cost coefficient of the Basis element in C_j row
C_j - Coefficient of the corresponding variable in the objective function. (j = 1, 2, ….. n)
x_j - Variable (for j = 1, 2, …. n)
X_B - Solution to the Basis elements.

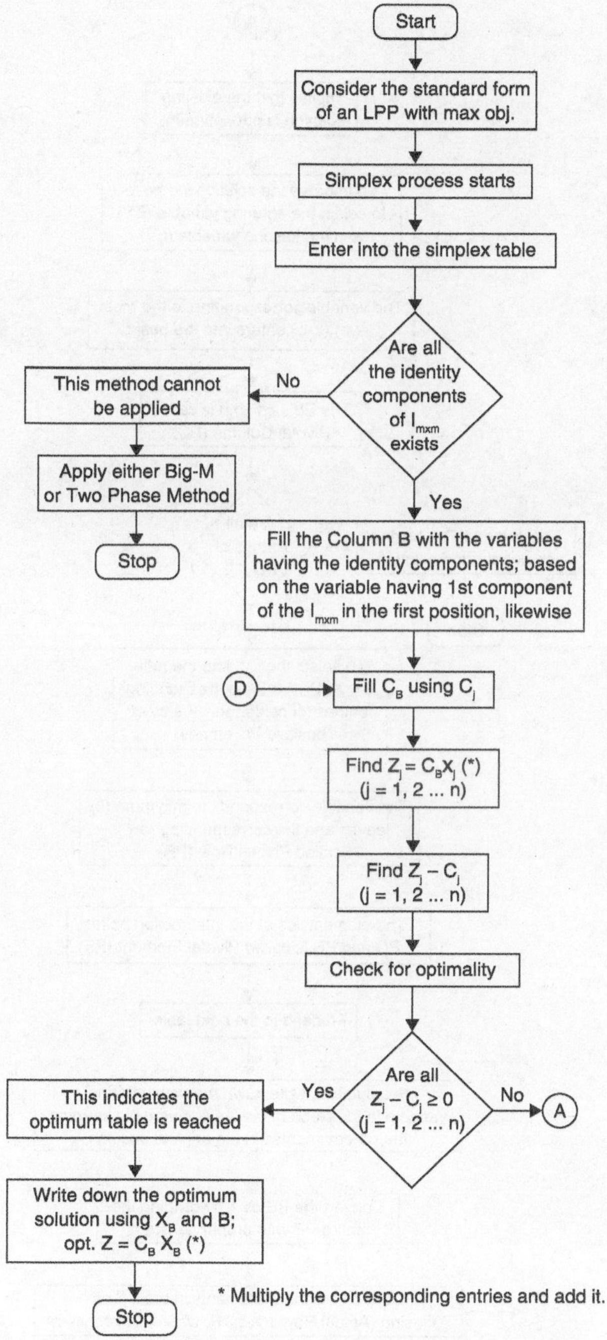

Flowchart 2.4 Simplex method-I

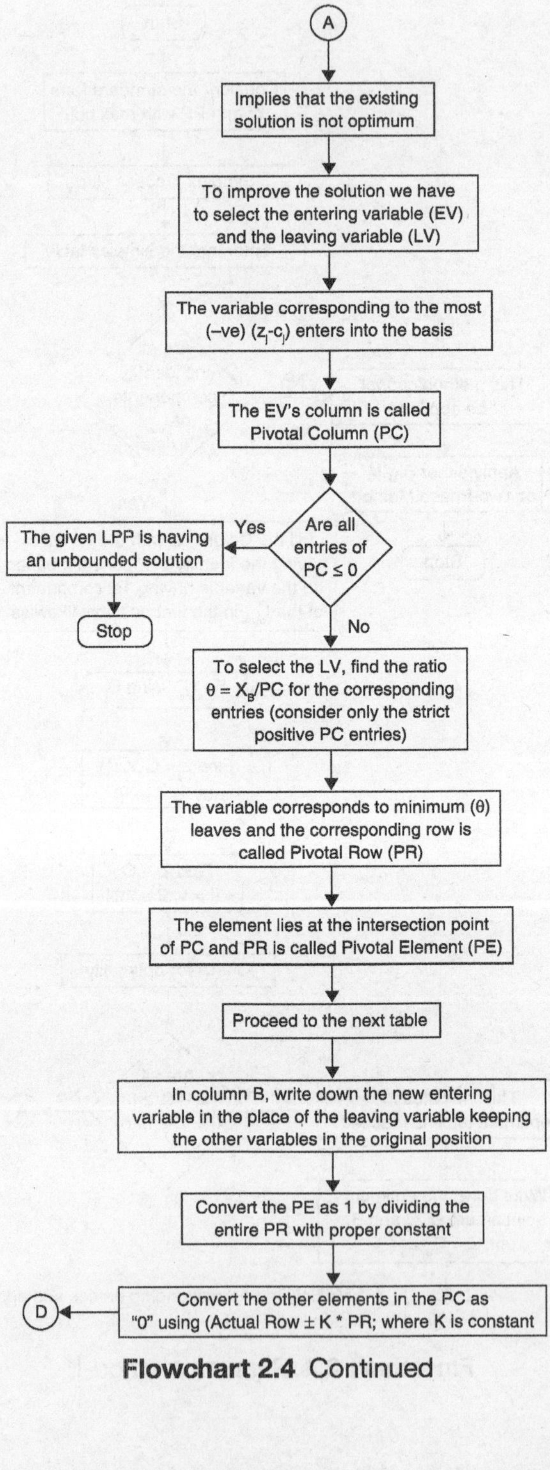

Flowchart 2.4 Continued

Z_j – $C_B X_j$ (j = 1, 2,...... n) (Multiply the corresponding entries & add it)
$Z_j - C_j$ – $C_B X_j - C_j$ (j = 1, 2 .... n) [called Net evaluation]
PC – Pivotal Column
PR – Pivotal Row
PE – Pivotal Element
θ – X_B/PC

The structure of a simplex table:

C_j	C_1, C_2 .... C_n	x_B	B	C_B	θ
x_i	x_1, x_2 .... x_n				
	Co-efficient matrix of the constraints	RHS values			
Z_j					
$Z_j - C_j$					

Example: 30

Solve the following LPP by simplex method

$$\text{Max } Z = 2x_1 + 4x_2;$$
$$\text{s.t.}\quad 2x_1 + 3x_2 \leq 48;$$
$$x_1 + 3x_2 \leq 42;$$
$$x_1 + x_2 \leq 21;$$
$$x_1, x_2 \geq 0$$

Step 1: Convert the given LPP into standard form with maximization objective (refer section 2.6). Here all the conditions are satisfied. Introduce the necessary slack variables (since all constraints having "≤" inequality)

Standard LPP

$$\text{Max } Z = 2x_1 + 4x_2 + 0x_3 + 0x_4 + 0x_5$$
$$\text{s.t.}\quad 2x_1 + 3x_2 + x_3 = 48;$$
$$x_1 + 3x_2 + x_4 = 42;$$
$$x_1 + x_2 + x_5 = 21;$$
$$x_1, x_2, x_3, x_4, x_5 \geq 0$$
(x_3, x_4 and x_5 are the slack variables)

Step 2: Enter into the simplex table. Since, the table-1 consists of the required three identity components of I_{3x3} (because number of constraints = 3) we can apply simplex method-I.

Step 3: Since x_3, x_4 and x_5 contains the 1st, 2nd and 3rd components ok I_{3x3}, write the variables x_3, x_4 & x_5 in the 1st position, 2nd position and 3rd position respectively in column B.

Step 4:

1. Fill C_B;
2. Evaluate $Z_j = C_B x_j$; (for j = 1, 2 .... n)
   Example: $Z_1 = C_B x_1 = 0 \times 2 + 0 \times 1 + 0 \times 1 = 0$
   $Z_2 = C_B x_2 = 0 \times 3 + 0 \times 3 + 0 \times 1 = 0$
3. Evaluate $(Z_j - C_j)$; (for j = 1, 2, ... n)
   Since, some of the $(Z_j - C_j)$ entries are negative,
   ⟹ The existing basic feasible solution in Table-1 (0, 0, 48, 42, 21) is not optimum.
   ⟹ Improvement is required.

Step 5: Selection of an entering variable
Consider the non-basic variables $(Z_j - C_j)$ value of table 1

Non-basic Variables	$(Z_j - C_j)$ Value	$-(Z_j - C_j)$	Unit Increment in the Z Value	Max.
x_1	–2	2	2	
x_2	–4	4	4	4

Note: Ties may be broken in arbitrary fashion.

We choose a non-basic variable as an entering variable based on most (–ve) $(Z_j - C_j)$ entry. Here the entering variable is x_2.

Alternatively, the most (–ve) entry of $(Z_j - C_j)$ is (–4) which corresponds to x_2; so, x_2 is the new basic variable which enters into the Basis. Then the column x_2 is called PC.

Step 6: Selection of a leaving variable

Since, we have decided to increase x_2, as x_2 increases, the values of the current basic variables (x_3, x_4, x_5) will change. This means that increasing x_2 may cause a basic variable to become negative. With this in mind, we see that increasing x_2 (while holding $x_j = 0$) changes the values of the current set of basic variables.

From table-1 we have, And $x_3, x_4, x_5 \geq 0$ $\Rightarrow 48 - 3x_2 \geq 0 \Rightarrow x_2 \leq 16$

$3x_2 + x_3 = 48 \Rightarrow x_3 = 48 - 3x_2$ $\quad 42 - 3x_2 \geq 0 \Rightarrow x_2 \leq 14$

$3x_2 + x_4 = 42 \Rightarrow x_4 = 42 - 3x_2$ $\quad 21 - x_2 \geq 0 \Rightarrow x_2 \leq 21$

$x_2 + x_5 = 21 \Rightarrow x_5 = 21 - x_2$

i.e., x_2 is min {16, 14, 21} = 14

Alternatively, Find the Ratio $\theta = X_B/PC$ for the corresponding strict (+ve) entries of PC. Here, $\theta = \min\{48/3, 42/3, 21/1\} = \min\{16, 14, 21\} = 14$ which corresponds to the second row, call it as PR; then the variable x_4 leaves. Then the element lying in (2nd row; 2nd column) is the Pivotal Element (PE).

Step 7: Proceed to the next table in which replace x_4 as x_2. Then convert the PE (which is 3) as 1 by dividing the entire 2nd row by 3 and call it as PR and convert the remaining elements of PC, (3 of Row1 and 1 of Row3) as Zero by using the row transformation technique as follows.

$R_1 ==> R_1 - 3PR$ and $R_3 ==> R_3 - PR$ and Goto Step(4) repeat the process till all $Z_j - C_j \geq 0$ (for j = 1, 2 … n)

C_j	2	4	0	0	0	X_B	B	C_B	θ
x_j	x_1	x_2	x_3	x_4	x_5				
	2	3	1	0	0	48	x_3	0	48/3 = 16
	1	(3)	0	1	0	42	x_4	0	42/3 = 14 <= min PR
	1	1	0	0	1	21	x_5	0	21/1 = 21
Z_j	0	0	0	0	0				
$Z_j - C_j$	–2	–4↑ PC	0	0	0				
R_1 – 3PR	(1) PE	0	1	–1	0	6	x_3	0	6/1 = 6 <= min PR
PR	1/3	1	0	1/3	0	14	x_2	4	14/0.33 = 42
R_3 – PR	2/3	0	0	–1/3	1	7	x_5	0	7/0.67 = 10.5
Z_j	4/3	4	0	4/3	0				
$Z_j - C_j$	–2/3↑ PC	0	0	4/3	0				
PR	1	0	1	–1	0	6	x_1	2	
R_2 – (1/3)PR	0	1	–1/3	2/3	0	12	x_2	4	
R_3 – (2/3)PR	0	0	– 2/3	1/3	1	3	x_5	0	
Z_j	2	4	2/3	2/3	0				
$Z_j - C_j$	0	0	2/3	2/3	0	60			

Since, all $Z_j - C_j$ Values are ≥ 0

==> The optimum table is reached. Hence the optimum solution is

$$x_1 = 6, x_2 = 12, x_3 = 0, x_4 = 0, x_5 = 3 \text{ and Max } Z = 60$$

Remark 1: All the variables presented in column B are called basic variables or basis elements.

Remark 2: The variables which are not presented in column B are called non-basic variables whose values are assumed to be zero.

***Example:* 31**

Use simplex method to solve the following LPP

$$\text{Max } Z = 4x_1+10x_2; \quad \text{s.t. } 2x_1 + x_2 \leq 50; \quad 2x_1 + 5x_2 \leq 100; \quad 2x_1 + 3x_2 \leq 90;$$

$$x_1, x_2 \geq 0$$

Introducing the slack variables x_3, x_4, & x_5 we have,

$$\text{Max } Z = 4x_1 + 10x_2 + 0x_3 + 0x_4 + 0x_5; \quad \text{s.t. } 2x_1 + x_2 + x_3 = 50; \quad 2x_1 + 5x_2 + x_4 = 100;$$
$$2x_1 + 3x_2 + x_5 = 90; \quad x_1, x_2, x_3, x_4, x_5 \geq 0$$

C_j	4	10	0	0	0	X_B	B	C_B	θ
x_j	x_1	x_2	x_3	x_4	x_5				
	2	1	1	0	0	50	x_3	0	50/1 = 50
	2	(5) PE	0	1	0	100	x_4	0	100/5 = 20 <= min PR
	2	3	0	0	1	90	x_5	0	90/3 = 30
Z_j	0	0	0	0	0				
$Z_j - C_j$	–4	–10↑ PC	0	0	0				
R_1 – PR	8/5	0	1	–1/5	0	30	x_3	0	
PR	2/5	1	0	1/5	0	20	x_2	10	
R_3 – 3PR	4/5	0	0	–3/5	1	30	x_5	0	
Z_j	4	10	0	2	0				
$Z_j - C_j$	0	0	0	2	0	200			

Since, all $Z_j - C_j \geq 0$; ==> The optimum table is reached.
Hence the optimum solution is $x_1 = 0$; $x_2 = 20$; $x_3 = 30$; $x_4 = 0$; $x_5 = 30$ and Max Z = 200

Example: 32

Using simplex method solve the problem

$$\text{Max } Z = 4x_1 + x_2 + 3x_3 + 5x_4;$$
$$\text{s.t. } 4x_1 - 6x_2 - 5x_3 + 4x_4 \geq -20;$$
$$3x_1 - 2x_2 + 4x_3 + x_4 \leq 10;$$
$$8x_1 - 3x_2 + 3x_3 + 2x_4 \leq 20;$$
$$x_1, x_2, x_3, x_4, \geq 0$$

In the given LPP, the RHS of constraint (1) is (–ve). So, multiply both sides of constraint (1) using (–1). Then the constraint (1) reduced to

$$-4x_1 + 6x_2 + 5x_3 - 4x_4 \leq 20$$

Introducing the slack variables x_5, x_6 & x_7 we have,

$$\text{Max } Z = 4x_1 + x_2 + 3x_3 + 5x_4 + 0x_5 + 0x_6 + 0x_7$$
$$\text{s.t. } -4x_1 + 6x_2 + 5x_3 - 4x_4 + x_5 = 20;$$
$$3x_1 - 2x_2 + 4x_3 + x_4 + x_6 = 10;$$
$$8x_1 - 3x_2 + 3x_3 + 2x_4 + x_7 = 20;$$
$$x_j \geq 0; \text{ (for } j = 1, 2, \ldots, 7)$$

C_j	4	1	3	5	0	0	0	X_B	B	C_B	θ
x_j	x_1	x_2	x_3	x_4	x_5	x_6	x_7				
	−4	6	5	−4	1	0	0	20	x_5	0	--
	3	−2	4	1	0	1	0	10	x_6	0	10/1 = 10
	8	−3	3	(2) PE	0	0	1	20	x_7	0	20/2 = 10 <= min PR
Z_j	0	0	0	0	0	0	0				
$Z_j - C_j$	−4	−1	−3	−5↑PC	0	0	0				
R_1 + 4PR	12	0	11	0	1	0	2	60	x_5	0	
R_2 − PR	−1	−1/2	5/2	0	0	1	−1/2	0	x_6	0	
PR	4	−3/2	3/2	1	0	0	1/2	10	x_4	5	
Z_j	20	−15/2	15/2	5	0	0	5/2				
$Z_j - C_j$	16	−17/2↑PC	9/2	0	0	0	5/2				

Since, all the entries in the PC are "≤ 0".
==> The given LPP has an unbounded solution.

Remark 1: In table (1), we have a tie in selection of min {θ}, whenever a tie occurs, we can select any one. This arbitrary selection may lead to CYCLING, i.e., the simplex process will become endless.

To avoid this kind of endless process, we have a method for breaking the tie in selection of min {θ}. Consider PC and 1st component of I_{3x3}.

B	PC	1st Component of I_{3x3}	Ratio
x_5	−4	1	*
x_6	1	0	0/1 = 0 } again tie occurs
x_7	2	0	0/2 = 0 }

Remark 2: Find the ratio only for the corresponding tied up variables. Here the tied up variables are x_6 & x_7. Then select the minimum.

Since, again tie occurs consider the 2nd component of I_{3x3}.

B	PC	2nd Component of I_{3x3}	Ratio
x_5	–4	0	*
x_6	1	1	1/1 = 1
x_7	2	0	0/2 = 0 <== min

The minimum value occurs, corresponding to x_7. Implies that, x_7 leaves the basis.

Example: 33

$$\text{Min } Z = x - 3y + 2z; \quad \text{s.t. } 3x - y + 2z \le 7; \quad -2x + 4y \le 12; \quad -4x + 3y + 8z \le 10; \quad x, y, z \ge 0$$

In the given LPP, the objective is minimization. So, convert into maximization objective. For which we have to multiply the given objective function by (–1).

$$\text{Max } Z^* = -x + 3y - 2z$$

Apply the simplex method with the actual constraints given. Then the optimum solution is max $Z^* = -11$; $x = 4$; $y = 5$ and $z = 0$. Hence, the optimum solution to the given problem is

$$\text{min } Z = -(-11) = 11; \quad x = 4; \quad y = 5 \text{ and } z = 0$$

****Example: 34***

$\text{Max } Z = 3x_1 + 5x_2 + 4x_3$

s.t. $2x_1 + 3x_2 \le 8$; $2x_1 + 5x_2 \le 10$; $3x_1 + 2x_2 + 4x_3 \le 15$; $x_1, x_2, x_3 \ge 0$

Introducing the slack variables x_4, x_5 and x_6, we have

$\text{Max } Z = 3x_1 + 5x_2 + 4x_3 + 0x_4 + 0x_5 + 0x_6$;

s.t. $2x_1 + 3x_2 + x_4 = 8$; $2x_1 + 5x_2 + x_5 = 10$; $3x_1 + 2x_2 + 4x_3 + x_6 = 15$; $x_j \ge 0$;
(for j = 1, 2, 3 …, 6)

C_j	3	5	4	0	0	0	X_B	B	C_B	θ
x_j	x_1	x_2	x_3	x_4	x_5	x_6				
	2	3	0	1	0	0	8	x_4	0	8/3 = 2.67
	2	(5) PE	0	0	1	0	10	x_5	0	10/5 = 2 <= min PR
	3	2	4	0	0	1	15	x_6	0	15/2 = 7.5
Z_j	0	0	0	0	0	0				
$Z_j - C_j$	–3	–5↑ PC	–4	0	0	0				

$R_1 - 3PR$	4/5	0	0	1	–3/5	0	2	x_4	0	--
PR	2/5	1	0	0	1/5	0	2	x_2	5	--
$R_3 - 2PR$	11/5	0	(4) PE	0	–2/5	1	11	x_6	0	11/4 <= min PR
Z_j	2	5	0	0	1	0				
$Z_j - C_j$	–1	0	↑	0	1	0				
R_1	4/5	0	0	1	–3/5	0	2	x_4	0	
R_2	2/5	1	0	0	1/5	0	2	x_2	5	
PR	11/20	0	1	0	–1/10	1/4	11/4	x_3	4	
Z_j	21/5	5	4	0	3/5	1				
$Z_j - C_j$	6/5	0	0	0	3/5	1	21			

Since, all $z_j - c_j \geq 0$; ==> The optimum table is reached. Hence the optimum solution is $x_1 = 0$; $x_2 = 2$; $x_3 = 11/4$; $x_4 = 2$; $x_5 = 0$; $x_6 = 0$; and max Z = 21

Remark: In the 2nd table the non-pivotal elements are zero. So, the next table rewrite the entries as it is.

Example: 35

$$\text{Min } Z = x_1 - x_2 + x_3 + x_4 + x_5 - x_6 \quad \text{s.t. } x_1 + x_4 + 6x_6 = 9; \quad 3x_1 + x_2 - 4x_3 + 2x_6 = 2;$$
$$x_1 + 2x_3 + x_5 + 2x_6 = 6; \; x_j \geq 0 \text{ (for } j = 1, 2, \ldots 6)$$

Converting the minimization objective into Maximization objective.

$$\text{Min } Z = -\text{Max } Z^*$$
$$\text{Max } Z^* = -x_1 + x_2 - x_3 - x_4 - x_5 + x_6$$

Here there is no necessity to introduce the slack variables, because all the given constraints are equations.

C_j	–1	1	–1	–1	–1	1	X_B	B	C_B	θ
x_j	x_1	x_2	x_3	x_4	x_5	x_6				
	1	0	0	1	0	6	9	x_4	–1	9/6 = 3/2
	3	1	–4	0	0	(2) PE	2	x_2	1	2/2 = 1 <= min PR
	1	0	2	0	1	2	6	x_5	–1	6/2 = 3
Z_j	1	1	–6	–1	–1	–6				
$Z_j - C_j$	2	0	–5	0	0	–7↑ PC				

$R_1 - 6PR$	–8	–3	(12) PE	1	0	0	3	x_4	–1	3/12 = 1/4 <= min PR
PR	3/2	1/2	–2	0	0	1	1	x_6	1	--
$R_3 - 2PR$	–2	–1	6	0	1	0	4	x_5	–1	4/6 = 2/3
Z_j	23/2	9/2	–20	–1	–1	1				
$Z_j - C_j$	25/2	7/2	–19↑ PC	0	0	0				
PR	–2/3	–1/4	1	1/12	0	0	1/4	x_3	–1	--
$R_3 + 2PR$	1/6	0	0	1/6	0	1	3/2	x_6	1	--
$R_3 - 6PR$	2	(1/2) PE	0	–1/2	1	0	5/2	x_5	–1	5 <= min PR
Z_j	–7/6	–1/4	–1	7/12	–1	1				
$Z_j - C_j$	–1/6	–5/4↑ PC	0	19/12	0	0				
$R_1 + (1/4)PR$	1/3	0	1	–1/6	1/2	0	3/2	x_3	–1	
R_2	1/6	0	0	1/6	0	1	3/2	x_6	1	
PR	4	1	0	–1	2	0	5	x_2	1	
Z_j	23/6	1	–1	–2/3	3/2	1				
$Z_j - C_j$	29/6	0	0	1/3	5/2	0	5			

Since, all $Z_j - C_j \geq 0$; the optimum table is reached.

==> The optimum solution is

$$x_1 = 0;\ x_2 = 5;\ x_3 = 3/2,\ x_4 = 0;\ x_5 = 0;\ x_6 = 3/2 \text{ and max } Z^* = 5$$

$$\min Z = -\max Z^* = -5$$

$$\min Z = -5$$

2.12 PENALTY METHOD/BIG-M METHOD/CHARNES METHOD

Consider the standard form of the given LPP. If all the required number of "m" identity components (if the LPP having m constraints) exists then we can apply simplex method-I. If not we have to apply penalty method. Introduce the required number of identity components under new variable names called artificial variables.

Add these variables in the LHS of the respective constraints. It is very clear that such artificial variables have no physical meaning with respect to the original problem given (so we call it as artificial), the process will be adequate only if we drive these variables to be zero when the optimality is reached.

To do the above stated things A. CHARNES has suggested a method to penalize the artificial variable in the maximization objective function. That is, each artificial variable will be charged a high penalty of M (where M is a very large positive constant). Then solve it using simplex process.

The various steps to be followed while applying penalty method are clearly explained in Flowchart (2.5).

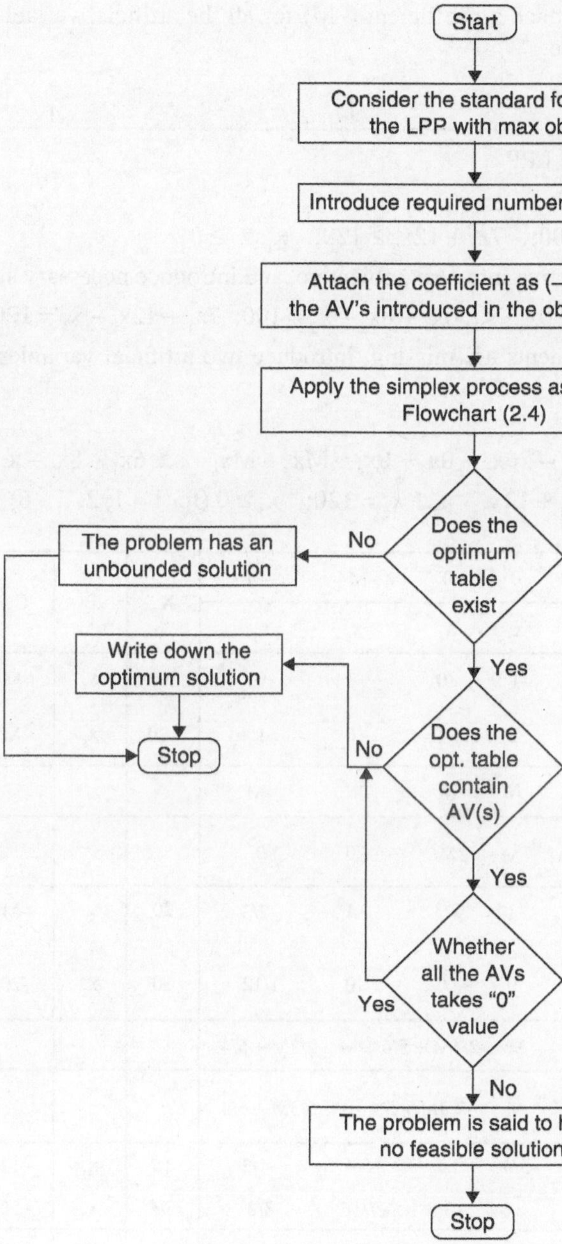

Flowchart 2.5 Big-M method

Remark: As a working rule, attach a coefficient (–M) for all the artificial variables introduced in the maximization objective function.

Example: 36

Use Big-M method to solve the LPP

$$\text{Min } Z = 12\,x_1 + 20x_2;$$

$$\text{s.t.}\quad 6x_1 + 8x_2 \geq 100;\quad 7x_1 + 12x_2 \geq 120;\quad x_1, x_2 \geq 0$$

Convert the minimization objective into Max. objective, and introduce necessary surplus variables.

$$\text{Max } Z = -12x_1 - 20x_2 + 0x_3 + 0x_4 \quad \text{s.t. } 6x_1 + 8x_2 - x_3 = 100;\ 7x_1 + 12x_2 - x_4 = 120;\ x_1, x_2, x_3, x_4 \geq 0$$

As both the identity components are missing, introduce two artificial variables one to each of the constraints.

Let it be x_5 & x_6

$$\Longrightarrow \text{Max } Z = -12x_1 - 20x_2 + 0x_3 + 0x_4 - Mx_5 - Mx_6 \quad \text{s.t. } 6x_1 + 8x_2 - x_3 + x_5 = 100;$$

$$7x_1 + 12x_2 - x_4 + x_6 = 120;\quad x_1 \geq 0 \text{ (for j = 1, 2, …. 6)}$$

C_j	–12	–20	0	0	–M	–M	X_B	B	C_B	θ
x_j	x_1	x_2	x_3	x_4	x_5	x_6				
	6	8	–1	0	1	0	100	x_5	–M	100/8 = 12.5
	7	(12) PE	0	–1	0	1	120	x_6	–M	120/12 = 10<= min PR
Z_j	–13M	–20M	M	M	–M	–M				
$Z_j - C_j$	–13M + 12	–20M + 20↑ PC	M	M	0	0				
R_1 – 8PR	(4/3) PE	0	–1	2/3	1	–2/3	20	x_5	–M	20/(4/3)= 15<= min PR
PR	7/12	1	0	–1/12	0	1/12	10	x_2	–20	$\frac{10}{(7/12)}$ = 17.14
Z_j	–4M/3 – 35/3	–20	M	–2/3M + 5/4	–M	2/3M + 5/4				
$Z_j - C_j$	–4/3M + 1/3↑ PC	0	M	–2/3M + 5/4	0	5/3M + 5/4				
PR	1	0	–3/4	1/2	3/4	–1/2	15	x_1	–12	
R_2 – (7/12)PR	0	1	7/16	–3/8	–7/16	3/8	5/4	x_2	–20	
Z_j	–12	–20	1/4	3/2	–1/4	–3/2				
$Z_j - C_j$	0	0	1/4	3/2	M–1/4	M–3/2	–205			

Since, all $Z_j - C_j \geq 0$, the optimum table is reached. Also, the artificial variables x_5 and x_6 are not in the Basis,

==> The optimum solution exists for the given problem.

$x_1 = 15;\ \ x_2 = 5/4;\ \ x_3 = 0;\ \ x_4 = 0;\ \ x_5 = 0;\ \ x_6 = 0$

$\text{Max } Z = -205$

i.e. $\text{Min } Z = 205$

Remark 1: $(M \pm K)$ is always ≥ 0; for any constant K.

Remark 2: $(-M \pm K)$ is always ≤ 0; for any constant K.

*Example: 37

$\text{Max } Z = 3x_1 + 2x_2;$ s.t. $2x_1 + x_2 \leq 2;\ \ 3x_1 + 4x_2 \geq 12;\ \ x_1, x_2 \geq 0$

Introducing the slack variable x_3 in constraint (1) and surplus variable x_4 in constraint (2), we have

$\text{Max } Z = 3x_1 + 2x_2 + 0x_3 + 0x_4;$ s.t. $2x_1 + x_2 + x_3 = 2;\ \ 3x_1 + 4x_2 - x_4 = 12;\ \ x_1, x_2, x_3, x_4 \geq 0$

Clearly, the 2nd identity component is missing. So, introduce the Artificial variable x_5 in constraint (2). Then the problem becomes,

$\text{Max } Z = 3x_1 + 2x_2 + 0x_3 + 0x_4 - Mx_5$

s.t. $2x_1 + x_2 + x_3 = 2$

$3x_1 + 4x_2 - x_4 + x_5 = 12;\ x_j \geq 0$ (for j = 1, 2, ….. 5)

C_j	3	2	0	0	–M	X_B	B	C_B	θ
x_j	x_1	x_2	x_3	x_4	x_5				
	2	① P E	1	0	0	2	x_3	0	21/2 = 2<= min PR
	3	4	0	–1	1	12	x_5	–M	12/4 = 3
Z_j	–3M	–4M	0	M	–M				
$Z_j - C_j$	–3M – 3	–4M – 2↑ P C	0	M	0				
PR	2	1	1	0	0	2	x_2	2	
R_2 – 4PR	–5	0	–4	–1	1	4	x_5	–M	
Z_j	5M + 4	2	4M + 2	M	–M				
$Z_j - C_j$	5M + 1	2	4M + 2	M	0				

Since, all $z_j - c_j \geq 0$, the optimum table is reached.
Note that in the Basis, x_5 is at "> 0" level.
Hence, the given LPP has "no feasible solution".

*Example: 38

$\text{Max } Z = 8x_2;$ s.t. $x_1 - x_2 \geq 0;\ \ 2x_1 + 3x_2 \leq 6;\ \ x_1, x_2$ unrestricted.

Given the variables x_1, x_2 are unrestricted, means that x_1 and x_2 can take negative value apart from positive/zero value. As it is we cannot apply the method.

Step 1: Remove the unrestrictedness of the variables. For this, we are going to express each unrestricted variable as a difference of two new variables, such that they are " > 0"

$$x_1 = x_3 - x_4; \quad x_3, x_4 \geq 0; \quad x_2 = x_5 - x_6; \quad x_5, x_6 \geq 0$$

substitute these values of x_1 & x_2 in the given problem. The revised problem is, Max $Z = 8x_5 - 8x_6$

$$\text{s.t. } x_3 - x_4 - x_5 + x_6 \geq 0; \quad 2x_3 - 2x_4 + 3x_5 - 3x_6 \leq 6; \quad x_3, x_4, x_5, x_6 \geq 0$$

Introduce the surplus and artificial variable x_7, & x_8 in constraint (1) and Slack variable x_9 in constraint (2)

$$\text{Max } Z = 8x_5 - 8x_6 + 0x_7 - Mx_8 + 0x_9$$

$$\text{s.t. } x_3 - x_4 - x_5 + x_6 - x_7 + x_8 = 0; \quad 2x_3 - 2x_4 + 3x_5 - 3x_6 + x_9 = 6; \quad x_j \geq 0 \ (j = 3,4, \ldots 9)$$

C_j	0	0	8	−8	0	−M	0	X_B	B	C_B	θ
x_j	x_3	x_4	x_5	x_6	x_7	x_8	x_9				
	① P E	−1	−1	1	−1	1	0	0	x_8	−M	0/1 = 0 <= min PR
	2	−2	3	−3	0	0	1	6	x_9	0	6/2 = 3
Z_j	−M	M	M	−M	M	−M	0				
$Z_j - C_j$	−M↑ P C	M	M − 8	−M + 8	M	0	0				
PR	1	−1	−1	1	−1	–	0	0	x_3	0	–
R_2 − 2PR	0	0	⑤ P E	−5	2	–	1	6	x_9	0	6/5 = 1.2 <= min PR
Z_j	0	0	0	0	0	*	0				
$Z_j - C_j$	0	0	−8↑ P C	+8	0	*	0				
R_1 + PR	1	−1	0	0	−3/5	–	0	6/5	x_3	0	--
PR	0	0	1	−1	2/5	–	1/5	6/5	x_4	8	
Z_j	8	0	8	−8	16/5	*	8/5				
$Z_j - C_j$	8	0	0	0	16/5	*	8/5				

Since, all $Z_j - C_j \geq 0$; the optimum table is reached.
Hence the optimum solution is

$$x_3 = 6/5; \quad x_4 = 0; \quad x_5 = 6/5; \quad x_6 = 0; \quad x_7 = 0; \quad x_8 = 0; \text{ and } x_9 = 0$$

Then,

$$x_1 = x_3 - x_4 = 6/5 - 0 = 6/5$$

$$x_2 = x_5 - x_6 = 6/5 - 0 = 6/5$$

$$\text{Max } Z = 48/5$$

Remark: In table-2, instead of the artificial variable x_8, the variable x_3 enters. Whenever if an A.V leaves the basis means there after no necessity to fill the column of the corresponding A.V. (This rule is applicable for the corresponding A.V. only)

Example: 39

Solve using Big-M method to

$$\text{Max } Z = 6x_1 + 4x_2; \quad \text{s.t. } 2x_1 + 3x_2 \le 30; \quad 3x_1 + 2x_2 \le 24; \quad x_1 + x_2 \ge 3; \quad x_1, x_2 \ge 0$$

Is the optimum solution unique? If not get the alternate solution.

Introducing the slack variables x_3 & x_4 in constraint (1) & (2); and introducing the surplus variable x_5 and artificial variable x_6 in constraint (3), we have

$$\text{Max } Z = 6x_1 + 4x_2 + 0x_3 + 0x_4 + 0x_5 - Mx_6$$

$$\text{s.t. } 2x_1 + 3x_2 + x_3 = 30$$

$$3x_1 + 2x_2 + x_4 = 24$$

$$x_1 + x_2 - x_5 + x_6 = 3$$

$$x_j \ge 0 \ (j = 1, 2, \ldots, 6)$$

C_j	6	4	0	0	0	−M				
x_j	x_1	x_2	x_3	x_4	x_5	x_6	X_B	B	C_B	θ
	2	3	1	0	0	0	30	x_3	0	30/2 = 15
	3	2	0	1	0	0	24	x_4	0	24/3 = 8
	① P E	1	0	0	−1	1	3	x_6	−M	3/1 = 3 <= min PR
Z_j	−M	−M	0	0	M	−M				
$Z_j - C_j$	−M − 6↑ P C	−M−4	0	0	M	0				
$R_1 - 2PR$	0	1	1	0	2	–	24	x_3	0	24/2 = 12
$R_2 - 3PR$	0	−1	0	1	③ P E	–	15	x_4	0	15/3 = 5 <= min PR
PR	1	1	0	0	−1	–	3	x_1	6	---
Z_j	6	6	0	0	−6↑ P C	*				
$Z_j - C_j$	0	2	0	0	−6↑ P C	*				
$R_1 - 2PR$	0	(5/3) P E	1	−2/3	0	–	14	x_3	0	42/5 <= min PR
PR	0	−1/3	0	1/3	1	–	5	x_5	0	--
$R_3 + PR$	1	2/3	0	1/3	0	–	8	x_1	6	8/0.67 = 12
Z_j	6	4	0	2	0	*				
$Z_j - C_j$	0	0↑ P C	0	2	0	*	48			
PR	0	1	3/5	−2/5	0	–	42/5	x_2	4	
$R_2 + (1/3)PR$	0	0	1/5	1/5	1	–	39/5	x_5	0	
$R_3 - (2/3)PR$	1	0	−2/5	3/5	2	–	12/5	x_1	6	

Z_j	6	4	0	2	0	*				
$Z_j - C_j$	0	0	0	2	0	*	48			

All $Z_j - C_j \geq 0$ in Table-3; then the optimum solution is attained. The solution is $x_1 = 8, x_2 = 0; x_3 = 14; x_4 = 0; x_5 = 0; x_6 = 0$ and max $Z = 48$.

In the optimum table-3, since the non-basic variable x_2's (which is not in basis) entry in $(Z_j - C_j)$ row is "0" ==> That there is an alternate solution for the given LPP. So, in table-3; select x_2 as an entering variable and proceed, we have another optimum solution in table-4.

The solution is

$$x_1 = 12/5; x_2 = 42/5; x_3 = 0; x_4 = 0; x_5 = 39/5; x_6 = 0 \text{ and Max } Z = 48.$$

2.13 TWO-PHASE METHOD

This method is a substitute for Big-M method. It is highly necessary because, while applying Big-M method in a computer system to solve an LPP, there is a chance for computational error (since, we make use of large value to the constant M).

Eventhough the artificial variables are added in the same manner that is used in Big-M method, the use of the constant M is eliminated by solving the problem in two phases, namely Phase-I and Phase-II (hence it is named as "Two-phase" method).

Phase-I

In Phase-I our main aim is to drive away the artificial variables. For which we construct a new dummy objective function in which penalty "1" is assigned to all the AVs and keep all other variables as zero. Solve by considering the new objective function and the given original constraints. If the new objective value is zero, it implies that the problem has a feasible solution space. Then proceed to Phase-II. If not the problem has no feasible solution, stop the process.

Phase II

There is a chance for the Artificial Variable not taking place and taking place at zero level in the basis. For the first type apply case (i) and for the second type apply case (ii).

Case (i): Consider the original objective function along with the optimum table entries of Phase-I. Then proceed with the simplex process.

Case (ii): Consider the original objective function and attach the co-efficient for the artificial variables as zero together with the optimum table entries of Phase-I. Then proceed with the simplex process. During the process, don't allow the Artificial Variable to take positive value (> 0) for which if in a PC, corresponding to the Artificial Variable the entry is strictly negative, then consider that value as a pivotal element (without finding the ratio) and proceed.

The various steps to be followed while applying penalty method are clearly explained in the Flowchart (2.6).

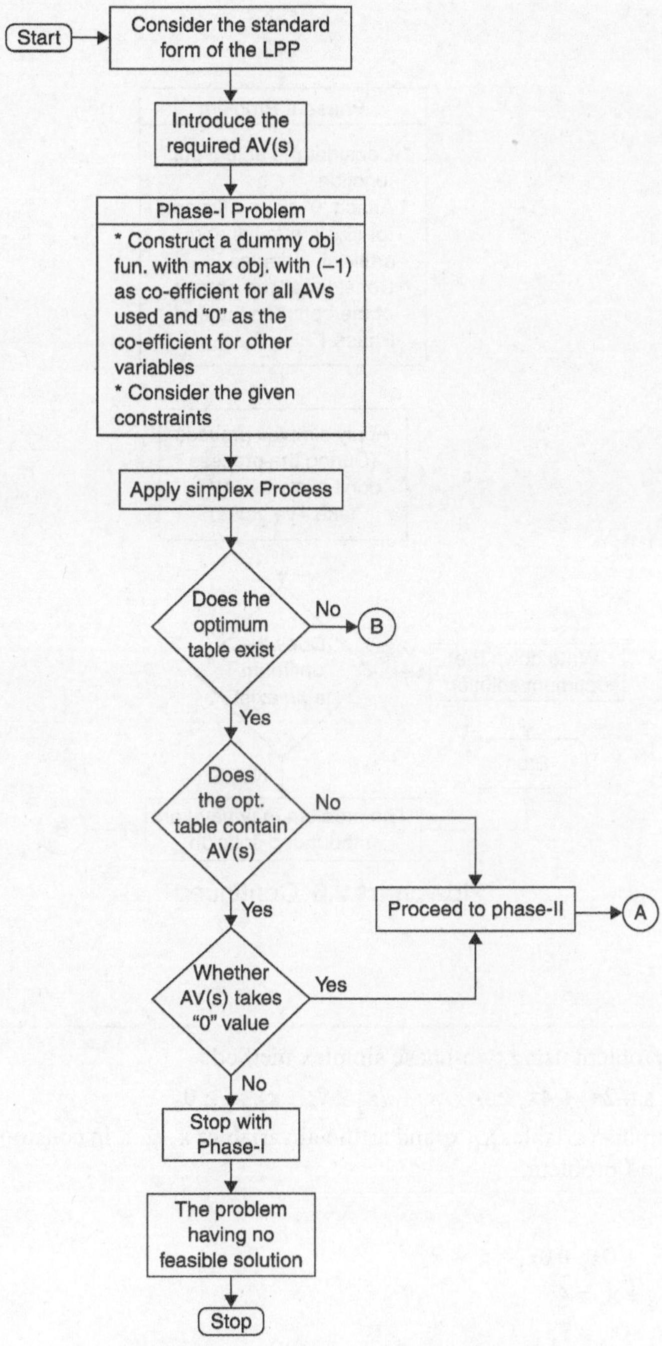

Flowchart 2.6 Two-phase method

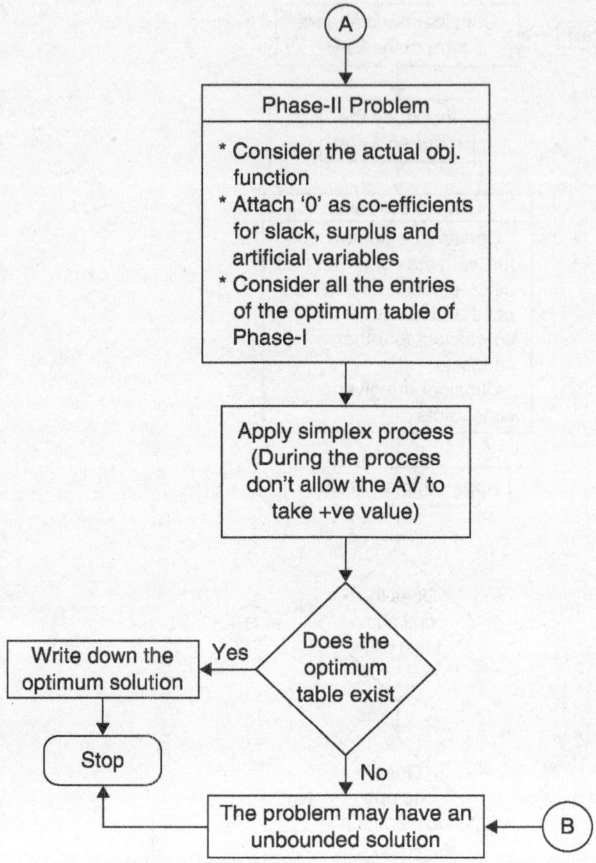

Flowchart 2.6 Continued

***Example:* 40**

Solve the following problem using two-phase simplex method:

Min $z = x_1 + x_2$; s.t. $2x_1 + 4x_2 \geq 4$; $x_1 + 7x_2 \geq 7$; $x_1, x_2 \geq 0$

Introducing the surplus variables x_3, x_4 and artificial variables x_5 & x_6 in constraint (1) and (2) respectively. Write the Phase-I problem.

Phase-I:

$$\text{Max } z^* = 0x_1 + 0x_2 + 0x_3 + 0x_4 - x_5 - x_6$$

$$\text{s.t. } 2x_1 + 4x_2 - x_3 + x_5 = 4;$$

$$x_1 + 7x_2 - x_4 + x_6 = 7;$$

$$x_1 \geq 0; (j = 1, 2, \ldots, 6)$$

C_j	0	0	0	0	−1	−1	X_B	B	C_B	θ
x_j	x_1	x_2	x_3	x_4	x_5	x_6				
	2	4	−1	0	1	0	4	x_5	−1	4/4 = 4
	1	(7) PE	0	−1	0	1	7	x_6	−1	7/7 = 1 <= min PR
Z_j	−3	−11	1	1	−1	−1				
$Z_j - C_j$	−3	−11↑ PC	1	1	0	0				
R_1 − 4PR	(10/17) PE	0	−1	4/7	1	–	0	x_5	−1	0 <= min PR
PR	1/7	1	0	−1/7	0	–	1	x_2	0	7
Z_j	−10/7	0	1	−4/7	−1	*				
$Z_j - C_j$	−10/7↑ PC	0	1	−4/7	0	*				
PR	1	0	−7/10	2/5	–	–	0	x_1	0	--
R_2 − (1/7)PR	0	1	1/10	−1/5	–	–	1	x_2	0	
Z_j	0	0	0	0	*	*				
$Z_j - C_j$	0	0	0	0	*	*				

Since, in Phase-I; $Z^* = 0$; proceed to Phase-II

Phase II: The actual objective is max $Z' = -x_1 - x_2 + 0x_3 + 0x_4$ and consider the optimum table of Phase-I.

C_j	−1	−1	0	0	X_B	B	C_B	θ
x_j	x_1	x_2	x_3	x_4				
	1	0	−7/10	(2/5) PE	0	x_1	−1	0<= min PR
	0	1	1/10	−1/5	1	x_2	−1	--
Z_j	−1	−1	6/10	−1/5				
$Z_j - C_j$	0	0	8/10	−1/5↑ PC				
PR	5/2	0	−7/4	1	0	x_4	0	
R_2 + (1/5)PR	1/2	1	1/4	0	1	x_2	−1	
Z_j	−1/2	−1	−1/4	0				
$Z_j - C_j$	1/2	0	1/4	0	−1			

Since, all $Z_j - C_j \geq 0$, implies that the Phase-II optimum exists. The optimum solution is $x_1 = 0$; $x_2 = 1$; $x_3 = 0$; $x_4 = 0$ and

$$\text{Max } Z' = -1 \Longrightarrow \text{Min } Z = 1$$

*Example: 41

Use Two-Phase simplex method to minimize $Z = -x_1 - 2x_2 - 3x_3$
subject to $-2x_1 + x_2 + 3x_3 = 2$; $2x_1 + 3x_2 + 4x_3 = 1$; $x_1, x_2, x_3 \geq 0$

Introducing the artificial variables x_4 and x_5 one each to the constraints, (1) & (2). Write the Phase-I problem.
$$\text{Max } Z^* = 0x_1 + 0x_2 + 0x_3 - x_4 - x_5$$
s.t. $-2x_1 + x_2 + 3x_3 + x_4 = 2$; $2x_1 + 3x_2 + 4x_3 + x_5 = 1$; $x_j \geq 0$ (j = 1, 2, ……., 5)

C_j	0	0	0	–1	–1	X_B	B	C_B	θ
x_j	x_1	x_2	x_3	x_4	x_5				
	–2	1	3	1	0	2	x_4	–1	2/3
	2	3	(4) PE	0	1	1	x_5	–1	1/4<= min PR
Z_j	0	–4	–7	–1	–1				
$Z_j - C_j$	0	–4	–7↑ PC	0	0				
R_1 – 3PR	–7/2	–5/4	0	1	–	5/4	x_4	–1	
PR	1/2	3/4	1	0	–	1/4	x_3	0	
Z_j	7/2	5/4	0	–1	*				
$Z_j - C_j$	7/2	5/4	0	0	*				

Since, all $Z_j - C_j \geq 0$, implies that Phase-I optimum table exists.

Also, the artificial variable x_4 exists in the basis at positive level, implies that the given LPP is not having any feasible solution.

Example: 42

Use Two-Phase method to maximize $Z = 2x_1 + x_2 + x$

s.t. $4x_1 + 6x_2 + 3x_3 \leq 8$;
$3x_1 - 6x_2 - 4x_3 \leq 1$;
$2x_1 + 3x_2 - 5x_3 \geq 4$;
$x_1, x_2, x_3 \geq 0$

Introducing the slack variables x_4, x_5, in constraint (1) and (2); a surplus variable x_6 and an artificial variable x_7 in constraint (3) write the Phase-I problem.

Max $Z^* = 0x_1 + 0x_2 + 0x_3 + 0x_4 + 0x_5 + 0x_6 - x_7$

s.t. $4x_1 + 6x_2 + 3x_3 + x_4 = 8;\quad 3x_1 - 6x_2 - 4x_3 + x_5 = 1;\quad 2x_1 + 3x_2 - 5x_3 - x_6 + x_7 = 4;$

$x_j \geq 0$ (for j = 1, 2, …., 7)

C_j	0	0	0	0	0	0	−1	X_B	B	C_B	θ
x_j	x_1	x_2	x_3	x_4	x_5	x_6	x_7				
	4	⑥ PE	3	1	0	0	0	8	x_4	0	8/6 = 4/3 <= min PR
	3	−6	−4	0	1	0	0	1	x_5	0	--
	2	3	−5	0	0	−1	1	4	x_7	−1	4/3
Z_j	−2	−3	5	0	0	1	−1				
$Z_j - C_j$	−2	−3↑ PC	5	0	0	1	0				
PR	2/3	1	1/2	1/6	0	0	0	4/3	x_2	0	
$R_2 + 6PR$	7	0	−1	1	1	0	0	9	x_5	0	
$R_3 - 3PR$	0	0	−6.5	−0.5	0	−1	1	0	x_7	−1	
Z_j	0	0	13/2	1/2	0	1	−1				
$Z_j - C_j$	0	0	13/2	1/2	0	1	0	0			

Since, the Phase-I optimum, $Z^* = 0$; proceed to Phase-II.

Remark: The artificial variable x_7 is in the basis at zero level.

Phase II

The actual objective is

Max $Z = 2x_1 + x_2 + x_3 + 0x_4 + 0x_5 + 0x_6 + 0x_7$ and consider the optimum table of Phase-I

C_j	2	1	1	0	0	0	0	X_B	B	C_B	θ
x_j	x_1	x_2	x_3	x_4	x_5	x_6	x_7				
	2/3	1	1/2	1/6	0	0	0	4/3	x_2	1	2
	⑦ PE	0	−1	1	1	0	0	9	x_5	0	9/7 <= min PR
	0	0	−13/2	−1/2	0	−1	1	0	x_7	0	--
Z_j	2/3	1	1/2	1/6	0	0	0				
$Z_j - C_j$	−4/3↑ PC	0	−1/2	1/6	0	0	0				

$R_1-(2/3)PR$	0	1	25/42	1/4	−2/21	0	0	10/21	x_2	1	
PR	1	0	−1/7	1/7	1/7	0	0	9/7	x_1	2	
R_3	0	0	(−13/2) PE	−1/2	0	−1	1	0	x_7	0	<= min PR
Z_j	2	1	13/42	5/14	4/21	0	0				
Z_j-C_j	0	0	−29/42↑ PC	5/14	4/21	0	0				
$R_1-(25/42)PR$	0	1	0	1/39	−2/21	−25/273	–	10/21	x_2	1	
$R_2+(1/7)PR$	1	0	0	2/13	1/7	2/91	–	9/7	x_1	2	
PR	0	0	1	1/13	0	2/13	–	0	x_3	0	
Z_j	2	1	1	16/39	4/21	29/273					
Z_j-C_j	0	0	0	16/39	4/21	29/273	64/21				

Since, all $Z_j - C_j \geq 0$, the optimum Table exists. The optimum solution is

$$x_1 = 9/7, \quad x_2 = 10/21; \quad x_3 = 0 \text{ and max } Z = 64/21$$

Example: 43

Solve the following system of linear equations by using simplex method.

$$x_1 + x_2 = 1; \; 2x_1 + x_2 = 3$$

Here, we are suppose to solve only the simultaneous equations. So, assume the objective function as

$$\text{Max } Z = 0x_1 + 0x_2 \quad \text{s.t. } x_1 + x_2 = 1; \quad 2x_1 + x_2 = 3$$

Since, x_1 & x_2 are unrestricted variables, replace $x_1 = x_3 - x_4$; $x_2 = x_5 - x_6$; where $x_3, x_4, x_5, x_6 \geq 0$. Then the problem becomes,

$$\text{Max } Z = 0x_3 - 0x_4 + 0x_5 - 0x_6$$

$$\text{s.t. } x_3 - x_4 + x_5 - x_6 = 1; \quad 2x_3 - 2x_4 + x_5 - x_6 = 3; \quad x_1 \geq 0 \; (j = 3, 4, 5, 6)$$

Introduce x_7 & x_8 as artificial variables, one each to the constraint and solve it using the Two-Phase method, we have the optimum solution as $x_3 = 2$; $x_4 = 0$; $x_5 = 0$; and $x_6 = 1$

i.e. $x_1 = x_3 - x_4 = 2 - 0 = 2$

$x_2 = x_5 - x_6 = 0 - 1 = -1$

2.14 THE DUALITY CONCEPT IN A LINEAR PROGRAMMING

2.14.1 Definition of the Dual problem

The dual problem is an auxiliary LPP defined directly and systematically from the actual LPP (called primal problem).

2.14.2 Standard Form of the Primal

$$\text{Max } Z_p = \sum_{j=1}^{n} c_j x_j$$

$$\text{s.t. } \sum_{j=1}^{n} a_{ij} x_j \le b_i; \quad x_j \ge 0 \quad \text{where } i = 1,2, \ldots m \text{ and } j = 1, 2, \ldots n \text{ and 'P' denotes primal objective.}$$

2.14.3 Standard Form of the Dual

$$\text{Min } Z_D = \sum_{i=1}^{m} b_i \, y_i$$

$$\text{s.t. } \sum_{i=1}^{m} a_{ij} y_j \ge c_i; \quad \text{where } j = 1,2, \ldots n \text{ and 'D' denotes Dual objective.}$$

2.14.4 Structural and Computational Relationship Between Primal and Dual Problems

1. If the primal is a maximization problem, the corresponding dual must be a minimization problem, and vice-versa.
2. The optimal values $Z_p = Z_D$
3. The RHS constants of the primal constraints correspond to the co-efficients of the dual objective function.
4. The co-efficients of the primal objective function correspond to RHS constants of the dual constraints.
5. For each decision variable in the primal we have to initiate a constraint in the dual problem.
6. If the constraints of the primal are of $\{\le\}$ type then, the constraints of the dual must be of $\{\ge\}$ type and vice-versa.
7. The structure of the dual is linear with respect to decision variables y_i's where $y_i \ge 0$ ($i = 1, 2, \ldots m$).
8. The nature of solution between primal and dual is

	Standard Primal	Standard Dual
1.	Optimum solution exists	Optimum solution exists
2.	Unbounded solution	No feasible solution
3.	No solution	Neither unbounded nor no solution

2.14.5 In Matrix Notation (Primal & Dual)

Whenever we want to convert the given LPP (Primal problem) into a Dual problem, we have to see to that the given primal is in anyone of the following form.

No.	Primal Form	Dual Form
1.	Max $Z_p = C^TX$ s.t. $AX \le b$ $X \ge 0$	Min $Z_D = b^TY$ s.t. $A^TY \ge C$ $Y \ge 0$
2.	Min $Z_p = C^TX$ s.t. $AX \ge b$ $X \ge 0$	Max $Z_D = b^TY$ s.t. $A^TY \le C$ $Y \ge 0$
3.	Max $Z_p = C^TX$ s.t. $AX \{\le, =\}\ b$ $X \ge 0$	Min $Z_D = b^TY$ s.t. $A^TY \ge C^T$ (*)

(*) Treat the dual variable Y introduced corresponding to the equality constraint of the primal as unrestricted and other dual variables as ≥ 0. Where $C^T, X^T \in R^n$; $b^T, Y^T \in R^m$.

Theorem: The dual of the dual is primal

Proof: The standard primal problem is

$$\max Z_p = c^Tx$$
$$\text{s.t. } Ax \le b$$
$$x \ge 0$$

Dual: $\min Z_D = b^Ty$

s.t. $A^Ty \ge c$

$y \ge 0$

Dual of dual: $\max Z_p = c^Tw$

s.t. $Aw \le b$

$w \ge 0$

This is precisely the actual primal problem.
Therefore the dual of the dual is primal.

Example: 44

Max $Z = 80x_1 + 90x_2$

s.t. $3x_1 + 5x_2 \le 30$; $3x_1 + 2x_2 \le 21$; $x_1, x_2 \ge 0$

Step 1: Primal in Matrix notation is, max $Z = [80\ 90]\begin{bmatrix} x_1 \\ x_2 \end{bmatrix}$

$$\text{s.t. } \begin{bmatrix} 3 & 5 \\ 3 & 2 \end{bmatrix}\begin{bmatrix} x_1 \\ x_2 \end{bmatrix} \le \begin{bmatrix} 30 \\ 21 \end{bmatrix}$$

$$x_1, x_2 \ge 0$$

Step 2: Introduce the dual variable one each to the primal constraints. Let it be $y = \begin{bmatrix} y_1 \\ y_2 \end{bmatrix}$

Step 3: Using Step 1;

$$c^T = [80\ 90];\ A = \begin{bmatrix} 3 & 5 \\ 3 & 2 \end{bmatrix}\ b = \begin{bmatrix} 30 \\ 21 \end{bmatrix}$$

then
$$c = (c^T)^T = \begin{bmatrix} 80 \\ 90 \end{bmatrix};\ A^T = \begin{bmatrix} 3 & 5 \\ 3 & 2 \end{bmatrix}^T = \begin{bmatrix} 3 & 3 \\ 5 & 2 \end{bmatrix}$$

$$b^T = \begin{bmatrix} 30 \\ 21 \end{bmatrix}^T = [30\ 21]$$

then the dual is, $\text{Min } z_D = [30\ 21]\begin{bmatrix} y_1 \\ y_2 \end{bmatrix}$

$$\text{s.t.} \begin{bmatrix} 3 & 3 \\ 5 & 2 \end{bmatrix}\begin{bmatrix} y_1 \\ y_2 \end{bmatrix} \geq \begin{bmatrix} 80 \\ 90 \end{bmatrix};\ y_1, y_2 \geq 0$$

i.e. $\text{Min } Z_D = 30y_1 + 21y_2$ s.t. $3y_1 + 3y_2 \geq 80$; $5y_1 + 2y_2 \geq 90$; $y_1, y_2 \geq 0$

Example: 45

$\text{Min } Z = 10x_1 + 8x_2$ s.t. $x_1 + 2x_2 \geq 5$; $2x_1 - x_2 \geq 12$; $x_1 + 3x_2 \geq 4$; $x_1 \geq 0$; x_2 unrestricted.

Since, the primal variable x_2 is unrestricted, it must be represented as a difference of two new variables. Let it be $x_2 = x_3 - x_4$; $x_3, x_4 \geq 0$; substituting the value of x_2 in the given problem,

$\text{Min } Z = 10x_1 + 8x_3 - 8x_4$; s.t. $x_1 + 2x_3 - 2x_4 \geq 5$; $2x_1 - x_3 + x_4 \geq 12$; $x_1 + 3x_3 - 3x_4 \geq 4$; $x_1, x_3, x_4 \geq 0$.

$$\text{Min } Z = [10\ \ 8\ \ -8]\begin{bmatrix} x_1 \\ x_3 \\ x_4 \end{bmatrix}$$

$$\text{s.t.} \begin{bmatrix} 1 & 2 & -2 \\ 2 & -1 & 1 \\ 1 & 3 & -3 \end{bmatrix}\begin{bmatrix} x_1 \\ x_3 \\ x_4 \end{bmatrix} \geq \begin{bmatrix} 5 \\ 12 \\ 4 \end{bmatrix}$$

$$x_1, x_3, x_4 \geq 0$$

Dual:

Let us introduce the dual variable one each to the primal constraints, let it be $y = \begin{bmatrix} y_1 \\ y_2 \\ y_3 \end{bmatrix}$

$$\text{Max } z_D = [5\ \ 12\ \ 4]\begin{bmatrix} y_1 \\ y_2 \\ y_3 \end{bmatrix}$$

$$\text{s.t.} \begin{bmatrix} 1 & 2 & 1 \\ 2 & -1 & 3 \\ -2 & 1 & -3 \end{bmatrix} \begin{bmatrix} y_1 \\ y_2 \\ y_3 \end{bmatrix} \leq \begin{bmatrix} 10 \\ 8 \\ -8 \end{bmatrix}; \quad y_1, y_2, y_3 \geq 0$$

i.e. The required dual problem is,

$$\text{Max } Z_D = 5y_1 + 12y_2 + 4y_3;$$

$$\text{s.t. } y_1 + 2y_2 + y_3 \leq 10; \quad 2y_1 - y_2 + 3y_3 \leq 8; \quad -2y_1 + y_2 - 3y_3 \leq -8; \quad y_1, y_2, y_3 \geq 0$$

Example: 46

Write the dual model of the primal

$$\text{Max } Z = 10x_1 + 20x_2; \quad \text{s.t.} \quad x_1 + x_2 = 4; \quad 2x_1 - 3x_2 \leq 7; \quad x_1, x_2 \geq 0$$

Remark: We know that an equality $a = b$ can be replaced by two inequalities, i.e., $a \leq b$ and $a \geq b$. Using this idea

$$x_1 + x_2 = 4 \text{ can be rewritten as } x_1 + x_2 \geq 4; \quad x_1 + x_2 \leq 4$$

So, the problem becomes Max $Z = 10x_1 + 20x_2$

$$\text{s.t.} \quad x_1 + x_2 \geq 4; \quad x_1 + x_2 \leq 4; \quad 2x_1 - 3x_2 \leq 7; \quad x_1, x_2 \geq 0$$

For the dual conversion, in the primal all the constraints must be "≤" type if the objective is maximum (In the primal all the constraints must be ≥ type if the objective is minimum). Hence, multiplying the first constraint by (–1).

We get $-x_1 - x_2 \leq -4$

$$\text{Max } Z_p = [10 \ 20] \begin{bmatrix} x_1 \\ x_2 \end{bmatrix}$$

$$\text{s.t.} \begin{bmatrix} -1 & -1 \\ 1 & 1 \\ 2 & -3 \end{bmatrix} \begin{bmatrix} x_1 \\ x_2 \end{bmatrix} \leq \begin{bmatrix} -4 \\ 4 \\ 7 \end{bmatrix}; \ x_1, x_2 \geq 0$$

Introducing the dual variable one each to the constraints,

$$y = \begin{bmatrix} y_1 \\ y_2 \\ y_3 \end{bmatrix}$$

we get
$$\text{Min } Z_D = [-4 \ \ 4 \ \ 7] \begin{bmatrix} y_1 \\ y_2 \\ y_3 \end{bmatrix}$$

$$\begin{bmatrix} -1 & 1 & 2 \\ -1 & 1 & -3 \end{bmatrix} \begin{bmatrix} y_1 \\ y_2 \\ y_3 \end{bmatrix} \geq \begin{bmatrix} 10 \\ 20 \end{bmatrix}$$

$$y_1, y_2, y_3 \geq 0$$

The required dual problem is,

$$\text{Min } Z_D = -4y_i + 4y_2 + 7y_3; \quad \text{s.t.} -y_1 + y_2 + 2y_3 \geq 10; \quad -y_1 + y_2 - 3y_3 \geq 20; \quad y_1, y_2, y_3 \geq 0$$

2.14.6 Shadow Price (Dual Price/Accounting Price)

In general, the shadow price of the ith constraint is the value of the change in the optimal value of the objective function. This occurs in a unit change in RHS value b_i of that constraint. Keep all other parameter values of the problem remain unchanged.

Remark: If the dimension of the objective function is in Rupees and the dimension of the b_i is in hours, then the shadow price is in Rupees per hour.

Example: 47

A furniture manufacturer makes use of two machines A and B to produce chairs and tables. The production formula and the profit are given in the following table:

Items	Machine hrs/Unit		Profit (₹)
	A	B	
Chair	2	6	1
Table	5	--	5
Max	16	30	

How many chairs and tables must be produced daily inorder to maximize the profit. Formulate it as an LPP and solve.

Primal: Let x_1, x_2 be the number of units produced in chair and table

$\text{Max } Z_p = x_1 + 5x_2;$

s.t. $2x_1 + 5x_2 \leq 16$ (Machine time in A)

$6x_1 \leq 30$ (Machine time in B); $x_1, x_2 \geq 0$

Dual: $\text{Min } Z_D = 16y_1 + 30y_2; \quad \text{s.t. } 2y_1 + 6y_2 \geq 1; \quad 5y_1 \geq 5; \quad y_1, y_2 \geq 0$

Primal Solution

Introduce the slack variables x_3 & x_4.

$$\text{Max } Z_p = x_1 + 5x_2 + 0x_3 + 0x_4 \quad \text{s.t. } 2x_1 + 5x_2 + x_3 = 16; \quad 6x_1 + x_4 = 30; \quad x_j \geq 0; (j = 1, 2, 3, 4)$$

C_j	1	5	0	0	X_B	B	C_B	θ
x_j	x_1	x_2	x_3	x_4				
	2	⑤ PE	1	0	16	x_3	0	16/5<= min PR
	6	0	6	1	30	x_4	0	--

Z_j	0	0	0	0			
$Z_j - C_j$	–1	$-5\uparrow^{P}_{C}$	0	0			
PR	2/5	1	1/5	0	16/5	x_2	5
R_2	6	0	0	1	30	x_4	0
Z_j	2	5	1	0			
$Z_j - C_j$	1	0	1	0	16		

The primal optimum solution is, $x_1 = 0$; $x_2 = 16/5$; and Max Z =16.
We can get the solution of the dual from the optimum table of primal and vice-versa.

Primal Variable Presented in the X_B Column of the Initial Table	j	The Constraint which it Belongs (i)	Dual Variable Related to it (y_i)	$Z_j - C_j$ value Corresponds to the Primal Variables in the Optimum Table	Optimum Solution	Min. Z_D
x_3	3	1	y_1	1	$y_1 = 1$	16 (1) + 30 (0)
x_4	4	2	y_2	0	$y_2 = 0$	= 16

Hence $y_1 = 1$; and $y_2 = 0$. Then min $Z_D = 16\ (1) + 30\ (0) = 16$.

Note that the optimal values of the dual variables $y_1 = 1$ and $y_2 = 0$ are equal to the shadow prices of (machine-A time) and (machine-B time) resources or constraints.

Accordingly $y_1 = 1$ is the profit contributed by each hour of the machine-A. This rate of ₹ 1, at which the objective function value will increase as the machine-A capacity is increased by 1 hour, is called shadow price of the machine-A resource or machine-A constraint.

2.14.7 Economic Importance of Shadow Price

Many LP applications in production functions, the shadow prices are as important as the solution of the problem itself, because they tell us whether or not certain changes in the use of production capacities affect the objective function.

Remark: Solving the dual problem instead of primal problem would reduce the solution time in the following cases:

1. When the primal having more constraints than the decision variables.
2. When the primal problem has all constraints in the form of ">".

2.15 DUAL SIMPLEX METHOD (DSM)

Solving by simplex method is complex for certain primal or dual problems. Also it takes much time in computing while using artificial variables. To overcome this difficulty, in 1954 LEMKE introduced a method called Dual Simplex Method. This method provides more efficient procedure for such problems.

The DSM method also constitutes an application of the simplex method in the opposite direction.

This method starts from one infeasible extreme point to another attempting to find a feasible solution. This feasible solution is also the optimal solution.

We cannot apply this method to all LPP unless the following conditions are satisfied.

1. The LPP must be in the canonical form
2. All the entries of $(Z_j - C_j)$ must be ≥ 0.

2.15.1 Canonical Form of an LPP

1. The objective function should be of maximization
2. All the constraints should have "≤" inequality
3. All variables must be "≥ 0"

If an LPP satisfies the conditions cited above, then it is called in canonical form.

***Example:* 48**

Rewrite the given LPP in the canonical form

$$\text{Max } Z = 3x_1 + x_2 \quad \text{s.t. } 2x_1 - x_2 \leq 10; \quad 3x_1 + 4x_2 \geq 24; \quad x_1 \geq 2; \quad x_1, x_2 \geq 0$$

In the given problem, the 2nd & 3rd constraints having "≥" inequality, hence, multiply both sides of the 2nd and 3rd constraints with (–1), we have

$$-3x_1 - 4x_2 \leq -24; -x_1 \leq -2$$

i.e. $\text{Max } Z = 3x_1 + x_2$

s.t. $2x_1 - x_2 \leq 10; \quad -3x_1 - 4x_2 \leq -24; \quad -x_1 \leq -2; \quad x_1, x_2 \geq 0$

Regarding the various steps to be followed while applying Dual simplex method to solve a given LPP are clearly explained in Flowchart (2.7).

***Example:* 49**

Use duality to solve the following linear programming problem

$$\text{Max } Z = 3x_1 + 2x_2 \quad \text{s.t.} \quad 2x_1 + x_2 \leq 5; \quad x_1 + x_2 \leq 3; \quad x_1, x_2 \geq 0$$

Use duality means, first we have to get the dual of the given primal and solve the dual. From the optimum table of the dual, get the primal solution.

Primal: $\text{Max } Z = [3\ 2]\begin{bmatrix} x_1 \\ x_2 \end{bmatrix}$ s.t. $\begin{bmatrix} 2 & 1 \\ 1 & 1 \end{bmatrix}\begin{bmatrix} x_1 \\ x_2 \end{bmatrix} \leq \begin{bmatrix} 5 \\ 3 \end{bmatrix}$ $x_1, x_2 \geq 0$

Dual:

$$\text{Min } Z_P = [5\ 3]\begin{bmatrix} y_1 \\ y_2 \end{bmatrix} \quad \text{s.t.} \quad \begin{bmatrix} 2 & 1 \\ 1 & 1 \end{bmatrix}\begin{bmatrix} y_1 \\ y_2 \end{bmatrix} \geq \begin{bmatrix} 3 \\ 2 \end{bmatrix} \quad y_1, y_2 \geq 0$$

$$\text{Min } Z_D = 5y_1 + 3y_2; \quad \text{s.t.} \quad 2y_1 + y_2 \geq 3; \quad y_1 + y_2 \geq 2; \quad y_1, y_2 \geq 0$$

This Dual problem can be solved by applying Big-M method or Two-Phase method or Dual Simplex method. Here, if we apply DSM, we can avoid the introduction of the artificial variable. Rewrite the dual as

$$\text{Max } Z_D^* = -5y_1 - 3y_2; \quad \text{s.t.} \quad -2y_1 - y_2 \leq -3; \quad -y_1 - y_2 \leq -2; \quad y_1, y_2 \geq 0$$

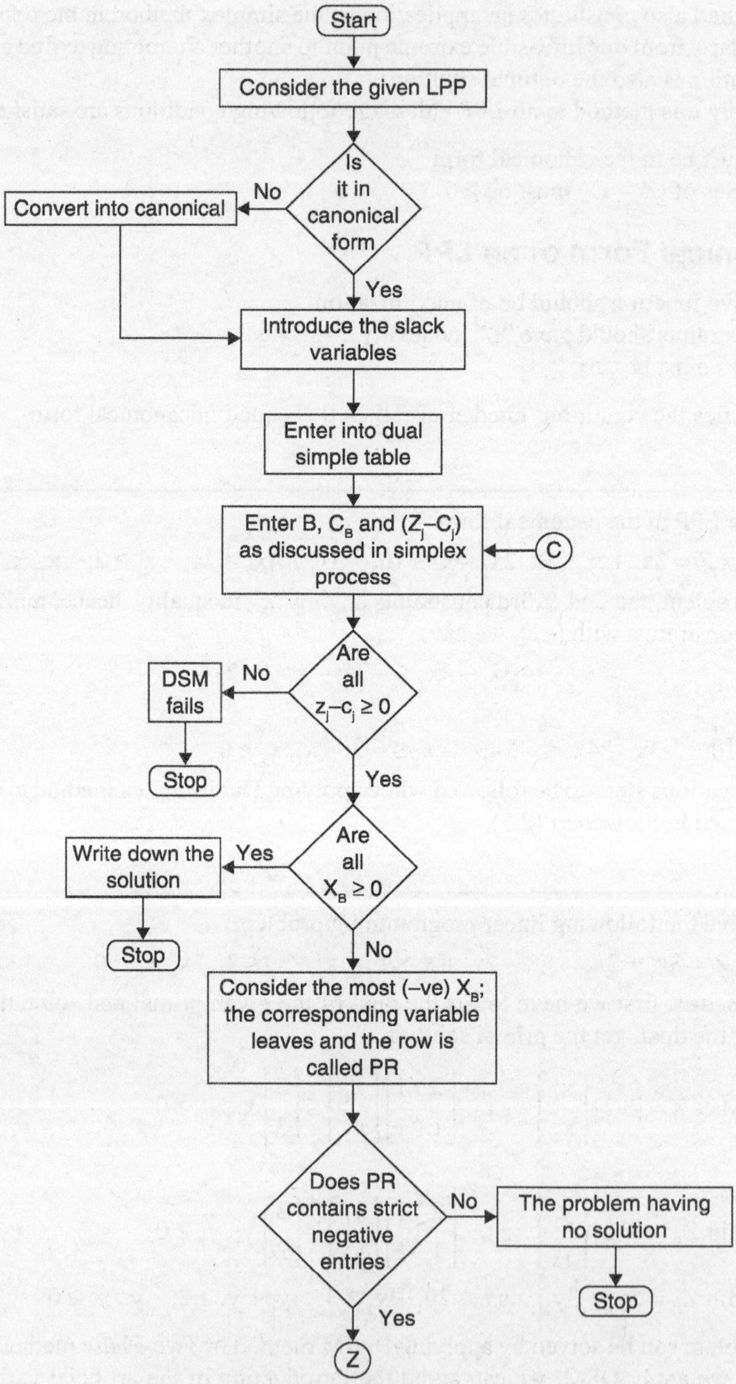

Flowchart 2.7 Dual simplex method

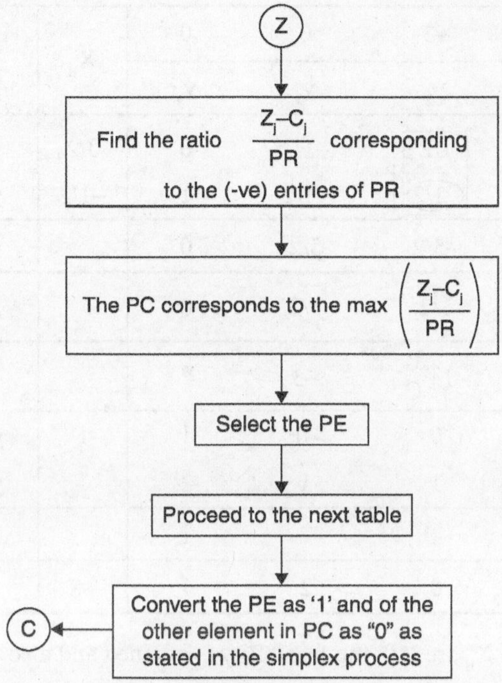

Flowchart 2.7 Continued

Introducing the slack variables
y_3 & y_4, we have

$$\text{Max } Z_0^* = -5y_1 - 3y_2; \quad \text{s.t.} \quad -2y_1 - y_2 + y_3 = -3; \quad -y_1 - y_2 + y_4 = -2$$

C_j	−5	−3	0	0	X_B	B	C_B
y_j	y_1	y_2	y_3	y_4			
	(−2) PE	−1	1	0	−3	y_3	0 <= PR most (−ve)
	−1	−1	0	1	−2	y_4	0
Z_j	0	0	0	0			
$Z_j - C_j$	5	3	0	0			
$(Z_j - C_j)$/PR	−5/2↑ PC	−3	*	*			

C_j	−5	−3	0	0	X_B	B	C_B
y_j	y_1	y_2	y_3	y_4			
PR	1	1/2	−1/2	0	3/2	y_1	−5
R_2 + PR	0	(−1/2) PE	−1/2	1	−1/2	y_4	0 <= PR most (−ve)
Z_j	−5	−5/2	5/2	0			
$Z_j - C_j$	0	1/2	5/2	0			
$(Z_j - C_j)$/PR	*	−1↑ PC	−5	*			
R_1 − (1/2)PR	1	0	−1	1	1	y_1	−5
PR	0	1	1	−2	1	y_2	−3
Z_j	−5	−3	2	1			
$Z_j - C_j$	0	0	2	1	−8		

Since, alt the elements in X_B are ≥ 0; the feasibility is attained and also optimal.

Dual Variable Presented in the X_B Column of the Initial Table	j	The Constraint which it Belongs (i)	Primal Variable Related to it (x_1)	$Z_j - C_j$ Value Corresponds to the Dual Variables in the Optimum Table	Optimum Solution	Max. Z_p
y_3	3	1	x_1	2	$x_1 = 2$	3(2) + 2(1)
y_4	4	2	x_2	1	$x_2 = 1$	= 8

Dual solution:

Max $Z_D^* = -8$ [i.e. Min $Z_D^* = 8$]

$y_1 = 1; y_2 = 1$

Primal solution:

$x_1 = 2; x_2 = 1;$ Max $Z_p = 8$

Remark 1: Note that Max Z_p = Min Z_n = 8

Remark 2: "*" in $(Z_j - C_j)$/PR indicates the ratio is not possible for the corresponding PR entries.

Example: 50

Solve the LPP by DSM

Max $Z = -2x_1 - 9x_2 - 24x_3 - 8x_4 - 5x_5$; s.t. $x_1 + x_2 + 2x_3 - x_5 - x_6 = 1$: $-2x_1 + x_3 + x_4 + x_5 - x_7 = 2$; $x_1 \geq 0$ (for j = 1, 2, ... 7)

Multiplying both sides of 1st and 2nd constraints by (−1).

We have

$$-x_1 - x_2 - 2x_3 + x_5 + x_6 = -1$$
$$2x_1 - x_3 - x_4 - x_5 + x_7 = -2$$

C_j	−2	−9	−24	−8	−5	0	0	X_B	B	C_B
x_j	x_1	x_2	x_3	x_4	x_5	x_6	x_7			
	−1	−1	−2	0	1	1	0	−1	x_6	0
	2	0	−1	−1	(−1) P E	0	1	−2	x_7	0<= most (−ve) PR
Z_j	0	0	0	0	0	0	0			
$Z_j - C_j$	2	9	24	8	5	0	0			
$(Z_j - C_j)$/PR	*	*	−24	−8	−5↑ P C	*	*			
R_1 − PR	1	−1	−3	(−1) P E	0	1	1	−3	x_6	0<= most (−ve) PR
PR	−2	0	1	1	1	0	−3	2	x_5	−5
Z_j	10	0	−5	−5	−5	0	5			
$Z_j - C_j$	12	9	19	3	0	0	5			
$(Z_j - C_j)$/PR	*	−9	−19/3	−3↑ P C	*	*	*			
PR	−1	1	3	1	0	−1	−1	3	x_4	−8
R_2 − PR	−1	−1	(−2) P E	0	1	1	0	−1	x_5	−5<= most (−ve) PR
Z_j	13	−3	−14	−8	−5	3	8			
$Z_j - C_j$	15	6	10	0	0	3	8			
$(Z_j - C_j)$/PR	−15	−6	−5↑ P C	*	*	*	*			
R_1 − 3PR	−5/2	−1/2	0	1	3/2	1/2	−1	3/2	x_4	−8
PR	1/2	1/2	1	0	−1/2	−1/2	0	1/2	x_3	−24
Z_j	8	−8	−24	−8	0	8	8			
$Z_j - C_j$	10	1	0	0	5	8	8	−24		

Since, all the entries in X_B are ≥ 0. The feasibility is attained.
The optimum solution is

$$x_1 = 0, \quad x_2 = 0, \quad x_3 = 1/2, \quad x_4 = 3/2, \quad x_5 = 0, \quad x_6 = 0, \quad x_7 = 0 \quad \text{and} \quad \text{Max } Z = -24.$$

2.16 THE REVISED SIMPLEX METHOD (RSM)

This method uses the same basic principles of the simplex method. Like simplex method no need to calculate the entire table. Instead, we can calculate the relevant information. It moves from one basic feasible solution to another directly using the original equations.

Notations Used

B	:	Basis column
B^{-1}	:	Inverse matrix of B. It is of order mxm.
b	:	Right hand side value of the given constraints
a_j	:	Coefficient matrix of the variable x_j both in the objective function and constraints. It is a column matrix of order $(m + 1) x_1$
$(Z_j - C_j)$	:	$z \times a_j$ = (z-row entries) (a_j) $j = 1, 2, \ldots n)$

For optimality, $(Z_j - C_j) \geq 0$ for all non-basic variables. If any one or more than one non-basic variables $(Z_j - C_j) < 0$, implies that the existing solution is not optimum and it must be improved further. Then select non-basic variable corresponds to the most negative $\{Z_j - C_j < 0\}$ as an entering variable. Find the entering variables current column entries using the relation.

$$y_j = (\text{extended } B^{-1}) * a_j$$

Enter the value of y into the table. Also to select the leaving variable, find min $\{X_B/y_j; y_j > 0\}$ for corresponding entries, then the minimum value corresponding to the variable leaves the basis. Then encircle the pivotal element. Convert the PE as 1 and the other non-pivotal entries as 0 including the Z-row using.

Matrix Row Transformation Technique

Structure of the table

Type 1:

	B	B^{-1}	X_B	Y_j	$\theta = X_B/Y_j$
	Z				

Type 2:

	B	Z	Z_D	B^{-1}		X_B	y_j	$\theta = x_B/y_j$
	Z							
	Z_D							

2.16.1 Type 1

***Example:* 51**

Use Revised Simplex Method (RSM) to solve

$$\text{Max } z = 3x_1 + 5x_2$$

$$\text{s.t.} \quad x_1 \le 4;\ x_2 \le 6; \quad 3x_1 + 2x_2 \le 18; \quad x_1, x_2 \ge 0$$

Step 1: Convert the given LPP into a standard form with maximization objective.

$$\text{Max } z = 3x_1 + 5x_2 + 0x_3 + 0x_4 + 0x_5$$

$$\text{s.t.} \quad x_1 + x_3 = 4;\ x_2 + x_4 = 6; \quad 3x_1 + 2x_2 + x_5 = 18; \quad x_j \ge 0\ (j = 1, \ldots .5)$$

where x_1 and x_2 be the decision variable, and x_3, x_4 and x_5 be the slack variables.

Step 2: Rewrite the LPP in step-1 as follows:

$$z - 3x_1 - 5x_2 - 0x_3 - 0x_4 - 0x_5 = 0$$

$$x_1 + x_3 = 4; \quad x_2 + x_4 = 6; \quad 3x_1 + 2x_2 + x_5 = 18 \tag{1}$$

Step 3: The basic variable matrix B can be defined as columns of z, x_3, x_4 & x_5.

$$B = [z \quad x_3 \quad x_4 \quad x_5]$$

$$= \begin{bmatrix} 1 & 0 & 0 & 0 \\ 0 & 1 & 0 & 0 \\ 0 & 0 & 1 & 0 \\ 0 & 0 & 0 & 1 \end{bmatrix}$$

Since B is an identity matrix of order 4×4. Naturally its inverse must be B itself. Hence

$$B^{-1} = \begin{bmatrix} 1 & 0 & 0 & 0 \\ 0 & 1 & 0 & 0 \\ 0 & 0 & 1 & 0 \\ 0 & 0 & 0 & 1 \end{bmatrix}; \quad b = \begin{bmatrix} 4 \\ 6 \\ 18 \end{bmatrix}$$

C_B = [Cost of the basis element] = [0 0 0]

$$a_1 = \begin{bmatrix} -3 \\ 1 \\ 0 \\ 3 \end{bmatrix}; \quad a_2 = \begin{bmatrix} -5 \\ 0 \\ 1 \\ 2 \end{bmatrix}; \quad a_3 = \begin{bmatrix} 0 \\ 1 \\ 0 \\ 0 \end{bmatrix}$$

$$a_4 = \begin{bmatrix} 0 \\ 0 \\ 1 \\ 0 \end{bmatrix}; \quad a_5 = \begin{bmatrix} 0 \\ 0 \\ 0 \\ 1 \end{bmatrix}$$

Initial objective value, $z = C_B \times b$

$$= [0 \quad 0 \quad 0]\begin{bmatrix} 4 \\ 6 \\ 18 \end{bmatrix} = 0$$

Step 4: Write down the extended B^{-1}, Basis and X_B by considering the given problem. Evaluate the $(z_j - c_j)$ values for all non-basic variables x_1 & x_2.

$$z_1 - c_1 = z * a_1 = [1 \quad 0 \quad 0 \quad 0]\begin{bmatrix} 3 \\ 1 \\ 0 \\ 3 \end{bmatrix} = -3$$

$$z_2 - c_2 = z * a_2 = [1 \quad 0 \quad 0 \quad 0]\begin{bmatrix} -5 \\ 0 \\ 1 \\ 2 \end{bmatrix} = -5$$

Since all $(z_j - c_j)$ values are ≤ 0, implies that the optimality is not attained. Most (– ve) = {– 3, – 5} = –5, corresponds to x_2. Hence x_2 enters the basis. Then evaluate x_2's current column elements using the relation,

$$y_2 = B^{-1} * a_2 = \begin{bmatrix} 1 & 0 & 0 & 0 \\ 0 & 1 & 0 & 0 \\ 0 & 0 & 1 & 0 \\ 0 & 0 & 0 & 1 \end{bmatrix}\begin{bmatrix} -5 \\ 0 \\ 1 \\ 2 \end{bmatrix} = \begin{bmatrix} -5 \\ 0 \\ 1 \\ 2 \end{bmatrix}.$$

Enter this y_2 column under y_j and find the ratio θ and proceed as like in the simplex method. Clearly x_4 leaves the basis.

$$z_1 - c_1 = [1 \quad 0 \quad 5 \quad 0]\begin{bmatrix} -3 \\ 1 \\ 0 \\ 3 \end{bmatrix} = -3$$

$$z_4 - c_4 = [1 \quad 0 \quad 5 \quad 0]\begin{bmatrix} 0 \\ 0 \\ 1 \\ 0 \end{bmatrix} = 5$$

Clearly x_1 enters,

$$y_1 = B^{-1} \times a_1 = \begin{bmatrix} 1 & 0 & 5 & 0 \\ 0 & 1 & 0 & 0 \\ 0 & 0 & 1 & 0 \\ 0 & 0 & -2 & 1 \end{bmatrix} \begin{bmatrix} -3 \\ 1 \\ 0 \\ 3 \end{bmatrix} = \begin{bmatrix} -3 \\ 1 \\ 0 \\ 3 \end{bmatrix}$$

Table 3

$$z_4 - c_4 = [1 \quad 0 \quad 3 \quad 3] \begin{bmatrix} 0 \\ 0 \\ 1 \\ 0 \end{bmatrix} = 3 \quad z_5 - c_5 = [1 \quad 0 \quad 3 \quad 3] \begin{bmatrix} 0 \\ 0 \\ 0 \\ 1 \end{bmatrix} = 3$$

		B	Z	B^{-1}			X_B	y_j	θ	
Table 1		Z	1	0	0	0	0	–5		
		x_3	0	1	0	0	4	0	–	
		x_4	0	0	1	0	6	(1) P E	6	<=PR
		x_5	0	0	0	1	18	2	9	
Table 2	$R_0 + 5PR$	Z	1	0	5	0	30	–3		
	R_1	x_3	0	1	0	0	4	1	4	
	PR	x_2	0	0	1	0	6	0	–	
	$R_3 - 2PR$	x_5	0	0	–2	1	6	(3) P E	2	<=PR
Table 3	$R_0 + 3PR$	z	1	0	3	3	36			
	$R_1 - PR$	x_3	0	1	2/3	–1	2			
	R_2	x_2	0	0	1	0	6			
	PR	x_1	0	0	–2/3	1	2			

Since all $(z_j - c_j)$ are ≥ 0, imply that the optimality is attained. Hence the optimum solution is, $x_1 = 2$; $x_2 = 6$ and Max $z = 36$

Example: 52

Max $z = 3x_1 + 2x_2 + 5x_3$

s.t. $x_1 + 2x_2 + x_3 \leq 430$; $3x_1 + 2x_3 \leq 460$;

$x_1 + 4x_2 \leq 420$; $x_1, x_2, x_3 \geq 0$ Use RSM.

Step 1: Convert the given LPP into a standard from with Maximization objective.

$$\text{Max } z = 3x_1 + 2x_2 + 5x_3 + 0x_4 + 0x_5 + 0x_6$$

$$\text{s.t. } x_1 + 2x_2 + x_3 + x_4 = 430; \quad 3x_1 + 2x_3 + x_5 = 460; \quad x_1 + 4x_2 + x_6 = 420;$$

$$x_j \geq 0 \ (j = 1....6)$$

Step 2: Rewrite the LPP in step-1 as follows:

$$z - 3x_1 - 2x_2 - 5x_3 - 0x_4 - 0x_5 - 0x_6 = 0$$

$$x_1 + 2x_2 + x_3 + x_4 = 430$$

$$3x_1 + 2x_3 + x_5 = 460$$

$$x_1 + 4x_2 + x_6 = 420$$

Step 3:

$$B = \begin{bmatrix} 1 & 0 & 0 & 0 \\ 0 & 1 & 0 & 0 \\ 0 & 0 & 1 & 0 \\ 0 & 0 & 0 & 1 \end{bmatrix}; \quad B^{-1} = \begin{bmatrix} 1 & 0 & 0 & 0 \\ 0 & 1 & 0 & 0 \\ 0 & 0 & 1 & 0 \\ 0 & 0 & 0 & 1 \end{bmatrix}; \quad b = \begin{bmatrix} 430 \\ 460 \\ 420 \end{bmatrix}$$

$$C_B = [0 \ \ 0 \ \ 0]$$

$$a_1 = \begin{bmatrix} -3 \\ 1 \\ 3 \\ 1 \end{bmatrix}; \quad a_2 = \begin{bmatrix} -2 \\ 2 \\ 0 \\ 4 \end{bmatrix}; \quad a_3 = \begin{bmatrix} -5 \\ 1 \\ 2 \\ 0 \end{bmatrix}; \quad a_4 = \begin{bmatrix} 0 \\ 1 \\ 0 \\ 0 \end{bmatrix};$$

$$a_5 = \begin{bmatrix} 0 \\ 0 \\ 1 \\ 0 \end{bmatrix}; \quad a_6 = \begin{bmatrix} 0 \\ 0 \\ 0 \\ 1 \end{bmatrix}$$

$$\text{initial value of the objective, } z = C_B * b = [0 \ \ 0 \ \ 0] \begin{bmatrix} 430 \\ 460 \\ 420 \end{bmatrix} = 0$$

Step 4:

Table 1:

Write down the extended B^{-1}, Basis and X_B, by considering the given problem in step-2. Evaluate the $(z_j - c_j)$ values for all the non-basic variables x_1, x_2 & x_3.

$$z_1 - c_1 = z * a_1 = [1 \ \ 0 \ \ 0 \ \ 0] \begin{bmatrix} -3 \\ 1 \\ 3 \\ 1 \end{bmatrix} = -3$$

$$z_2 - c_2 = z^* a_2 = [1 \quad 0 \quad 0 \quad 0]\begin{bmatrix}-2\\2\\0\\4\end{bmatrix} = -2$$

$$z_3 - c_3 = z^* a_3 = [1 \quad 0 \quad 0 \quad 0]\begin{bmatrix}-5\\1\\2\\0\end{bmatrix} = -5$$

Since all $(z_j - c_j) \leq 0$, imply that the optimality is not attained. Most (–ve) = {–3, –2, –5} = –5, corresponds to x_3. Hence x_3 enters the basis.

		B	Z	B^{-1}			X_B	y_j	θ	
Table 1		Z	1	0	0	0	0	–5		
		x_4	0	1	0	0	430	1	430	
		x_5	0	0	1	0	460	(2) PE	230	<=PR
		x_6	0	0	0	1	420	0	–	
Table 2	$R_0 + 5PR$	Z	1	0	5/2	0	1150	–2		
	$R_1 - PR$	x_4	0	1	–1/2	0	200	(2) PE	100	<=PR
	PR	x_3	0	0	1/2	0	230	0	–	
	R_3	x_6	0	0	0	1	420	4	105	
Table 3	$R_0 + 2PR$	Z	1	1	2	0	1350			
	PR	x_2	0	1/2	–1/4	0	100			
	R_2	x_3	0	0	1/2	0	230			
	$R_3 - 4PR$	x_6	0	–1/2	1	1	20			

Then evaluate x_3's current column elements using the relation.

$$y_3 = B^{-1*} a_3 = \begin{bmatrix}1 & 0 & 0 & 0\\0 & 1 & 0 & 0\\0 & 0 & 1 & 0\\0 & 0 & 0 & 1\end{bmatrix}\begin{bmatrix}-5\\1\\2\\0\end{bmatrix} = \begin{bmatrix}-5\\1\\2\\0\end{bmatrix}$$

Enter this y_3 column under y_j and find the ratio θ and proceed like in the simplex process. Clearly x_5 leaves the basis.

Table 2:

$$z_1 - c_1 = z^* a_1 = [1 \quad 0 \quad 5/2 \quad 0] \begin{bmatrix} -3 \\ 1 \\ 3 \\ 1 \end{bmatrix} = 9/2$$

$$z_2 - c_2 = z^* a_2 = [1 \quad 0 \quad 5/2 \quad 0] \begin{bmatrix} -2 \\ 2 \\ 0 \\ 4 \end{bmatrix} = -2$$

$$z_5 - c_5 = z^* a_5 = [1 \quad 0 \quad 5/2 \quad 0] \begin{bmatrix} 0 \\ 0 \\ 1 \\ 0 \end{bmatrix} = 5/2$$

Implies that the variable x_2 enters the basis. Then evaluate x_2's current column elements using the relation.

$$y_2 = B^{-1*} a_2 = \begin{bmatrix} 1 & 0 & 5/2 & 0 \\ 0 & 1 & -1/2 & 0 \\ 0 & 0 & 1/2 & 0 \\ 0 & 0 & 0 & 1 \end{bmatrix} \begin{bmatrix} -2 \\ 2 \\ 0 \\ 4 \end{bmatrix} = \begin{bmatrix} -2 \\ 2 \\ 0 \\ 4 \end{bmatrix}$$

Enter this y_2 column under y_j and find the ratio θ and proceed. The variable x_4 leaves the basis.

Table 3:

$$z_1 - c_1 = z^* a_1 = [1 \quad 1 \quad 2 \quad 0] \begin{bmatrix} -3 \\ 1 \\ 3 \\ 1 \end{bmatrix} = 4$$

$$z_4 - c_4 = z^* a_4 = [1 \quad 1 \quad 2 \quad 0] \begin{bmatrix} 0 \\ 1 \\ 0 \\ 0 \end{bmatrix} = 1$$

$$z_5 - c_5 = z^* a_5 = [1 \quad 1 \quad 2 \quad 0] \begin{bmatrix} 0 \\ 0 \\ 1 \\ 0 \end{bmatrix} = 2$$

Since all $(z_j - c_j)$ are ≥ 0, imply that the optimality is attained. Hence the optimum solution is,

$$x_1 = 0; \quad x_2 = 100; \quad x_3 = 230 \text{ and}$$

$$\text{Max } z = 1350$$

2.16.2 Type-II

In the previous type, we have seen how to include slack variables along with the decision variables and to proceed. In this type, we take into consideration the introduction of surplus and artificial variable, and the procedure to adopt. For each artificial variable introduced we attach a penalty cost of ₹M in the objective (where M is a very large value) and apply the procedure of type-I.

Example: 53

Use RSM to solve the following LPP.

$$\text{Min } z = x_1 + x_2; \quad \text{s.t. } x_1 + 2x_2 \geq 7; \quad 4x_1 + x_2 \geq 6; \quad x_1, x_2 \geq 0$$

Step 1: Convert the given LPP into a standard form with maximization objective.

$$\text{Max } z^* = -x_1 - x_2 + 0x_3 + 0x_4$$

$$\text{s.t. } \quad x_1 + 2x_2 - x_3 = 7; \quad 4x_1 + x_2 - x_4 = 6; \quad x_j \geq 0 \ (j = 1, ..., 4)$$

Step 2: Introduce the artificial variables, x_5 and x_6 one each to the constraints that is,

$$\text{Max } z^* = -x_1 - x_2 + 0x_3 + 0x_4 - Mx_5 - Mx_6$$

$$\text{s.t. } \quad x_1 + 2x_2 - x_3 + x_5 = 7; \quad 4x_1 + x_2 - x_4 + x_6 = 6; \quad x_j \geq 0 \ (j = 1, ..., 6)$$

Step 3: Rewrite the problem in step-2 as,

$$\left.\begin{array}{l} z^* + x_1 + x_2 - 0x_3 - 0x_4 + Mx_5 + Mx_6 = 0 \\ \text{s.t. } \quad x_1 + 2x_2 - x_3 + x_5 = 7; \quad 4x_1 + x_2 - x_4 + x_6 = 6 \end{array}\right\} \quad (1)$$

The basic variable matrix B can be defined as

$$B = [z \quad x_5 \quad x_6] \begin{bmatrix} 1 & M & M \\ 0 & 1 & 0 \\ 0 & 0 & 1 \end{bmatrix}$$

Note: In any square matrix, if the following conditions are satisfied then we can define its inverse in an easy way.

1. Diagonal elements must be "1"
2. Below diagonal elements must be "0"
3. Above the diagonal elements may be 0 or any +ve entry.

Case(i): Along with (1) & (2); if all entries of above diagonal elements are '0', then clearly that matrix under consideration is an Identity. Clearly inverse of any identity matrix is itself.

Case(ii): Along with (1) & (2); if some or all above diagonal elements are +ve (> 0); for the inverse, change the sign of above diagonal elements sign and keep all other elements as "0".

$$B = [z \quad x_5 \quad x_6]\begin{bmatrix} 1 & M & M \\ 0 & 1 & 0 \\ 0 & 0 & 1 \end{bmatrix}$$

Clearly

$$B^{-1} = \begin{bmatrix} 1 & -M & -M \\ 0 & 1 & 0 \\ 0 & 0 & 1 \end{bmatrix}$$

$$C_B = [-M \quad -M]; \qquad b = \begin{bmatrix} 7 \\ 6 \end{bmatrix}$$

Initial objective value is, $z = C_B b = [-M \quad -M]\begin{bmatrix} 7 \\ 6 \end{bmatrix} = -13M$

$$a_1 = \begin{bmatrix} 1 \\ 1 \\ 4 \end{bmatrix}; \quad a_2 = \begin{bmatrix} 1 \\ 2 \\ 1 \end{bmatrix}; \quad a_3 = \begin{bmatrix} 0 \\ -1 \\ 0 \end{bmatrix}; \quad a_4 = \begin{bmatrix} 0 \\ 0 \\ -1 \end{bmatrix};$$

$$a_5 = \begin{bmatrix} M \\ 1 \\ 0 \end{bmatrix}; \quad a_6 = \begin{bmatrix} M \\ 0 \\ 1 \end{bmatrix}$$

Step 4:

Table 1: Write down the extended B^{-1} Basis and X_B by considering the given problem. Evaluate

$$z_1 - c_1 = z * a_1 = [1 \quad -M \quad -M]\begin{bmatrix} 1 \\ 1 \\ 4 \end{bmatrix} = -5M + 1$$

$$z_2 - c_2 = z * a_2 = [1 \quad -M \quad -M]\begin{bmatrix} 1 \\ 2 \\ 1 \end{bmatrix} = -3M + 1$$

$$z_3 - c_3 = z * a_3 = [1 \quad -M \quad -M]\begin{bmatrix} 0 \\ -1 \\ 0 \end{bmatrix} = M$$

$$z_4 - c_4 = z * a_4 = [1 \quad -M \quad -M]\begin{bmatrix} 0 \\ 0 \\ -1 \end{bmatrix} = M$$

Clearly –5M + 1 being the most negative item, which corresponds to the variable x_1. Hence x_1 enters into the basis. The current column of x_1 can be defined as,

$$y_1 = B^{-1} * a_1 = \begin{bmatrix} 1 & -M & -M \\ 0 & 1 & 0 \\ 0 & 0 & 1 \end{bmatrix} \begin{bmatrix} 1 \\ 1 \\ 4 \end{bmatrix} = \begin{bmatrix} -5M+1 \\ 1 \\ 4 \end{bmatrix}$$

Enter y_1 in Table-1 and proceed.

Table 2:

$$z_2 - c_2 = z^* a_2 = [1 \quad -M \quad (1/4)(M-1)] \begin{bmatrix} 1 \\ 2 \\ 1 \end{bmatrix} = (1/4)(-7M+3)$$

$$z_3 - c_3 = z^* a_3 = [1 \quad -M \quad (1/4)(M-1)] \begin{bmatrix} 0 \\ -1 \\ 0 \end{bmatrix} = M$$

$$z_4 - c_4 = z^* a_4 = [1 \quad -M \quad (1/4)(M-1)] \begin{bmatrix} 0 \\ 0 \\ -1 \end{bmatrix} = (1/4)(M+1)$$

That is, variable x_2 enters into the basis. Then the current column of x_2 can be given as,

$$y_2 = B^{-1} * a_2 = \begin{bmatrix} 1 & -M & (1/4)[M-1] \\ 0 & 1 & -1/4 \\ 0 & 0 & 1/4 \end{bmatrix} \begin{bmatrix} 1 \\ 2 \\ 1 \end{bmatrix} = \begin{bmatrix} (1/4)[-7M+1] \\ 7/4 \\ 1/4 \end{bmatrix}$$

Table 3:

$$z_3 - c_3 = [1 \quad -3/7 \quad -1/7] \begin{bmatrix} 0 \\ -1 \\ 0 \end{bmatrix} = 3/7$$

$$z_4 - c_4 = [1 \quad -3/7 \quad -1/7] \begin{bmatrix} 0 \\ 0 \\ -1 \end{bmatrix} = 1/7$$

Since all $z_j - c_j \geq 0$ for the non-basic variables, the optimality is attained. The required optimum solution is,

$$x_1 = 5/7; \; x_2 = 22/7; \quad \text{Max } Z^* = -27/7 \quad \text{and} \quad \text{Min } Z = 27/7.$$

		B	Z*	B^{-1}		X_B	y_j	θ
Table 1:		Z	1	–M	–M	–13M	–5M + 1	
		x_5	0	1	0	7	1	7/1 = 7
		x_6	0	0	1	6	(4) P E	6/4 = 3/2 <= PR
Table 2:	R_0 – [–5M + 1] *PR	Z	1	–M	(1/4)(M–1)	(–1/2)[11M+3]	(1/4)[–7M+3]	
	R_1 – PR	x_5	0	1	–1/4	11/2	(7/4) P E	22/7 <= PR
	PR	x_1	0	0	1/4	3/2	1/4	6
	R_0 – (1/4) [–7M + 3]PR	Z	1	–3/7	–1/7	–27/7		
Table 3:	PR	x_2	0	4/7	–1/7	22/7		
	R_2 – (1/4)PR	x_1	0	–1/7	2/7	5/7		

2.16.3 Type-III

In this section, we are going to solve the type-ll model using two-phase. As usual in phase-I concentration is based on artificial variables, then decide the nature of solution and proceed to phase-II.

Z_D : The dummy maximization objective based on artificial variables.
Z : The actual maximization objective.

The structure of the table is also modified as follows.

B	z	z_D	B^{-1}	X_B	y_j	θ
Z						
z_D						

Example: 54

Max $Z = 2x_1 + 3x_2$
s.t. $x_1 \le 4$; $x_1 - x_2 \ge 0$; $x_1, x_2 \ge 0$

Step 1: Convert the given LPP into a standard form with maximization objective.

Max $Z = 2x_1 + 3x_2 + 0x_3 - 0x_4$
s.t. $x_1 + x_3 = 4$; $x_1 - x_2 - x_4 + x_5 = 0$; $x_j \ge 0$ (j = 1, ... 4)

x_3 : slack variable
x_4 : surplus variable
x_5 : artificial variable

Construct the dummy maximization objective function based on the artificial variables.

Phase I

$$\text{Max } Z_D = -x_5, \text{ subject to the constraints given.}$$

Step 2: Rewrite the LPP in step-1 as follows:

$$z - 2x_1 - 3x_2 - 0x_3 - 0x_4 = 0; \quad z_D + x_5 = 0; \quad x_1 + x_3 = 4; \quad x_1 - x_2 - x_4 + x_5 = 0 \qquad (1)$$

Step 3: Write down a_j (j = 1, ... 5)

Current objective value; $z = [0 \quad 0]\begin{bmatrix} 4 \\ 0 \end{bmatrix} = 0$

Current Dummy objective value; $z_D = [0 \quad -1]\begin{bmatrix} 4 \\ 0 \end{bmatrix} = 0$

$$B = \begin{bmatrix} 1 & 0 & 0 & 0 \\ 0 & 1 & 0 & 1 \\ 0 & 0 & 1 & 0 \\ 0 & 0 & 0 & 1 \end{bmatrix}; \quad B^{-1} = \begin{bmatrix} 1 & 0 & 0 & 0 \\ 0 & 1 & 0 & -1 \\ 0 & 0 & 1 & 0 \\ 0 & 0 & 0 & 1 \end{bmatrix}$$

$$a_1 = \begin{bmatrix} -2 \\ 0 \\ 1 \\ 1 \end{bmatrix}; \quad a_2 = \begin{bmatrix} -3 \\ 0 \\ 0 \\ -1 \end{bmatrix}; \quad a_3 = \begin{bmatrix} 0 \\ 0 \\ 1 \\ 0 \end{bmatrix}; \quad a_4 = \begin{bmatrix} 0 \\ 0 \\ 0 \\ -1 \end{bmatrix}; \quad a_5 = \begin{bmatrix} 0 \\ 1 \\ 0 \\ 1 \end{bmatrix}$$

Step 4:

Table 1: Write down the B^{-1}, B and X_B, by considering the given problem in Step-2. Evaluate the $z_j - c_j$ values for all non-basic variables x_1, x_2, & x_4.

$$z_1 - c_1 = z_D * a_1 = [0 \quad 1 \quad 0 \quad -1]\begin{bmatrix} -2 \\ 0 \\ 1 \\ 1 \end{bmatrix} = -1$$

$$z_2 - c_2 = z_D * a_2 = [0 \quad 1 \quad 0 \quad -1]\begin{bmatrix} 0 \\ -3 \\ 0 \\ -1 \end{bmatrix} = 1$$

$$z_4 - c_4 = z_D * a_4 = [0 \quad 1 \quad 0 \quad -1]\begin{bmatrix} 0 \\ 0 \\ 0 \\ -1 \end{bmatrix} = 1$$

This implies that the variable x_1 enters the basis.

$$y_1 = B^{-1} * a_1 = \begin{bmatrix} 1 & 0 & 0 & 0 \\ 0 & 1 & 0 & -1 \\ 0 & 0 & 1 & 0 \\ 0 & 0 & 0 & 1 \end{bmatrix} \begin{bmatrix} -2 \\ 0 \\ 1 \\ 1 \end{bmatrix} = \begin{bmatrix} -2 \\ -1 \\ 1 \\ 1 \end{bmatrix}$$

Enter this y_1 column under y_j and proceed.

Table 2:

$$z_1 - c_1 = z_D * a_1 = [0 \quad 1 \quad 0 \quad 0] \begin{bmatrix} -2 \\ 0 \\ 1 \\ 1 \end{bmatrix} = 0$$

$$z_2 - c_2 = z_D * a_2 = [0 \quad 1 \quad 0 \quad 0] \begin{bmatrix} -3 \\ 0 \\ 0 \\ -1 \end{bmatrix} = 0$$

$$z_5 - c_5 = z_D * a_5 = [0 \quad 1 \quad 0 \quad 0] \begin{bmatrix} 0 \\ 1 \\ 0 \\ 1 \end{bmatrix} = 0$$

Since all $z_j - c_j \geq 0$, imply that the phase-I optimum is attained. Since the phase-I optimum doesn't contain any artificial variable, implies that the given problem said to have a feasible solution. Hence, we can proceed to phase-II.

For phase-II table 1, write down the entries of the optimum table of phase-I as it is.

Phase-I

Table 1:

	B	Z	Z_D	B^{-1}		X_B	y_j	
	Z	1	0	0	0	0	–2	θ
	Z_D	0	1	0	– 1	0	–1	
	x_3	0	0	1	0	4	1	4
	x_5	0	0	0	1	0	(1) PE	0 <= PR
$R_0 + 2PR$	Z	1	0	0	2	0		

Table 2:

$Z_D + PR$	Z_D	0	1	0	0	0		
$R_1 - PR$	x_3	0	0	1	–1	4		
PR	x_1	0	0	0	1	0		

Phase-II

Table 1:

	B	Z	Z_D	B^{-1}		X_B	y_j	
	Z	1	0	0	2	0	–5	θ
	Z_D	0	1	0	0	0	0	
	x_3	0	0	1	–1	4	(1) P E	4 <= PR
	x_1	0	0	0	1	0	–1	–
$R_0 + 5PR$	Z	1	0	5	–3	20		
R_D	Z_D	0	1	0	0	0		
PR	x_2	0	0	1	–1	4		
R_2	x_1	0	0	0	1	0		

Table 1:

$$z_2 - c_2 = z^* a_2 = [1 \quad 0 \quad 0 \quad 2]\begin{bmatrix} -3 \\ 0 \\ 1 \\ -1 \end{bmatrix} = -5$$

$$z_4 - c_4 = z^* a_4 = [1 \quad 0 \quad 0 \quad 2]\begin{bmatrix} 0 \\ 0 \\ 0 \\ -1 \end{bmatrix} = -2$$

Clearly x_2 enters the basis. Find the current column of x_2, using

$$y_2 = B^{-1*} a_2 = \begin{bmatrix} 1 & 0 & 0 & 2 \\ 0 & 1 & 0 & 0 \\ 0 & 0 & 1 & -1 \\ 0 & 0 & 0 & 1 \end{bmatrix}\begin{bmatrix} -3 \\ 0 \\ 0 \\ -1 \end{bmatrix} = \begin{bmatrix} -5 \\ 0 \\ 1 \\ -1 \end{bmatrix}$$

Enter this y_2 column under y_j and find the ratio θ and proceed. Clearly, x_3 leaves the basis.

Table 2:

$$z_3 - c_3 = z^* a_3 = [1 \quad 0 \quad 5 \quad -3] \begin{bmatrix} 0 \\ 0 \\ 1 \\ 0 \end{bmatrix} = 5$$

$$z_4 - c_4 = z^* a_4 = [1 \quad 0 \quad 5 \quad -3] \begin{bmatrix} 0 \\ 0 \\ 0 \\ -1 \end{bmatrix} = 3$$

Since all $Z_j - C_j \geq 0$ in the phase-II, imply that the phase-II optimality is attained. Hence the optimum solution is $x_1 = 0$; $x_2 = 4$ and Max $z = 20$.

EXERCISE PROBLEMS

*1. Three products are processed through three different operations. The following table gives the time required, profit per unit of each product and operation capacity:

Operations	Time per Unit (Minutes)			Operation Capacity Minutes/Day
	Product I	Product II	Product III	
1	3	4	3	43
2	5	0	4	46
3	3	6	2	42
Profit/Unit (₹)	2	1	3	

The objective is to find product-mix which maximizes the profit. Formulate this as an LP. Problem.

*2. A manufacturer produces two types of models M_1 and M_2. Each M_1 model requires 4 hours of grinding and 2 hours of polishing, whereas each M_2 model requires 2 hours of grinding and 6 hours of polishing. There are 2 grinders and 3 polishers. Each grinder works for 40 hours a week and each polisher works for 60 hours a week. Profits on M_1 model is ₹3.00 and M_2 model is ₹4.00. Whatever produced is sold in the market. How does the manufacturer allocate his production capacity to two models so that he maximizes the profit? Set up the L.P. Problem.

*3. A manufacturer of cylindrical containers receives tin sheets in widths of 30 cm and 60 cm respectively. For these containers, the sheets are to be cut to three different widths of 15 cm, 21 cm and 27 cm respectively. The number of rolls required in these three widths are 400, 200, 300 respectively. There is no limit on the length of standard tin sheets. Formulate the L.P. Model for the production schedule that minimizes the trim losses.

*4. A farmer has 1000 acres of land in which he can grow corn, wheat or soyabeans. Each acre of corn costs ₹100 for preparation, requires 7 man-day of work and yields a profit of ₹30. An acre of wheat costs ₹120 to prepare, requires 10 man-days of work and yields a profit of ₹40. An acre of soyabeans costs ₹70 to prepare, requires 8 man-days of work and yields a profit of ₹20. The farmer has ₹1,00,000 for preparation and 8,000 man-days of work. Set-up the LPP.

*5. Two products A and B are processed on three machines M_1, M_2 and M_3. The processing times per unit, machine availability and profit per unit are:

Machine	Processing Time (Hours)		Availability (Hours)
	A	B	
M_1	2	3	1,500
M_2	3	2	1,500
M_3	1	1	1,000
Profit	₹10	₹12	

Any unutilized time on machine M_3 can be given on rental basis to another firm at an hourly rate of ₹1.50. Formulate the Mathematical Model.

*6. Old hens can be bought at ₹30 each and young ones at ₹50 each. The old hens lay 3 eggs per week, and the young ones lay 5 eggs per week, each egg worth ₹2. A hen costs ₹1.00 per week to feed. Veeramani has only ₹8000 to spend on purchasing hens. How many of each kind should Veeramani buy to give a profit of more than ₹600 per week, assuming that Veeramani cannot house more than 100 hens. Formulate this as an LPP.

*7. A person requires 10, 12 and 12 units of chemicals A, B and C respectively for his garden. A liquid product contains 5, 2, and 1 units of A, B and C respectively per Jar. A dry product contains 1, 2, and 4 units of A, B and C per carton. If the liquid product is sold for ₹3 per Jar and the dry product is ₹2 per carton, how many of each should be purchased to minimize the cost and meet the requirements. Set up an L.P. Problem to minimize the cost. Do not solve it.

*8. Three grades of coal A, B and C contains phosphorus and ash as impurities. In a particular Industrial process, fuel upto 100 ton (Maximum) is required which should contain ash not more than 3% and phosphorus not more than 0.03%. It is desired to maximize the profit while satisfying these conditions. There is an unlimited supply of each grade. The percentage of impurities and the profits of grades are given below:

Coal	Phosphorus/Ton (%)	Ash/Ton (%)	Profit in ₹/Ton
A	0.02	2.0	12.00
B	0.04	3.0	15.00
C	0.03	5.0	14.00

Find the proportion in which the three grades be used.

*9. The Indian Bank, a full-service facility, is in the process of formulating a loan policy involving a maximum of ₹100 million. The following table provides the pertinent data about the different types of loans the bank deals with

Loan Type	Rate of Interest	Probability of Bad Debt
Personal	0.140	0.10
Car	0.130	0.07
Home	0.120	0.03
Farm	0.125	0.05
Commercial	0.100	0.02

Bad debts are unrecoverable and hence produce no interest revenue. Competition with other financial institutions in the area requires the bank to allocate atleast 40% of the funds to farm and commercial loans. To assist the housing industry in the region, home loans must equal atleast 50% of the personal, car and home loans. The bank also has a stated policy specifying that the overall ratio for bad debts on all loans may not exceed 0.04.

*10. The Asset Management Company (AMC) of Canbank Mutual fund Ltd. is considering to determine a best investment portfolio with different proportions of six highly traded blue chip scrips. The details of the said scrips are as follows:

Particulars	Blue-chip Scrips					
	1	2	3	4	5	6
Current Market Price (₹)	80	100	160	120	150	200
Projected Annual Growth Rate	.08	.07	.10	.12	.09	.15
Projected Annual Dividend per share	4.0	6.5	1.0	0.5	2.75	--
Projected Risk	.05	.03	.10	.20	.06	.08

i. Total amount of proposed investment: ₹2.5 crores
ii. For the sake of diversity, at least 100 shares of each stock should be purchased.
iii. Atleast 10 per cent of total investment should be in stocks 1 and 2 combined.
iv. The maximum amount to be invested in stock 6 is ₹2.5 lakhs.
v. Not more than ₹5.0 lakhs should be invested in stock 1 and 2 combined.
vi. Total weighted risk should not be greater than 0.10.
vii. Dividends for the year should be atleast ₹1 lakh.

If the objective is to maximize total return (both in terms of growth and dividend) formulate LP model for determining the optimal number of shares to purchase from each of above said six categories. (Only formulate the problem. No solution to be worked out)

*11. Trichy city is studying the feasibility of introducing a mass transit bus system that will alleviate the smog problem by reducing in-city driving. The initial study seeks the determinations of the minimum number of buses that can handle the transportation needs. After gathering necessary information, the city engineer noticed that the minimum number of buses needed fluctuates with the time of the day. Studying the data further, it became evident that the required number of buses can be approximated by constant values and successive intervals of 4 hours each. The following table summarizes the engineer's findings. To carry out the required daily maintenance, each bus could operate only 8 successive hours a day.

Time	Number of Buses Needed
12.00 m.n – 4.00 a.m	4
4.00 a.m – 8.00 a.m	8
8.00 a.m – 12.00 noon	10
12.00 noon – 4.00 p.m	7
4.00 p.m – 8.00 p.m	12
8.00 p.m – 12.00 m.n	4

*12. M/s Funschool Toys Ltd. manufactures two models of toys T_1 and T_2. Each of these products must go through two manufacturing processes A and B. Each unit of T_1 requires one hour in Process A and 1½ hours in Process B; each unit of T_2 requires 2 hours in Process A and 3/4 hour in process B. The machinery supporting Process A currently has a weekly machine hour capacity of 100 hours, and the weekly capacity of process B machinery is 90 machine hours. Per unit profits on T_1 and T_2 are ₹100 and ₹150 respectively. Determine the optimal product mix for Funschool Ltd.

*13. The emergency Ward of M/s. Appolo Hospitals Ltd., schedules Nurses to come on duty every 3 hours for a 6 hours shift. The shift times and the minimum number of nurses required during each 3 hours time interval are given below. Determine the schedule that will minimize the total number of nurses employed in said Emergency ward.

Time	Number of Nurses Needed
12.00 – 3.00 a.m	1
3.00 a.m – 6.00 a.m	2
6.00 a.m – 9.00 a.m	3
9.00 a.m – 12.00 noon	7
12.00 noon – 3.00 p.m	9

Time	Number of Nurses Needed
3.00 p.m – 6.00 p.m	5
6.00 p.m – 9.00 p.m	4
9.00 p.m – 12.00	3

14. **A work scheduling problem:** A Post Office requires different number of full-time employees on different days of the week. The number of full-time employees required on each day is given in Table-1. Union rules state that each full-time employee must work five consecutive days then receive two days off. For example, an employee who worked Monday to Friday must be off on Saturday & Sunday. The Post Office wants to meet its daily requirements using only full-time employees. Formulate an LP that the Post Office can use to minimize the number of full-time employees that must be hired.

Table: 1

Day1: Monday – 17
Day2: Tuesday – 13
Day3: Wednesday – 15
Day4: Thursday – 19
Day5: Friday – 14
Day6: Saturday – 16
Day7: Sunday – 11

*15. An electronic company manufactures two TV models, each on a separate production line. The daily capacity of the first line is 60 TVs and that of the second is 75 TVs. Each unit of the first model uses 10 pieces of a certain electronic component, whereas each unit of the second model requires 8 pieces of the same component. The maximum daily availability of the special component is 800 pieces. The profit per unit of models 1 and 2 is ₹6000 and ₹4000 respectively. Formulate the above model as an LPP.

*16. Rewrite the LPP in the standard form.

Max $Z = 2x_1 + x_2 + 4x_3$

s.t. $-2x_1 + 4x_2 \leq 4$; $x_1 + 2x_2 + x_3 \geq 5$; $2x_1 + 3x_3 \leq 2$; $x_1, x_2 \geq 0$; x_3 is unrestricted.

(Hint: First replace $x_3 = x_4 - x_5$; such that $x_4, x_5 \geq 0$)

*17. Compute all the basic feasible solutions for the following LPP.

Max $Z = 2x_1 + 3x_2 + 4x_3 + 7x_4$;

st. $2x_1 + 3x_2 - x_3 + 4x_4 = 8$

$x_1 - 2x_2 + 6x_3 - 7x_4 = -3$; $x_1, x_2, x_3, x_4 \geq 0$

*18. Show that the following system of Linear equations have a degenerate solution.
$2x_1 + x_2 - x_3 = 2; \quad 3x_1 + 2x_2 + x_3 = 3$

*19. Find all the basic feasible solutions of the equations
$2x_1 + 6x_2 + 2x_3 + x_4 = 3$
$6x_1 + 4x_2 + 4x_3 + 6x_4 = 2$

Solve the following using Graphical Method:

*20. Max $Z = 3x_1 + 2x_2$; s.t. $-2x_1 + x_2 \le 1$; $x_1 \le 2$; $x_1 + x_2 \le 3$; $x_1, x_2 \ge 0$

21. Max $Z = 2x_1 + x_2$; s.t. $x_1 - x_2 \le 10$; $2x_1 \le 40$; $x_1, x_2 \ge 0$
(Hint: Similar to example: 3)

*22. Min $Z = 4x_1 + 2x_2$; s.t. $x_1 + x_2 \ge 2$; $3x_1 + x_2 \ge 3$; $4x_1 + 3x_2 \ge 6$; $x_1, x_2 \ge 0$

*23. Min $Z = 5x_1 + 3x_2$; s.t. $3x_1 + 5x_2 \le 15$; $5x_1 + 2x_2 \le 10$; $x_1, x_2 \ge 0$

*24. Min $Z = 2x - y$; s.t. $x + y \le 5$; $x + 2y \le 8$; $x, y \ge 0$

*25. Min $Z = 3x_1 + 2x_2$; s.t. $5x_1 + x_2 \ge 10$; $x_1 + x_2 \ge 6$; $x_1 + 4x_2 \ge 12$; $x_1, x_2 \ge 0$

*26. Min $Z = 3x_1 + 2x_2$; s.t. $x_1 + x_2 \ge 1$; $100x_1 + 10x_2 \ge 50$; $10x_1, + 100x_2 \ge 10$; $x_1, x_2 \ge 0$

*27. Max $Z = 5x_1 + 7x_2$; s.t. $x_1 + x_2 \le 4$; $3x_1 + 8x_2 \le 24$; $10x_1 + 7x_2 \le 35$; $x_1, x_2 > 0$

*28. Max $Z = x_1 + 3x_2$; s.t. $3x_1 + 6x_2 \le 8$; $5x_1 + 2x_2 \le 10$; $x_1, x_2 \ge 0$

29. Min $Z = 40x_1 + 60x_2$; s.t. $0 \le x_1 \le 8$; $0 \le x_2 \le 10$; $5x_1 + 3x_2 \ge 45$

*30. Max $Z = 3x_1 + 5x_2$; s.t. $3x_1 + 2x_2 \le 18$; $0 \le x_1 \le 4$; $0 \le x_2 \le 6$

31. Max $Z = 3x_1 + 2x_2$; s.t. $2x_1 + x_2 \le 2$; $3x + 4x_2 > 12$; $x_1, x_2 \ge 0$

*32. Max $Z = 3x_1 + 4x_2$; s.t. $4x_1 + 2x_2 \le 80$; $2x_1 + 5x_2 \le 180$; $x_1, x_2 \ge 0$

*33. Min $Z = 3x_1 + 2x_2$; s.t. $2x_1 - 2x_2 \le 2$; $x_1 + x_2 \ge 3$; $x_2 \le 4$; $x_1, x_2 \ge 0$

34. Max $Z = 3x_1 + 5y$; s.t. $x - 2y \le 5$; $-x + 3y \le 2$; $x, y \ge 0$

*35. A table gives a contribution of ₹80, while a chair gives a contribution of ₹60. A table needs 4 units of material X and 2 units of material Y, and a chair needs 2 units of material X and 4 units of material Y. There are 60 units of X and 48 units of Y. Formulate the problem as an LPP and solve.

*36. Max $Z = 6x_1 + 9x_2$; s.t. $x_1 + 9x_2 \le 44$; $x_1 + x_2 \le 12$; $x_1, x_2 \ge 0$

37. Min $Z = 8x_1 - 2x_2$; s.t. $5x_1 - 4x_2 \le 3$; $4x_1 - 2x_2 \ge -1$; $x_1, x_2 \ge 0$

*38. Min $Z = 3x_1 + 5x_2$; s.t. $x_1 + 2x_2 \le 2000$; $x_1 + x_2 \le 1500$; $x_2 \le 600$; $x_1, x_2 \ge 0$

*39. Min $Z = 2x_1 + x_2$; s.t. $5x_1 + 10x_2 \le 50$; $x_1 + x_2 \ge 1$; $x_1 \le 4$; $x_1, x_2 \ge 0$

40. Max $Z = 2x_1 - x_2$; s.t. $x_1 + x_2 \le 1$; $2x_1 + x_2 \ge 6$; $x_1, x_2 \ge 0$

*41. Use simplex method to solve Max $Z = 400x_1 + 100x_2$
s.t. $4x_1 + 2x_2 \le 1600$; $5/2x_1 + x_2 \le 1200$; $9/2x_1 + 3/2x_2 \le 1600$; $x_1, x_2 \ge 0$

*42. Solve the following LPP by simplex method: Max $Z = 12x_1 + 3x_2 + x_3$
s.t. $10x_1 + 2x_2 + x_3 \le 100$; $7x_1 + 3x_2 + 2x_3 \le 77$ $2x_1 + 4x_2 + x_3 \le 80$; $x_1, x_2, x_3 \ge 0$

*43. Max $Z = 20x_1 + 6x_2 + 8x_3$ s.t. the constraints $3x_1 + 2x_2 + 3x_3 \le 250$;
$4x_1 + 3x_2 \le 150$; $2x_1 + 3x_3 \le 90$; $x_1, x_2, x_3 \ge 0$

*44. Solve by simplex method
Max $Z = 50x_1 + 20x_2 + 30x_3$ s.t. $5x_1 + x_2 + 7x_3 \le 5$; $5x_1 + x_2 + 6x_3 \le 6$;
$3x_1 - x_2 - 9x_3 \le 3$; $x_1, x_2, x_3 \ge 0$

*45. Use simplex method to solve Max $Z = 2x_1 + 5x_2$
s.t. $x_1 + 4x_2 \le 24$; $3x_1 + x_2 \le 21$; $x_1 + x_2 \le 9$; $x_1, x_2 \ge 0$

*46. Solve the following problem by simplex method Max $Z = 4x_1 + 5x_2 + 9x_3 + 11x_4$
s.t. $x_1 + x_2 + x_3 + x_4 \le 15$; $7x_1 + 5x_2 + 3x_3 + 2x_4 \le 120$;
$3x_1 + 5x_2 + 10x_3 + 15x_4 \le 100$; $x_1, x_2, x_3, x_4 \ge 0$

*47. Solve the following LPP. Max $Z = 3x_1 + 4x_2 + x_3 + 7x_4$ s.t. $8x_1 + 3x_2 + 4x_3 + x_4 \le 7$;
$2x_1 + 6x_2 + x_3 + 5x_4 \le 3$; $x_1 + 4x_2 + 5x_3 + 2x_4 \le 8$; $x_j \ge 0$ $(j = 1, \ldots, 4)$

*48. Use simplex method to maximize $Z = 4x_1 + 5x_2$ s.t. $x_1 + 3x_2 \le 12$;
$4x_1 + 3x_2 \le 24$, $x_1, x_2 \ge 0$

*49. Solve using simplex method Max $Z = x_1 + 2x_2$ s.t. $-x_1 + 2x_2 \le 8$;
$x_1 + 2x_2 \le 12$; $x_1 - 2x_2 \le 3$; $x_1, x_2 \ge 0$

*50. Use simplex method to solve the following: Max $Z = 50x_1 + 70x_2$
s.t. $120x_1 + 120x_2 \le 8400$; $x_1 + 2x_2 \le 100$; $2x_1 + x_2 \le 120$; $x_1, x_2 \ge 0$

*51. Solve using simplex method Max $Z = 3x_1 + 5x_2$
s.t. $3x_1 + 2x_2 \le 18$; $x_1 \le 4$; $x_2 \le 6$; $x_1, x_2 \ge 0$

*52. Use simplex method to solve the following LPP: Max $Z = x_1 - x_2 + 3x_3$
s.t. $x_1 + x_2 + x_3 \le 10$; $2x_1 - x_1 \le 2$; $2x_1 - 2x_2 + 3x_3 \le 6$; $x_1, x_2, x_3 \ge 0$

*53. Max $Z = 3x_1 + 4x_2 + 6x$ s.t. $2x_1 + 2x_2 + x_3 \le 400$; $3x_1 + 2x_3 \le 440$;
$x_1 + 3x_2 < 300$; $x_1, x_2, x_3 > 0$

*54. Use simplex method to solve: Max $Z = 5x_1 + 3x_2$;
s.t. $x_1 + x_2 \le 2$; $5x_1 + 2x_2 \le 10$; $3x_1 + 8x_2 \le 12$; $x_1, x_2 \ge 0$

55. Use simplex method to solve the following:

$\text{Min } Z = -10x_1 + x_2 + 2x_3$ s.t. $x_1 + x_2 - 2x_3 \le 10$; $4x_1 + x_2 + x_3 \le 20$; $x_1, x_2, x_3 \ge 0$

56. Use the simplex algorithm to find two optimal solutions to the following LPP. How many optimal solutions do this LP have? Find the "alternate optimal solution.

$\text{Max } Z = 4x_1 + x_2$; s.t. $2x_1 + 3x_2 \le 4$; $x_1 + x_2 \le 1$; $4x_1 + x_2 \le 2$; $x_1, x_2 \ge 0$.

57. An auto company manufactures cars and trucks. Each vehicle must be processed in the paint shop and body assembly shop. If the paint shop where only painting truck, takes place 40 per day could be painted. If the paint shop where only painting car, 60 per day could be painted. If the body shop where only producing cars, it could process 50 per day. If the body shop where only producing trucks, it could process 50 per day. Each truck contributes ₹10000 to profit, and each car contributes ₹8000 to profit. Use linear programming to determine a daily production schedule that will maximize the company's profits.

[Hint: 1 day = 24hrs]

*58. Minimize $Z = 5x_1 + 6x_2 + 3x_3$ s.t. $5x_1 + 5x_2 + 3x_3 \ge 50$; $x_1 + x_2 - x_3 \ge 20$;

$7x_1 + 6x_2 - 9x_3 \ge 30$; $x_1, x_2, x_3 \ge 0$

*59. $\text{Max } Z = x_1 + 6x_2$; s.t. $x_1 + x_2 \ge 2$; $x_1 + 3x_2 \le 3$; $x_1, x_2 \ge 0$

*60. $\text{Min } Z = 3x_1 + 2x_2$; s.t. $5x_1 + x_2 \ge 10$; $x_1 + x_2 \ge 6$; $x_1 + 4x_2 \ge 12$; $x_1, x_2 \ge 0$

*61. $\text{Min } Z = x_1 + x_2 + x_3$; s.t. $x_1 + 2x_2 \ge 3$; $x_2 + 7x_3 \le 6$; $x_1 - 3x_2 + 5x_3 = 5$; $x_1, x_2, x_3 \ge 0$

*62. Use Big-M method to maximize $Z = 2x_1 + 3x_2$ s.t. $x_1 + 2x_2 \le 4$; $x_1 + x_2 = 3$; $x_1, x_2 \ge 0$

63. $\text{Min } Z = 5x + 6y$; s.t. $2x + 5y \ge 1500$; $3x + y \ge 1200$; $x, y \ge 0$

*64. Use penalty (or Big-M) method to maximize $Z = 2x_1 + x_2 + x_3$

s.t. $x_1 + x_2 + 2x_3 \le 5$; $2x_1 + 3x_2 + 4x_3 = 12$; $x_1, x_2, x_3 \ge 0$

65. Use Big-M method to solve the LPP $\text{Max } Z = 2x_1 + x_2$;

s.t. $x_1 - x_2 \le 10$; $x_1 \le 20$; $x_1, x_2 \ge 0$

*66. $\text{Max } Z = x_1 + 2x_2 + 3x_3 - x_4$ s.t. $x_1 + 2x_2 + x_3 = 15$; $2x_1 + x_2 + 5x_3 = 20$;

$x_1 + 2x_2 + x_3 + x_4 = 10$; $x_1, x_2, x_3, x_4 \ge 0$

*67. $\text{Max } Z = 3x_1 + 2x_2 + x_3$ s.t. $2x_1 + 5x_2 + x_3 = 12$; $3x_1 + 4x_2 = 11$

$x_2, x_3 \ge 0$ and x_1 is unrestricted.

*68. $\text{Min } Z = 3x_1 + 2x_2$ s.t. $2x_1 - 2x_2 \le 2$; $x_1 + x_2 \ge 3$; $x_2 \le 4$; $x_1, x_2 \ge 0$

*69. Use Big-M method to $\text{Min } Z = 2x_1 + x_2$; s.t. $3x_1 + x_2 = 3$; $4x_1 + 3x_2 \ge 6$;

$x_1 + 2x_2 \le 3$; $x_1, x_2 \ge 0$

*70. Solve the following by simplex algorithm: Min $Z = 4x_1 + 3x_2 + x_3$;

s.t. $2x_1 + 5x_2 + x_3 \geq 30$; $4x_1 + x_2 + 2x_3 = 25$; $x_1, x_2 \geq 0$ x_3 is unrestricted

*71. Min $Z = 3x_1 + 6x_2 + x_3$; s.t. $x_1 + x_2 + x_3 \geq 6$; $x_1 - 5x_2 - x_3 \geq 4$;

$x_1 + 5x_2 + x_3 \geq 24$; $x_1, x_2, x_3 \geq 0$

*72. Max $Z = 3x_1 - x_2$; s.t. $2x_1 + x_2 \geq 2$; $x_1 + 3x_2 \leq 3$; $x_2 \leq 4$; $x_1, x_2 \geq 0$

*73. Max $Z = 2x_1 + x_2 + x_3$ s.t. $2x_1 + 3x_2 - x_3 \leq 9$; $2x_2 + x_3 \geq 4$; $x_1 + x_3 = 6$; $x_1, x_2, x_3 \geq 0$

*74. Use simplex method to solve the following system of linear equations

$x_1 - x_3 + 4x_4 = 3$; $2x_1 - x_2 = 3$; $3x_1 - 2x_2 - x_4 = 1$; $x_1, x_2, x_3, x_4 \geq 0$

*75. Max $Z = 2x_1 + x_2$ s.t. $3x_1 + x_2 \geq 3$; $4x_1 + 3x_2 \geq 6$; $x_1 + 2x_2 \geq 2$; $x_1, x_2 \geq 0$

76. Max $Z = 2x_1 + x_2 + 3x_3$ s.t. $x_1 + x_2 + 2x_3 \leq 15$; $2x_1 + 3x_2 + 4x_3 = 12$;

$4x_1 + 6x_2 + 8x_3 = 24$; $x_1, x_2, x_3 \geq 0$

*77. Max $Z = -8x_1 + 3x_2 - 6x_3$ s.t. $x_1 - 3x_2 + 5x_3 = 4$; $5x_1 + 3x_2 - 4x_3 \geq 6$; $x_1, x_2, x_3 \geq 0$

78. Max $Z = 4x_1 + 3x_2$ s.t. $x_1 + x_2 \leq 50$; $x_1 + 2x_2 \geq 80$; $3x_1 + 2x_2 \geq 140$; $x_1 \geq 0$ & x_2 is unrestricted.

79. Min $X_0 = -x_1 + x_2$ s.t. $x_1 - 2x_2 - x_3 = 1$; $-2x_1 + 2x_2 - x_4 = 1$; $x_1, x_2, x_3, x_4 \geq 0$

*80. Use Two-Phase simplex method to solve

Max $Z = 5x_1 - 4x_2 + 3x_3$ s.t. $2x_1 + x_2 - 6x_3 = 20$; $6x_1 + 5x_2 + 10x_3 \leq 76$;

$8x_1 - 3x_2 + 6x_3 \leq 50$; $x_1, x_2, x_3 \geq 0$

*81. Solve by Two-Phase simplex method:

Min $Z = 2x_1 + x_2 + 3x_3$ s.t. $5x_1 - 2x_2 \geq -5$; $3x_1 + 2x_2 + 4x_3 \geq 7$; $2x_1 + 5x_3 \leq 3$; $x_1, x_2, x_3 \geq 0$

82. Solve: Max $Z = 5x_1 - x_2$ s.t. $2x_1 + x_2 = 6$; $x_1 + x_2 \leq 4$; $x_1 + 2x_2 \leq 5$; $x_1, x_2 \geq 0$

83. Solve: Min $Z = -3x_1 + x_2$ s.t. $x_1 - 2x_2 \geq 2$; $-x_1 + x_2 \geq 3$; $x_1, x_2 \geq 0$

*84. Find the dual of the following LPP:

Max $Z = x_1 - x_2 + 3x_3$; s.t. $x_1 + x_2 + x_3 \leq 10$; $2x_1 - x_3 \leq 2$; $2x_1 - 2x_2 + 3x_3 \leq 6$; $x_1, x_2, x_3 \geq 0$

*85. Write the dual of the LPP:

Min $Z = 4x_1 + 6x_2 + 18x_3$; s.t. $x_1 + x_2 \geq 3$; $x_2 + 2x_3 \geq 5$; $x_1, x_2, x_3 \geq 0$

*86. Obtain the dual of the following LPP:

Max $Z = x_1 - 2x_2 + 3x_3$; s.t. $-2x_1 + x_2 + 3x_3 = 2$; $2x_1 + 3x_2 + 4x_3 = 1$; $x_1, x_2, x_3 \geq 0$

[Hint: Equality constraints in the primal can be dealt in the following alternate method also. The combination of [≤; =] symbols in primal can be converted as [≥] symbol in dual and the corresponding dual variable to the equality constraint must be treated as an unrestricted variable. Here y_1 & y_2 are the two dual variables introduced one each to the primal constraints, then y_1 & y_2 must be unrestricted].

*87. Solve using duality principle

Max $Z = 2x_1 + x_2$; s.t. $x_1 + 2x_2 \le 10$; $x_1 + x_2 \le 6$; $x_1 - x_2 \le 2$; $x_1 - 2x_2 \le 1$; $x_1, x_2 \ge 0$

*88. Using dual simplex method solve the LPP,

Min $Z = 3x_1 + x_2$; s.t. $x_1 + x_2 \ge 1$; $2x_1 + 3x_2 \ge 2$; $x_1, x_2 \ge 0$

*89. Solve the following LPP by DSM:

Min $Z = x_1 + 2x_2 + 3x_3 + 4x_4$; s.t. $x_1 + 2x_2 + 2x_3 + 3x_4 \ge 30$; $2x_1 + x_2 + 3x_3 + 2x_4 \ge 20$; $x_j \ge 0$ (for $j = 1, 2, 3, 4$)

*90. Use DSM to solve the following LPP:

Min $Z = 6x_1 + 7x_2 + 3x_3 + 5x_4$; s.t. $5x_1 + 6x_2 - 3x_3 + 4x_4 \ge 12$; $x_2 + 5x_3 - 6x_4 \ge 10$; $2x_1 + 5x_2 + x_3 + x_4 \ge 8$; $x_j \ge 0$ (for $j = 1,\dots.4$)

*91. Use DSM to solve the following LPP:

Max $Z = -3x_1 - 2x_2$; s.t. $x_1 + x_2 \ge 1$; $x_1 + x_2 \le 7$; $x_1 + 2x_2 \ge 10$; $x_2 \le 3$; $x_1, x_2 \ge 0$

92. Solve the following problem using DSM:

Min $Z = x_1 + 2x_2 + 5x_3$; s.t. $|-x_1 + 10x_2 - 3x_3| \le 15$; $2x_1 + x_2 + x_3 \ge 10$; $x_1, x_2, x_3 \ge 0$

[Hint: $|-x_1 + 10x_2 - 3x_2| \le 15$; means that $-x_1 + 10x_2 - 3x_3 \le 15$; $-x_1 + 10x_2 - 3x_3 \ge -15$]

*93. Apply the principle of duality to solve the following LPP

Min $Z = 2x_1 + 2x_2$; s.t. $2x_1 + 4x_2 \ge 1$; $x_1 + 2x_2 \ge 1$; $2x_1 + x_2 \ge 1$; $x_1, x_2 \ge 0$

*94. Write the dual of min $Z = x_1 + x_2$; s.t. $2x_1 + x_2 \ge 4$; $x_1 + 7x_2 \ge 7$; $x_1, x_2 \ge 0$

Solve the dual problem and hence obtain the solution of the given primal problem.

*95. Obtain the dual of the problem Max $Z = 2x_1 + 5x_2 + 3x_3$; s.t. $2x_1 + 4x_2 - x_3 \le 8$; $-2x_1 + 2x_2 + 3x_3 \ge 7$; $x_1 + 3x_2 - 5x_3 \ge -2$; $4x_1 + x_2 + 3x_3 \le 4$; $x_1, x_2, x_3 \ge 0$

*96. Use dual simplex method to solve the following LPP

Max $Z = -3x_1 - 2x_2$; s.t. $x_1 + x_2 \ge 1$; $x_1 + x_2 \le 7$; $x_1 + 2x_2 \ge 10$; $x_2 \le 3$; $x_1, x_2 \ge 0$

*97. Write the dual of the following:

Max $Z = 3x_1 - 2x_2 + 4x_3 - 6x_4$; s.t. $x_1 + x_2 + x_3 + x_4 = 25$; $2x_1 - 3x_2 - x_3 + x_4 \le 40$; $7x_1 - 3x_2 + 4x_3 - x_4 \le 35$; $6x_1 + 6x_2 - x_3 + 3x_4 \ge 28$; $x_1, x_3 \ge 0$; $x_2, x_4 \le 0$;

*98. Write the dual of the problem min $Z = 4x_1 + 3x_2 + 6x_3$; s.t. $x_1 + x_3 \ge 2$; $x_2 + x_3 \ge 5$; $x_1, x_2, x_3 \ge 0$

*99. Solve the following LPP using its dual

Max $Z = 5x_1 - 2x_2 + 3x_3$; s.t. $2x_1 + 2x_2 - x_3 \ge 2$; $3x_1 - 4x_2 \le 3$; $x_2 + 3x_3 \le 5$; $x_1, x_2, x_3 \ge 0$

Use RSM to solve the following LPP:

100. Max $Z = 60x_1 + 30x_2 + 20x_3$ s.t. $8x_1 + 6x_2 + x_3 \le 48$; $4x_1 + 2x_2 + 3/2x_3 \le 20$; $2x_1 + 3/2x_2 + 1/2x_3 \le 8$; $x_j \ge 0$ (j = 1, 2, 3)

101. Max $z = 4x_1 + x_2$ s.t. $x_1 + x_2 \le 4$; $2x_1 + x_2 \ge 6$; $3x_2 \ge 6$; $x_1, x_2 \ge 0$

102. Max $z = x_1 + 2x_2$ s.t. $x_1 + x_2 \le 3$; $x_1 + 2x_2 \le 5$; $3x_1 + x_2 \le 6$; $x_1, x_2 \ge 0$

103. Max $z = x_1 + x_2$ s.t. $2x_1 + 5x_2 \ge 6$; $x_1 + x_2 \ge 2$; $x_1, x_2 \ge 0$

104. Max $z = 30x_1 + 23x_2 + 29x_3$ s.t. $6x_1 + 5x_2 + 3x_3 \le 26$; $4x_1 + 2x_2 + 5x_3 \le 7$; $x_1, x_2, x_3 \ge 0$

ANSWERS TO THE EXERCISE PROBLEMS

1. Max $Z = 2x_1 + x_2 + 3x_3$

   s.t. $3x_1 + 4x_2 + 3x_3 \le 43$, $5x_1 + 4x_3 \le 46$, $3x_1 + 6x_2 + 2x_3 \le 42$, $x_j \ge 0$ (j = 1, 2, 3)
2. Max $Z = 3x_1 + 4x_2$ s.t. $4x_1 + 2x_2 \le 80$, $2x_1 + 6x_2 \le 180$, $x_1, x_2 \ge 0$
3. Min $Z = 9x_2 + 3x_3 + 18x_5 + 6x_6 + 9x_7 + 3x_8 + 12x_9$ s.t. $2x_1 + 4x_4 + 2x_7 + 2x_8 = 400$;

   $x_2 + 2x_5 + x_7 + x_9 = 200$; $x_3 + 2x_6 + x_8 + x_9 = 300$, $x_j \ge 0$ (j = 1, ... 9)
4. Max $Z = 30x_1 + 40x_2 + 20x_3$, s.t. $x_1 + x_2 + x_3 \le 1000$, $100x_1 + 120x_2 + 70x_3 \le 100000$;

   $7x_1 + 10x_2 + 8x_3 \le 8000$, $x_j \ge 0$ (j = 1, 2, 3)
5. Max $Z = 10x_1 + 12x_2 + (3/2)\, x_3$

   s.t. $2x_1 + 3x_2 \le 1500$, $3x_1 + 2x_2 \le 15000$, $x_1 + x_2 + x_3 = 1000$, $x_j \ge 0$ (j = 1, 2, 3)
6. Max $Z = 5x_1 + 9x_2$

   s.t. $30x_1 + 50x_2 \le 8000$, $x_1 + x_2 \le 100$, $5x_1 + 9x_2 \ge 600$, $x_1, x_2 \ge 0$
7. Min $Z = 3x_1 + 2x_2$

   s.t. $5x_1 + x_3 \ge 10$; $2x_1 + 2x_2 \ge 12$; $x_1 + 4x_2 \ge 12$, $x_1, x_2 \ge 0$
8. Max $Z = 12x_1 + 15x_2 + 14x_4$

   st. $x_1 \le 100d_1$; $x_2 \le 100d_2$; $x_3 \le 100d_3$; $x_1 + x_2 + x_3 \le 100$

   $0.02d_1 + 0.04d_2 + 0.03d_3 \le 0.03$; $2d_1 + 3d_2 + 5d_3 \le 3$; $x_j \ge 0$ & d_1, d_2, d_3 can lake either 0 or 1.
9. Max $Z = (0.14)(0.9)x_1 + (0.13)(0.93)x_2 + (0.12)(0.97)x_3 + (0.125)(0.95)x_4 + (0\ 1)(0.98)x_5$

   s.t. $x_1 + x_2 + x_3 + x_4 + x_5 \le 100$ million

   $x_4 + x_5 \ge 0.4^* \ 100$ million

   $x_3 \ge (x_1 + x_2 + x_3)(0.5)$

   $$\frac{0.1x_1 + 0.07x_2 + 0.03x_3 + 0\ 05x_4 + 0.02x_5}{x_1 + x_2 + x_3 + x_4 + x_5} \le 0.04 \qquad x_j \ge 0 \ (j = 1, \ldots 5)$$

10. Max $Z = 4.08x_1 + 6.57x_2 + 1.1x_3 + 0.62x_4 + 2.84x_5 + 0.15x_6$

s.t. $80x_1 + 100x_2 + 160x_3 + 120x_4 + 150x_5 + 200x_6 \le 2.5$ crores

$x_j \ge 100$; $(j = 1, \ldots 6)$

$80x_1 + 100x_2 \ge 0.25$ crores

$200x_6 \le 2.5$ lakhs

$80x_1 + 100x_2 \le 5$ lakhs

$4x_1 + 6.5x_2 + 1.0x_3 + 0.5x_4 + 2.75x_5 \ge 1$ lakh

$$\frac{0.05x_1 + 0.03x_2 + 0.04x_3 + 0.2x_4 + 0.06x_5 + 0.08x_6}{0.46} \le 0.1 \qquad x_j \ge 0 \ (j = 1, \ldots. 6)$$

11. Min $Z = x_1 + x_2 + x_3 + x_4 + x_5 + x_6$

s.t. $x_1 + x_6 \ge 4$; $x_1 + x_2 \ge 8$; $x_2+x_3 \ge 10$; $x_3 + x_4 \ge 7$; $x_4 + x_5 \ge 12$; $x_5 + x_6 \ge 4$; $x_j \ge 0$ $(j = 1, \ldots 6)$

12. Max $Z = 100x_1 + 150x_2$

s.t. $x_1 + 2x_2 \le 100$; $(3/2)\ x_1 + (3/4)\ x_2 \le 90$, $x_1, x_2 \ge 0$

13. Min $Z = \sum_{j=1}^{8} x_j$

s.t. $x_8 + x_1 \ge 1$; $x_1 + x_2 \ge 2$; $x_2 + x_3 \ge 3$; $x_3 + x_4 \ge 7$; $x_4 + x_5 \ge 9$; $x_5 + x_6 \ge 5$; $x_6 + x_7 \ge 4$; $x_7 + x_8 \ge 3$; $x_j \ge 0$ $(j = 1, \ldots. 8)$

14. Min $Z = \sum_{j=1}^{7} x_j$

s.t. $x_1 + x_4 + x_5 + x_6 + x_7 \ge 17$; $x_1 + x_2 + x_5 + x_6 + x_7 \ge 13$; $x_1 + x_2 + x_3 + x_6 + x_7 \ge 15$; $x_1 + x_2 + x_3 + x_4 + x_7 \ge 19$; $x_1 + x_2 + x_3 + x_4 + x_5 \ge 14$; $x_2 + x_3 + x_4 + x_5 + x_6 \ge 16$; $x_3 + x_4 + x_5 + x_6 + x_7 \ge 11$; $x_j \ge 0$ $(j = 1, \ldots 7)$

15. Max $Z = 6000x_1 + 4000x_2$; s.t. $10x_1 + 8x_2 \le 800$; $x_1 \le 60$; $x_2 \le 75$; $x_1. x_2 \ge 0$

16. Max $Z = 2x_1 + x_2 + 4\ (x_4 - x_5)$

st. $2x_1 + 4x_2 + x_6 = 4$, $x_1 + 2x_2 + (x_4 - x_5) - x_7 = 5$; $2x_1 + 3\ (x_4 - x_5) + x_6 = 2$; $x_j \ge 0$ $(j = 1, 2, 4, \ldots 8)$

17. [0, 45/16. 7/16, 0); [22/9, 0, 0, 7/9]; [1, 2, 0, 0]

18. [1, 0, 0]

19. (0, 0, 2, –1); (8/3, 0, 0, –7/3); etc.

No.	x_1	x_2	Objective	Remark
20.	2	1	Max Z = 8	-
21.	-	-	-	Unbounded
22.	1/2	3/2	Min Z = 5	-

No.	x_1	x_2	Objective	Remark
23.	0	0	Min Z = 0	-
24.	0	4	Min Z = – 4	-
25.	1	5	Min Z = 13	-
26.	0.4444	0.5556	Min Z = 2.4444	-
27.	1.6	2.4	Max Z = 24.8	-
28.	0	1.3333	Max Z = 4	-
29.	8	1.6667	Min Z = 420	-
30.	2	6	Max Z = 36	-
31.	-	-	-	No feasible solution
32.	2.5	35	Max Z = 147.5	-
33.	0	3	Min Z = 6	-
34.	19	7	Max Z = 92	-
35.	12	6	Max Z= 1320	-
36.	8	4	Max Z = 84	-
37.	0	1/2	Min Z = – 1	-
38.	0	0	Min Z = 0	-
39.	0	1	Min Z = 1	-
40.	-	-	-	No feasible solution

No.	x_1	x_2	x_3	x_4	Objective	Remark
41.	355.5555	0	*	*	Max Z = 142222.22	-
42.	9.1250	4.3750	0	*	Max Z = 122.63	-
43.	0	30	63.333	*	Max Z = 686.67	-
44.	0	5	0	*	Max Z = 100	-
45.	4	5	*	*	Max Z = 33	-
46.	7.1429	0	7.8571	0	Max Z = 99.29	-

No.	x_1	x_2	x_3	x_4	Objective	Remark
47.	0.8421	0	0	0.2632	Max Z = 4.3684	-
48.	4	2.667	*	*	Max Z = 29.33	-
49.	2	5	*	*	Max Z = 12	-
50.	40	30	*	*	Max Z = 4100	-
51.	2	6	*	*	Max Z = 36	-
52.	0	6	4	*	Max Z = 6	-
53.	0	90	220	*	Max Z =1680	-
54.	2	0	*	*	Max Z = 10	-
55.	5	0	0	*	Min Z = – 50	-
56.	1/2	0	*	*	Max Z = 2	Only one alternate
	0.3333	0.6667	*	*	Max Z = 2	Solution exists
57.	11.5218	0	*	*	Max Z = 115218	-
58.	20	0	0	*	Min Z = 100	-
59.	1.5	0.5	*	*	Max Z = 4.5	-
60.	1	5	*	*	Min Z = 13	-
61.	2.2	0.4	0.8	*	Min Z = 3.4	-
62.	2	1	*	*	Max Z = 7	-
63.	346.1539	161.5385	*	*	Min Z = 2700	-
64.	3	2	0	*	Max Z = 8	-
65.	15	5	*	*	Max Z = 35	-
66.	-	-	-	-	-	No feasible solution
67.	3.667	0	4.6667	*	Max Z = 15.6667	-
68.	0	3	*	*	Min Z = 6	-
69.	0.6	1.2	*	*	Min Z = 2.4	-
70.	0	3.8889	10.5556	*	Min Z = 22.222	-
71.	14	0	10	*	Min Z = 52	-
72.	3	0	*	*	Max Z = 9	-
73.	4	1	2	*	Max Z = 11	-
74.	5	7	2	0	-	-

No.	x_1	x_2	x_3	x_4	Objective	Remark
75.	-	-	-	-	-	Unbounded solution
76.	6	0	0	*	Max Z = 12	-
77.	0	15	10	*	Max Z = – 14	-
78.	-	-	-	-	-	No feasible solution
79.	-	-	-	-	-	No feasible solution
80.	7.8571	4.2587	0	*	Max Z = 22.1429	-
81.	0.25	3.125	0	*	Min Z = 3.625	-
82.	3	0	*	*	Max Z = 15	-
83.	-	-	-	-	-	No feasible solution

84. Min $Z_D = 10y_1 + 2y_2 + 6y_3$

    s.t. $y_1 + 2y_2 + 2y_3 \geq 1$; $y_1 - 2y_3 \geq -1$; $y_1 - y_2 + 3y_3 \geq 3$; $y_1, y_2, y_3 \geq 0$

85. Max $Z_D = 3y_1 + 5y_2$

    s.t. $y_1 \leq 4$; $y_1 + y_2 \leq 6$; $2y_2 \leq 18$; $y_1, y_2 \geq 0$

86. Min $Z_D = 2y_1 + y_2$

    s.t. $-2y_1 + 2y_2 \geq 1$; $y_1 + 3y_2 \geq -2$; $3y_1 + 4y_2 \geq 3$; y_1, y_2 are unrestricted

No	x_1	x_2	x_3	x_4	Objective	Remark
87.	4	2	*	*	Max Z = 10	-
88.	0	1	*	*	Min Z = 1	-
89(a).	30	0	0	0	Min Z = 30	-
89(b).	10/3	40/3	0	0	Min Z = 30	Alternate solution
90.	0	2.7273	1.4545	0	Min Z = 23.4545	-
91.	4	3	*	*	Max Z = – 18	-
92.	5	0	0	*	Min Z = 5	-
93.	1/3	1/3	*	*	Min Z = 4/3	-
94.	1.62	0.77	*	*	Min Z = 2.38	-

95. Min $Z_D = 8y_1 - 7y_2 + 2y_3 + 4y_4$

    s.t. $2y_1 + 2y_2 - y_3 + 4y_4 \geq 2$; $4y_1 + 2y_2 - 3y_3 + y_4 \geq 5$; $-y_1 - 3y_2 + 5y_3 + 3y_4 \geq 3$, $y_1, y_2, y_3, y_4 \geq 0$

No	x_1	x_2	x_3	x_4	Objective	Remark
96.	4	3	*	*	Max Z = −18	-

97. Min $Z_D = 25y_1 + 40y_2 + 35y_3 - 28y_4$

    s.t. $y_1 + 2y_2 + 7y_3 - 6y_4 \geq 3$; $y_1 - y_2 + 4y_3 + y_4 \geq 4$; $y_1 - 3y_2 - 3y_3 - 6y_4 \geq -2$;

    $-y_1 + 3y_2 + 3y_3 + 6y_4 \geq 2$

    $y_1 + y_2 - y_3 - 3y_4 \geq -6$; $-y_1 - y_2 + y_3 + 3y_4 \geq 6$; $y_2, y_3, y_4 \geq 0$ and y_1 is unrestricted

98. Max $Z = 2y_1 + 5y_2$

    s.t. $y_1 \leq 4$; $y_2 \leq 3$; $y_1 + y_2 \leq 6$, $y_1, y_2 \geq 0$

No	x_1	x_2	x_3	x_4	Objective	Remark
99.	7.67	5.0	0	*	Max Z = 28	-
100.	2	0	8	*	Max Z = 280	-
101.	2	2	*	*	Max Z = 10	-
102.	1	2	*	*	Max Z = 5	-
103.	-	-	-	-	-	Unbounded solution
104.	0	7/2	0	*	Max Z = 161/2	-

REVIEW QUESTIONS

Short Questions

1. Mention and describe the three elements that constitute a decision model. (Section 2.2)
2. When is a mathematical programming problem called a Linear one? (Section 2.4)
3. Define i) Slack variables ii) Surplus variables (Section 2.6)
4. Write a general LPP in the standard form. (Section 2.6)
5. What do you mean by a simplex technique?

   The process of selecting and testing the corner points of the feasible region by modified algebraic process.

6. While solving an LPP using simplex method, how will you know that the problem has an alternate optimum solution? (Section 2.12)
7. If an LPP has two optimum solutions, how many more optimum solutions can it have?

   If there are two/more optimum solutions which exist for an LPP, then there exists an infinite number of optimal solutions. That is, if we have $S_1, S_2, \ldots S_x$ as different basic feasible solutions giving the same optimum value, then any convex combination of $S_1, S_2, \ldots S_x$ is also an optimum solution.

8. State the areas of applications of O.R. 1. Production Management 2. Marketing Management
9. Write an LPP in matrix form. (Section 2.14.5)
10. Define an unrestricted variable.

    If a variable is permitted to take any real value then it is called an unrestricted variable.
11. Define a restricted variable.

    If a variable is restricted to take 0 and +ve values it is known as restricted variable
12. Define a convex set. (Section 2.9)
13. State all the assumptions of an LPP (Section 2.4)
14. For what purpose are the artificial variables used? (Section 2.12)
15. Define an Extreme point.

    The corner point of feasible region is known as an Extreme point
16. Define a constraint in an LPP?

    In order to maximize or minimize the given objective function, we have to take care of the availability of resources. This kind of restriction is called a constraint in an LPP.
17. What is degeneracy?

    In a simplex process, if any tie occurs in the selection of the leaving variable (the ratios are equal), then we can select any one variable to leave the basis, then the corresponding X_B value of the other tied up variable lakes the value "0" in the subsequent iteration. This process is called degeneracy.
18. Define duality. (Section 2.15)
19. Define a basic feasible solution and an optimal solution. (Section 2.8)
20. What is meant by objective function?

    Each and every business problem is going to have a specific aim. That aim can be expressed in the form of a function using the defined variable(s) known as objective function.
21. What are the methods to solve LPP with ≥ constraints?

    (i) Graphical (ii) Big-M (iii) Two-phase
22. Write the dual of the following primal problem.

    Min $Z = 3x_1 + x_2$

    s.t. $2x_1 + 3x_2 \geq 2;\ x_1 + x_3 \geq 1;\ x_1, x_2 \geq 0$

    Max $Z_D = 2y_1 + y_2$

    s.t. $2y_1 + y_2 \leq 3;\ 3y_1 + y_2 \leq 1;\ y_1, y_2 \geq 0$
23. Explain the term 'nonnegative restrictions in an LPP'.

In the real life situation, the variables used in the LPP can not take any strict negative value. That is, the variables can take either 0 or strict positive value. This situation is referred to as constraints in an LPP.

24. What are all the limitations of Linear Programming? (Section 2.4.1)

25. State the principle of duality. (Section 2.15)

26. Name the methods to resolve degeneracy in an LPP.

    (i) Perturbation Method (ii) Lexicographic Method

27. Define an unbounded solution.

    There is no bound for the variables defined in the given LPP.

28. Draw the structure of the simplex table (Section 2.11.1)

29. State the rules to be followed while selecting the entering and leaving variables (Section 2.11.1)

30. Define cycling. (Section 2.11.1)

31. Define the Canonical form of an LPP (Section 2.15.1)

32. What is the difference between the slack variable and the surplus variable.

Slack	Surplus
i. Used to convert the "≤" inequality constraint	Used to convert the "≥" inequality into an equality one
ii. Add it from RHS of the constraint	Subtract it from RHS of the constraint

33 Express the canonical form in matrix notation.

Max $Z = C^T X$

s.t $AX \le b$ where $C^T, X^T \in R^n$

$X \ge 0$ and $b^T \in R^m$, A is a mxn matrix.

34. What do you mean by a degenerate solution?

    If one/more decision variables are taking the value "0" in an optimum solution, we call it degenerate

35. What is the maximum number of basic feasible solution we can think of, if the given system is having 3 equality constraints and 6 variables?

    Maximum number of basic feasible solutions = ${}^6C_3 = 20$

36. If a given LPP is having n variables and m equality constraints, then how many basic solutions exist? (${}^nC_m$)

37. State the optimality condition for the simplex method

Objective	Condition
i. Maximization	All $z_j - c_j \ge 0$
ii. Minimization	All $z_j - c_j \le 0$

38. Bring out the difference between the resource and the constraint.

    Resource: Refers to the availability of raw material

    Constraint: The combination required for the corresponding raw material

39. What do you mean by key row and key column? (Write about PR and PC). (2.52 & 2.53)

    The solution for the given LPP cannot be fixed, because even the optimum value may be nearer to infinity.

40. Define the artificial variable. (Section 2.12)

41. When is an LPP going to have a Pseudo optimum solution?

    In the optimum table if one or more artificial variable exists at the +ve level (> 0), then we can say the given LPP is said to have a Pseudo Optimum solution.

State True or False

42. All equality constraints can be replaced equivalents by two inequality constraints (using ≤ and ≥). (T)
43. If an LPP has m constraints then the number of basic variables can exceed m. (F)
44. In the simplex method, all variables should be ≥ 0. (T)
45. All basic feasible solutions of the simplex iteration need not be an extreme point of the corresponding feasible region. (F)
46. In the simplex process the optimality condition for a maximization and minimization models are different. (T)
47. In the simplex process the feasibility condition for a maximization and minimization model is different. (F)
48. The pivotal element can be zero in a simplex process. (F)
49. An artificial variable leaves the basis means, there is no chance for the artificial variable to enter once again. (T)
50. Instead of selecting the most (–ve) $(Z_j - C_j)$ value for selecting PC, if we select any one (–ve) value in the $(Z_j - C_j)$ as a PC, then we cannot get the optimum solution. (F)
51. Instead of selecting the most (–ve) $(Z_j - C_j)$ value for PC, if we select any one negative $(Z_j - C_j)$, then it will take more number of iterations to reach the optimum table (if it exists). (T)
52. The unbounded feasible region implies that the solution is unbounded (F)
53. The optimality of the solution is affected if there is a change in the availability of resource. (T)
54. If the primal has no solution then it implies that the dual also has no solution. (F) [ ]
55. In a simplex process, if we have an unbounded indication together with some artificial variable in the basis at +ve level it implies that the LPP has no solution. (T)
56. We can apply DSM for a given LPP directly. (F)

Fill in the Blanks

57. Simplex method was designed by _________. (Dantzig)
58. _________ has suggested a method to penalize the artificial variable. (A. Charnes)
59. Penalty method is otherwise called _________ and _________. (Big-M, Charnes Method)
60. Big-M method is a substitute method for _________. (Two-Phase)
61. In phase-l of a Two-phase method our main aim is to _________. (drive away the artificial variables)
62. Dual of the Dual is _________. (Primal)
63. The Dual variable corresponding to the equality constraini of the primal must be treated as _________ in the dual model. (unrestricted)
64. Shadow price is otherwise called _________ or _________. (dual price/accounting price)
65. DSM was introduced by _________ in _________. (Lemke, 1954)
66. Canonical form and the standard forms are _________. (different)
67. RSM stands for _________. (Revised simplex method)
68. An LPP having more than one optimal solution referred to as having _________ or _________ optimal solutions. (alternate, multiple).
69. The _________ is the intersection of the PR and PC (PE)
70. For a maximization model the coefficient of an artificial variable is _________. (–M)

3

Sensitivity Analysis (or) Post-Optimal Analysis

3.1 INTRODUCTION

We now explore how changes in LP's parameters (objective function co-efficients, right hand sides and technological co-efficients) change the optimal solution. The study of how an LP's optimal solution depends on its parameters is called **Sensitivity Analysis**.

Due to change of parameters in the existing model can result in one of four cases

1. The optimality is not affected.
2. The latest solution becomes infeasible.
3. The latest solution may be feasible but the optimality is affected.
4. The latest solution becomes infeasible and also non-optimal.

There is no problem in case-1 because the optimality is retained. Regarding case-2, the solution is infeasible. To recover from this we can apply the Dual Simplex Method to get the new optimum solution. In case-3, to get the optimal solution we can make use of a simplex process. Since case-4 is the contribution of cases-2 and 3, a suitable measure can be taken to overcome this situation.

We now discuss how six types of changes in LP's parameter change the optimal solution.

Change-1: Changing the objective function co-efficient of a non-basic variable.
Change-2: Changing the objective function co-efficient of a basic variable.
Change-3: Changing the right hand side of a constraint.
Change-4: Changing the column of a non-basic variable.
Change-5: Adding a new variable.
Change-6: Adding a new constraint.

3.2 CHANGE IN THE OBJECTIVE FUNCTION CO-EFFICIENT OF A NON-BASIC VARIABLE

***Example:* 1**

Consider the given LPP along with its optimum table 3.1.

$$\text{Max } Z = 6x_1 + 4x_2; \quad \text{s.t.} \quad 2x_1 + 6x_2 \le 12; \quad 4x_1 + 4x_2 \le 16; \quad x_1, x_2 \ge 0$$

Optimum Table 3.1

C_j	6	4	0	0			
x_j	x_1	x_2	x_3	x_4	X_B	B	C_B
	0	4	1	–1/2	4	x_3	0
	1	1	0	1/4	4	x_1	6
$Z_j - C_j$	0	2	0	3/2			

As per the above problem, the only non-basic decision variable is x_2 (refer optimum table 3.1). Currently, the objective function co-efficient of x_2 is $c_2 = 4$. Now we want to discuss for what values of c_2 the optimality will be retained.

Let us take the change in c_2 as Δc_2, then Δc_2 can be given as $\Delta c_2 \le z_2 - c_2$ (i.e.) $\Delta c_2 \le 2$

Hence the range for c_2 (non-basic variables co-efficient) without affecting the optimality can be given by

$$-\infty \le c_2 \le c_2 + \Delta c_2$$
$$-\infty \le c_2 \le 4 + 2$$
$$-\infty \le c_2 \le 6$$

⇒ The co-efficient of x_2 can be varied from $-\infty$ to 6.

3.3 CHANGE IN THE OBJECTIVE FUNCTION CO-EFFICIENT OF A BASIC VARIABLE

Example: 2

Consider the given LPP along with its optimum table 3.2.

$$\text{Max } Z = 3x_1 + 5x_2; \quad \text{s.t.} \quad x_1 + x_2 \le 1; \quad 2x_1 + 3x_2 \le 1; \quad x_1, x_2 \ge 0$$

Optimum Table 3.2

C_j	3	5	0	0			
x_j	x_1	x_2	x_3	x_4	X_B	B	C_B
	1/3	0	1	–1/3	2/3	x_3	0
	2/3	1	0	1/3	1/3	x_2	5
$Z_j - C_j$	1/3	0	0	5/3	5/3		

As per the above problem, the only basic decision variable is x_2 (refer optimum table 3.2). Currently, the objective function co-efficient of x_2 is $c_2 = 5$. Now we want to discuss for what range values of c_2 the optimality will be retained.

Let us take the change in c_2 as Δc_2, to decide the +ve and –ve change for Δc_2 evaluate the following.

$Z_j - C_j$	1/3	0	0	5/3
x_2	2/3	1	0	1/3
$(Z_j - C_j)/x_2$	1/2	0	*	5

Note: The table entries are taken from the optimum table 3.2 entries.

For a max +ve change: |Min{(–ve) ratios}|
For a max –ve change: min {+ve ratios}

Here, –ve ratios: {0}: min {0}: 0
+ve ratios: {1/2, 5}: min {1/2, 5}: 1/2

$c_2 - 1/2 \le c_2 \le c_2 + 0$
$5 - 1/2 \le c_2 \le 5$
$9/2 \le c_2 \le 5$

⇒ The co-efficient of x_2 can be varied from 9/2 to 5.

3.4 CHANGE IN THE RIGHT-HAND SIDE OF A CONSTRAINT

***Example:* 3**

Consider the given LPP along with its optimum table 3.3.

Max $Z = 4x_1 + 5x_2$; s.t. $x_1 + 2x_2 \le 10$; (Labour) $6x_1 + 6x_2 \le 36$; (Material)
$8x_1 + 4x_2 \le 40$; (Storage space) $x_1, x_2 \ge 0$

Optimum Table 3.3

C_j	4	5	0	0	0			
x_j	x_1	x_2	x_3	x_4	x_5	X_B	B	C_B
	0	1	1	–1/6	0	4	x_2	5
	1	0	–1	+1/3	0	2	x_1	4
	0	0	4	–2	1	8	x_5	0
$Z_j - C_j$	0	0	1	1/2	0			

Def: Matrix B

The m × m matrix B [m-constrained LPP] is the Matrix whose j[th] column is the column of the jth positions basic variable in the optimum table 3.3.

Here, the optimum table 3.3 contains {x_2 x_1 x_5} as Basic variables.

Hence the corresponding matrix B is nothing but the column of the variables x_2, x_1 and x_5 in the initial simplex table.

$$B = \begin{array}{c} \begin{matrix} x_2 & x_1 & x_5 \end{matrix} \\ \begin{bmatrix} 2 & 1 & 0 \\ 6 & 6 & 0 \\ 4 & 8 & 1 \end{bmatrix} \end{array}$$

To get the inverse of B (i.e.) B^{-1}, consider the initial table Basic variable, $\{x_3, x_4, x_5\}$ column entry in the optimum table 3.3.

$$B^{-1} = \begin{array}{c} \begin{matrix} x_3 & x_4 & x_5 \end{matrix} \\ \begin{bmatrix} 1 & -1/6 & 0 \\ -1 & 1/3 & 0 \\ 4 & -2 & 1 \end{bmatrix} \end{array}$$

Evaluation of right-hand side values of the optimum table 3.3;

RHS of the optimum table's constraints = B^{-1} **b** (1)

Where b is the actual given column vector of the RHS values.

RHS of the optimum table 3.3 for the given example problem is $= -1\begin{bmatrix} 1 & -1/6 & 0 \\ 1/3 & 0 & 36 \\ 4 & -2 & 1 \end{bmatrix}\begin{bmatrix} 10 \\ \\ 40 \end{bmatrix} = \begin{bmatrix} 4 \\ 2 \\ 8 \end{bmatrix}$

Clearly $\begin{bmatrix} 4 \\ 2 \\ 8 \end{bmatrix}$ is the X_B column entries of the optimum table 3.3.

Suppose we want to determine how changing the amount of Labour (b_1) not affects the optimal solution. Currently, $b_1 = 10$. If we change b_1, to $b_1 + \Delta$, then from (1) the right-hand side of the constraints in the optimum table 3.3 will become,

$$B^{-1} * \begin{bmatrix} 10+\Delta \\ 36 \\ 40 \end{bmatrix} = \begin{bmatrix} 1 & -1/6 & 0 \\ -1 & 1/3 & 0 \\ 4 & -2 & 1 \end{bmatrix}\begin{bmatrix} 10+\Delta \\ 36 \\ 40 \end{bmatrix} = \begin{bmatrix} 4+\Delta \\ 2-\Delta \\ 8+4\Delta \end{bmatrix}$$

For the current basis to remain optimal, we require that the RHS of each constraint in the optimal table 3.3 remain non negative (≥ 0). This means that the current basis will remain optimal if the following conditions are satisfied.

$$\begin{aligned} 4+\Delta \geq 0 &\quad \Rightarrow \quad \Delta \geq -4 \\ 2-\Delta \geq 0 &\quad \Rightarrow \quad \Delta \leq 2 \\ 8+4\Delta \geq 0 &\quad \Rightarrow \quad \Delta \geq -2 \end{aligned}$$

$$\Delta \le 2; \quad \Delta \ge -4; \quad \Delta \ge -2$$
$$\Rightarrow \Delta \le 2 \text{ and } \Delta \ge -2$$
$$-2 \le \Delta \le 2$$
$$10 - 2 \le b_1 \le 10 + 2$$
$$8 \le b_1 \le 12$$

Thus, if the value of b_1 lies in the interval [8, 12] the current basis remains feasible and also optimal.

3.5 CHANGE IN THE COLUMN OF A NON-BASIC VARIABLE

Example: 4

Consider the LPP

Max $Z = 5x_1 + 12x_2 + 4x_3$ s.t. $x_1 + 2x_2 + x_3 \le 5$; $2x_1 - x_2 + 3x_3 = 2$; $x_1, x_2, x_3 \ge 0$

Optimum Table 3.4

C_j	5	12	4	0	–M			
x_j	x_1	x_2	x_3	x_4	x_5	X_B	B	C_B
	0	1	–1/5	2/5	–1/5	8/5	x_2	12
	1	0	7/5	1/5	2/5	9/5	x_1	5
$Z_j - C_j$	0	0	3/5	29/5	M – 2/5	41/5		

Here we are changing elements of column for x_3 in the original problem.

Clearly; $C_3 = 4$

$$B^{-1} = \begin{bmatrix} 2/5 & -1/5 \\ 1/5 & 2/5 \end{bmatrix} \quad \text{and} \quad a_3 = \begin{bmatrix} -5 \\ 2 \end{bmatrix}$$

$$Z_3 - C_3 = C_B{}^T \cdot B^{-1} \cdot a_3 - C_3$$
$$= [12 \quad 5] \begin{bmatrix} 2/5 & -1/5 \\ 1/5 & 2/5 \end{bmatrix} \begin{bmatrix} -5 \\ 2 \end{bmatrix} - 4$$
$$= [29/5 \quad -2/5] \begin{bmatrix} -5 \\ 2 \end{bmatrix} - 4$$
$$= \frac{-145-4}{5} - 4 = \frac{-169}{5} < 0$$

Since $z_3 - c_3 < 0$, the current basis is no longer optimal. The fact that $z_3 - c_3 = -169/5$ means that for each unit increase of the variable x_3 increases the revenue by 169/5. Introduce x_3 into the basis.

To find the new optimal table for the given problem, we re-create the table for the Basic variable $\{x_2, x_1\}$ and then apply simplex algorithm.

Column for x_3 in the optimal table's constraints $= B^{-1}a_3$

$$= \begin{bmatrix} 2/5 & -1/5 \\ 1/5 & 2/5 \end{bmatrix} \begin{bmatrix} -5 \\ 2 \end{bmatrix}$$

$$= \begin{bmatrix} -12/5 \\ -1/5 \end{bmatrix}$$

Since all entries of the column of x_3 in the optimum table 3.4 are –ve (<0), it is not possible for us to proceed. That is the indication of unbounded solution.

Note: If there is no unbounded indication, we have to introduce the new column corresponding to the variable in the existing optimum table 3.4 and apply the simplex algorithm to get the optimum solution.

3.6 ADDING A NEW CONSTRAINT

In many situations, opportunities arise to undertake new constraints.

Example: 5

Consider an LPP along with its optimum solution.

$$\text{Max } Z = 6x_1 + 4x_2; \quad \text{s.t.} \quad 2x_1 + 6x_2 \le 12; \quad 4x_1 + 4x_2 \le 16; \quad x_1, x_2 \ge 0$$

Optimal Solution:

$$x_1 = 4 \quad x_3 = 4 \quad \text{Max } Z = 24$$
$$x_2 = 0 \quad x_4 = 0$$

Check whether the addition of a new constraint will have any impact on the optimum solution. The additional constraint is, $4x_1 + 2x_2 \le 20$ (new resource).

Substitute the values for the variables and verify whether the condition is satisfied.

Here, $x_1 = 4$ & $x_2 = 0$ $\quad 4(4) + 2(0) \le 20 \quad 16 \le 20$

Clearly the condition is satisfied. This implies that, we can include the constraint along with the existing problem. Induction of this new constraint not causing any problem with the existing optimal solution.

Note: If the condition is violated, it is not possible for us to include the constraint.

3.7 ADDING A NEW VARIABLE

At various situations, opportunities arise to undertake new variables. Consider an LPP along with its dual solution.

Max $Z = 6x_1 + 4x_2$; s.t. $2x_1 + 6x_2 \le 12$ (Labour) $\quad 4x_1 + 4x_2 \le 16$ (Machine hr) $\quad x_1, x_2 \ge 0$

Dual Solution: $y_1 = 0 \quad y_2 = 3/2$

Dual Problem: Min $Z^* = 12y_1 + 16y_2$; $\quad$ s.t. $2y_1 + 4y_2 \ge 6$; $\quad 6y_1 + 4y_2 \ge 4$; $\quad y_1, y_2 \ge 0$

In the above LP problem, the company may be presented with the opportunity to manufacture additional type of a product.

Let x_3 be the number of units to be produced in the new product. Its contribution to the objective is ₹3 per unit and one unit of the new product requires one man hour and 2 machine hours.

Then the new LPP with the additional new product can be given as

$$\text{Max } Z = 6x_1 + 4x_2 + 3x_3; \quad \text{s.t.} \quad 2x_1 + 6x_2 + x_3 \leq 12; \quad 4x_1 + 4x_2 + 2x_3 < 16; \quad x_1, x_2, x_3 \geq 0$$

Convert the above problem into a dual one:

$$\text{Min } Z^* = 12y_1 + 16y_2; \quad \text{s.t} \quad 2y_1 + 4y_2 \geq 6; \quad 6y_1 + 4y_2 \geq 4; \quad y_1 + 2y_2 \geq 3; \quad y_1, y_2 \geq 0$$

The addition of the new variable in the primal becomes addition of a new constraint in the dual. Hence, substitute the value of the dual variables and verify the condition.

$$0 + 2\,(3/2) \geq 3 \qquad \text{(i.e.) } 3 \geq 3$$

Clearly the condition is true. This implies that the inclusion of the third variable for a new additional product in the existing model not affect the optimality situation. Hence the additional constraint is redundant and the optimum solution of the given LPP is also optimum for the new LPP.

Example: 6

Given the LPP

$$\text{Min } Z = 3x_1 + 6x_2 + x_3; \quad \text{s.t. } x_1 + x_2 + x_3 \geq 6; \quad x_1 - 5x_2 - x_3 \geq 4; \quad x_1 + 5x_2 + x_3 \geq 24; \quad x_1, x_2, x_3 \geq 0$$

(i) Solve the LPP.

(ii) Discuss the effect of changing the requirement vector from [6 4 24] to [6 2 12] on the optimum solution.

Apply the simplex method and get the optimum solution.

C_j	−3	−6	−1	0	0	0			
x_j	x_1	x_2	x_3	x_4	x_5	x_6	X_B	B	C_B
	0	4	0	1	0	−1	18	x_4	0
	1	0	0	0	−1/2	−1/2	14	x_1	−3
	0	5	1	0	1/2	−1/2	10	x_3	−1
$Z_j - C_j$	0	1	0	0	1	2	−52		

$$B^{-1} = \begin{bmatrix} 1 & 0 & -1 \\ 0 & -1/2 & -1/2 \\ 0 & 1/2 & -1/2 \end{bmatrix}$$

*These questions have appeared in Indian university examination papers.

$$\begin{bmatrix} x_4 \\ x_1 \\ x_3 \end{bmatrix} = X_B = B^{-1}\,[\text{new vector}]$$

$$= \begin{bmatrix} 1 & 0 & -1 \\ 0 & -1/2 & -1/2 \\ 0 & 1/2 & -1/2 \end{bmatrix} \begin{bmatrix} -6 \\ -2 \\ -12 \end{bmatrix}$$

$$= \begin{bmatrix} 6 \\ 7 \\ 5 \end{bmatrix}$$

Since all entries of X_B are ≥ 0, the existing solution is feasible. Hence the solution is

$$x_4 = 6; \quad x_1 = 7; \quad x_3 = 5; \quad \text{and} \quad \text{Min } Z = 26$$

Note 1: Since the objective is minimum we have to multiply the new vector by (–1). If the object is maximum don't multiply by (–1).

Note 2: Apply the method stated above for the combined effect.

*Example: 7

For the following LPP, the optimal solution is shown in the table 3.5.

$$\text{Max } Z = -x_1 + 2x_2 - x_3; \quad \text{s.t.} \quad 3x_1 + x_2 - x_3 \leq 10; \quad -x_1 + 4x_2 + x_3 \geq 6;$$
$$x_2 + x_3 \leq 4; \quad x_1, x_2, x_3 \geq 0$$

Determine the ranges for discrete changes in the components b_2 and b_3 of the requirement vectors so as to maintain the optimality of the current optimum solution.

Optimum Table 3.5

			–1	2	–1	0	0	0	–M
C_B	Y_B	X_B	x_1	x_2	x_3	S_1	S_2	S_3	R_1
0	S_1	6	3	0	–2	1	0	–1	0
2	X_2	4	0	1	1	0	0	1	0
0	S_2	10	1	0	3	0	1	4	–1
		8	1	0	3	0	0	2	M

$$\text{Here,} \quad B^{-1} = [S_1 \quad R_1 \quad S_3] = \begin{bmatrix} 1 & -1 & 0 \\ 0 & 1 & 0 \\ 0 & 4 & -1 \end{bmatrix}$$

Change in b_2.

At present $b_2 = 6$, Let it be $6 + \Delta$.

$$B^{-1}\begin{bmatrix}10\\6+\Delta\\4\end{bmatrix} = \begin{bmatrix}1 & -1 & 0\\0 & 1 & 0\\0 & 4 & -1\end{bmatrix}\begin{bmatrix}10\\6+\Delta\\4\end{bmatrix}$$

$$= \begin{bmatrix}10-6-4\\6+\Delta\\24+4\Delta-4\end{bmatrix}$$

For optimality, all the entries must be ≥ 0.

$$4-\Delta \geq 0 \Rightarrow \Delta \leq 4$$
$$6+\Delta \geq 0 \Rightarrow \Delta \geq -6$$
$$4\Delta + 20 \geq 0 \Rightarrow \Delta \geq -5$$
$$\Delta \leq 4 \text{ and } \Delta \geq -5$$
$$-5 \leq \Delta \leq 4$$
$$6-5 \leq b_2 \leq 6+4$$
$$1 \leq b_2 \leq 10$$

Change in b_3: $b_3 = 4$

Let it be $4 + \Delta$

$$B^{-1} * \begin{bmatrix}10\\6\\4+\Delta\end{bmatrix} = \begin{bmatrix}1 & -1 & 0\\0 & 1 & 0\\0 & 4 & -1\end{bmatrix}\begin{bmatrix}10\\6\\4+\Delta\end{bmatrix}$$

$$= \begin{bmatrix}10\\6\\24-4-\Delta\end{bmatrix}$$

for optimality,

$$24-4-\Delta \geq 0$$
$$20-\Delta \geq 0$$
$$\Delta \leq 20$$
$$b_3 \leq 4+20$$
$$b_3 \leq 24$$

Example: 8

Find the optimum solution to the LPP Max $Z = 15x_1 + 45x_2$;

s.t. $x_1 + 16x_2 \leq 240$; $5x_1 + 2x_2 \leq 162$; $x_2 \leq 50$; $x_1, x_2 \geq 0$

If maximum $Z = \Sigma\, c_j\, x_j$; $j = 1, 2$, and c_2 is kept fixed at 45, determine how much can c_1 be j changed without affecting the above optimum solution.

Consider the given LPP and get the optimum table 3.6 by applying the simplex method.

Optimum Table 3.6

C_j	15	45	0	0	0			
x_j	x_1	x_2	x_3	x_4	x_5	X_B	B	C_B
	0	1	5/78	–1/78	0	173/13	x_2	45
	1	0	–1/39	8/39	0	352/13	x_1	15
	0	0	–5/78	1/78	1	477/13	x_5	0
	0	0	5/2	5/2	0	1005		

Clearly from the optimum table 3.6 x_1 is the basic variable.

Currently, the objective function coefficient of x_1 is $c_1 = 15$. Now we want to discuss for what range values of c_1, the optimality will be retained.

Let us take the change in c_1, as Δc_1, to decide the +ve and –ve change for Δc_1. Evaluate the following.

$Z_j - C_j$	0	0	5/2	5/2	0
x_1	1	0	–1/39	8/39	0
$Z_j - C_j/x_1$	0	*	–195/2	195/16	*

For a Max +ve change: |Min [0, –195/2]| : 195/2

For a Min –ve change: Min {195/16} : 195/16

$15 - (195/16) \le c_1 \le 15 + (195/2)$

$45/16 \le c_1 \le 225/2$

EXERCISE PROBLEMS

*1. Solve the following LPP.

Max $Z = 3x_1 + 4x_2 + x_3 + 7x_4$; s.t. $8x_1 + 3x_2 + 4x_3 + x_4 \le 7$;

$2x_1 + 6x_2 + x_3 + 5x_4 \le 3$; $x_1 + 4x_2 + 5x_3 + 2x_4 \le 8$; $x_1, x_2, x_3, x_4 \ge 0$

Also check whether there is any change in the optimum solution of the primal problem if the constraint $2x_1 + 3x_2 + x_3 + 5x_4 \le 4$ is added.

*2. Given the LPP. Max $Z = 3x_1 + 5x_2$; s.t. $3x_1 + 2x_2 \le 18$; $x_1 \le 4$; $x_2 \le 6$; $x_1, x_2 \ge 0$

Discuss the effect on optimality of the solution when the objective function is changed to $[3x_1 + x_2]$

[**Hint:** Find the range for c_2, then compare with the coefficient of c_2 in the modified one]

*3. Consider the LPP.

Max $Z = 5x_1 + 12x_2 + 4x_3$; s.t. $x_1 + 2x_2 + x_3 \le 5$; $2x_1 - x_2 + 3x_3 = 2$; $x_1, x_2, x_3 \ge 0$

The optimum solution to this problem is contained in the following simplex table:

y_B	x_B	y_1	y_2	y_3	y_4	y_5
y_2	8/5	0	1	–1/5	2/5	–1/5
y_1	9/5	1	0	7/5	1/5	2/5
$Z_j - C_j$	281/5	0	0	3/5	29/5	M –2/5

For each of the discrete parameter changes listed below, make the necessary corrections in the optimum table and solve the resulting problem:

(a) Change b to {3, 10} from {5, 2}
(b) Change c to (4, 10, 4) from (5, 12, 4)

[**Hint:**

(a) Find $B^{-1} * \begin{bmatrix} 3 \\ 10 \end{bmatrix} = \begin{bmatrix} -4/5 \\ 23/5 \end{bmatrix}$ Since one entry is –ve implies that it is not optimal.

To get the new optimal solution introduce $\begin{bmatrix} -4/5 \\ 23/5 \end{bmatrix}$ as the X_B value in the existing optimum table and apply Dual Simplex method.

(b) Replace the c_j entries as the new values [4 10 4] in the optimum table and find $(z_j - c_j)$ if the optimality condition is satisfied stop the process. If not apply the simplex process to get the optimum solution.]

ANSWERS TO THE EXERCISE PROBLEMS

1. $x_1 = 16/19$; $x_2 = 0$; $x_3 = 0$; $x_4 = 5/19$ and Max $Z = 83/19$
   No change in the optimum solution.
2. $x_1 = 2$; $x_2 = 6$; Max $Z = 36$
   range for c_1 is $0 \le c_1 \le 7.5$
   range for c_2 is $2 \le c_2 \le$ infinity
3. (a) No feasible solution to the problem.
   (b) $x_1 = 0$; $x_2 = 13/7$; $x_3 = 9/7$ and Max $Z = 166/7$

REVIEW QUESTIONS

Short Questions

1. Define the term 'Sensitivity Analysis'. (Section 3.1)
2 State the four cases possible in the Sensitivity Analysis. (Section 3.1)
3. Can the optimum value change by introducing an additional constraint? (No)
4. What is the use of sensitivity analysis.
   i. To know the range of the resources for which the optimality is retained.
   ii. To make sure of the possibility of introducing the new variable/new constraints with the existing model.

State True or False

5. Sensitivity analysis decides how sensitive the optimum value is with respect to each coefficients of the LPP in objective and constraints. (T)
6. Sensitivity analysis can be adopted only after getting the optimum solution of a given LPP. (T)
7. Addition of a new constraint with the existing LPP will always improve the optimum solution. (F)
8. Addition of a new constraint can retain the optimality. (T)
9. The addition of a new variable can improve the value of the objective. (T)

Fill in the Blanks

10. Sensitivity analysis is otherwise called _________. (Post-optimal analysis)
11. Change in resource level may affect the _________. (Optimality)
12. To verify the nature of inclusion of a new variable with the existing model, we seek the help of _________. (Dual, principle)
13. Sensitivity analysis spells out the _________ of resource levels for which the optimality is not affected. (range)

4

Transportation Problem

4.1 INTRODUCTION

The term "transportation" is somewhat deceptive it appears to be restricted only to transportation systems, but the case is different. In fact many of the resource allocation problems arising in production systems can be treated as transportation problems. Typical examples are production scheduling, transportation scheduling, etc.

This problem was first introduced by HITCHCOCK (1941) and KOOPMANS (1947), and was solved by DANTZIG (1947) using revised simplex method.

4.1.1 The Transportation Problem can be Described as Follows

Consider m origins (production centres/warehouses) and n destinations (market places) in a transportation system.

The Origin i (i = 1, 2, … m) has s_i ($s_i > 0$) units of single product of supply available to be transported/shipped, whereas the destination j (j = 1, 2, … n) requires the demand ($d_j > 0$) d_j units of the same product.

Assume that the cost of transporting from the origin i to the destination j is directly proportional to the amount of units transported and that the unit transportation cost from each origin to each destination c_{ij} is known and constant.

Then the model requires to determine the number of units of the product to be transported from origin i to destination j, for all i and j, so that the total transportation cost can be minimized, subject to the supply and demand constraints.

Let x_{ij} ($x_{ij} \geq 0$) denotes the number of units transported from the ith origin to j^{th} destination for all i = 1, 2, … m and j = 1, 2, … n.

The relationship between the different elements of a transportation problem can be given in a tabular form (Table 4.1). The box (i, j) will be called cell (i, j).

Table 4.1

Destination / Origin	1	2	....	j	....	n	Supply/ Availability
1	x_{11} c_{11}	x_{12} c_{12}	....	x_{1j} c_{1j}	....	x_{1n} c_{1n}	s_1
2	x_{21} c_{21}	x_{22} c_{22}	....	x_{2j} c_{2j}	....	x_{2n} c_{2n}	s_2
:							
i	x_{i1} c_{i1}	x_{i2} c_{i2}	....	x_{ij} c_{ij}	....	x_{in} c_{in}	s_i
:							
m	x_{m1} c_{m1}	x_{m2} c_{m2}	....	x_{mj} c_{mj}	....	x_{mn} c_{mn}	s_m
Demand/ Requirement	d_1	d_2	....	d_j	....	d_n	

4.2 CONVERSION OF A TP INTO AN EQUIVALENT LPP FORM

Based on the description cited above, the general model of the transportation problem can be formulated as an LPP.

$$\text{Min } Z = C_{11}x_{11} + c_{12}x_{12} + \ldots\ldots\ldots\ldots + c_{mn}x_{mn}$$

$$\text{subject to } x_{11} + x_{12} + \ldots\ldots\ldots\ldots + x_{1n} = s_1$$

$$\ldots\ldots\ldots\ldots\ldots\ldots\ldots\ldots\ldots\ldots\ldots\ldots$$

$$x_{m1} + x_{m2} + \ldots\ldots\ldots\ldots + x_{mn} = s_m$$

$$x_{11} + x_{21} + \ldots\ldots\ldots\ldots + x_{m1} = d_1$$

$$\cdot \quad \cdot \quad \cdot \quad \cdot \quad \cdot \quad \cdot \quad \cdot \quad \cdot$$

$$x_{1n} + x_{2n} + \ldots\ldots\ldots\ldots + x_{mn} = d_n$$

Where $x_{ij} > 0$ (for $i = 1, 2, \ldots m; j = 1, 2, \ldots n$)

Inorder to get a feasible solution for the problem, we must have the total supply $(\sum_{i=1}^{m} S_i)$ equal to the total demand $(\sum_{j=1}^{n} d_i)$

$$\text{i.e. } \sum_{i=1}^{m} s_i = \sum_{j=1}^{n} d_i$$

If the condition is satisfied for a transportation model, then it is called **balanced transportation problem**. Otherwise, the problem is called **unbalanced transportation problem**. Even an unbalanced model can be converted to a balanced one. That we can see at a later stage of this unit.

4.3 FORMULATION OF A TRANSPORTATION PROBLEM

***Example:* 1**

A company has 3 warehouses and 4 marketing outlets. The warehouses' capacities are 30, 70 and 50. The market demands are 40, 30, 40 and 40. The unit cost of shipping is given by the following table.

		Marketing outlets			
		1	2	3	4
warehouses	1	2	2	2	1
	2	10	8	5	4
	3	7	6	6	8

Formulate the above problem as a transportation problem. The equivalent transportation model is,

		Marketing outlets				
		1	2	3	4	supply
warehouses	1	2	2	2	1	30
	2	10	8	5	4	70
	3	7	6	6	8	50
demand		40	30	40	40	150 / 150

Also, the model is a balanced one.

4.4 CONCEPTS OF FEASIBILITY BASICNESS, AND DEGENERACY IN THE SOLUTION

A solution for a given transportation problem is a planned shipment indicating the number of units to be transported from certain origins to certain destinations in such a way as to satisfy the supply and demand constraints.

4.4.1 Basic and Non-basic Cells

If x_{ij} be the allocation to the cell (i, j) and $x_{ij} > 0$ is called positive allocation and the cell (i, j) is called occupied/basic cell.

When $x_{ij} = 0$ for any cell (i, j), it is named as non-basic cell (the value need not be represented in that cell).

Feasible Solution

In a transportation problem, a solution is called feasible if the set of x_{ij} values ($x_{ij} > 0$) satisfies all the supply and demand requirements.

Basic Feasible Solution

If a feasible solution, containing exactly $(m + n - 1)$ basic/occupied cells then it is called basic feasible solution (non-degenerate solution).

Degenerate Basic Feasible Solution

If a feasible solution, containing less than $(m + n - 1)$ basic cells, then it is called degenerate basic feasible solution (degenerate solution).

Optimal Solution

A basic feasible solution that minimizes (or maximizes the profit) the total transportation cost is called optimal solution.

4.5 METHODS USED TO FIND THE SOLUTION TO A TRANSPORTATION PROBLEM

The next step after the formulation of a transportation problem is to find the solution to the same. We have so many ways to find the solution that are clearly stated in the Flowchart 4.1.

Notations Used

TP : Transportation Problem
S : Supply corresponding to the origin
TS : Total Supply
D : Demand corresponding to the column
TD : Total Demand
NWCR : North West Corner Rule
RMM : Row Minima Method
CMM : Column Minima Method
LCM : Least Cost Method
VAM : Vogel's Approximation Method
MODI : Modified Distributive Method
m : Number of origins (rows)
n : Number of destinations (columns)
A : Actual allocation of units
TC : Total transportation cost
RS : Remaining supply
RD : Remaining Demand
VN : Vogel Number
$\in$: The least positive quantity (10^{-6})
K : Number of basic cells

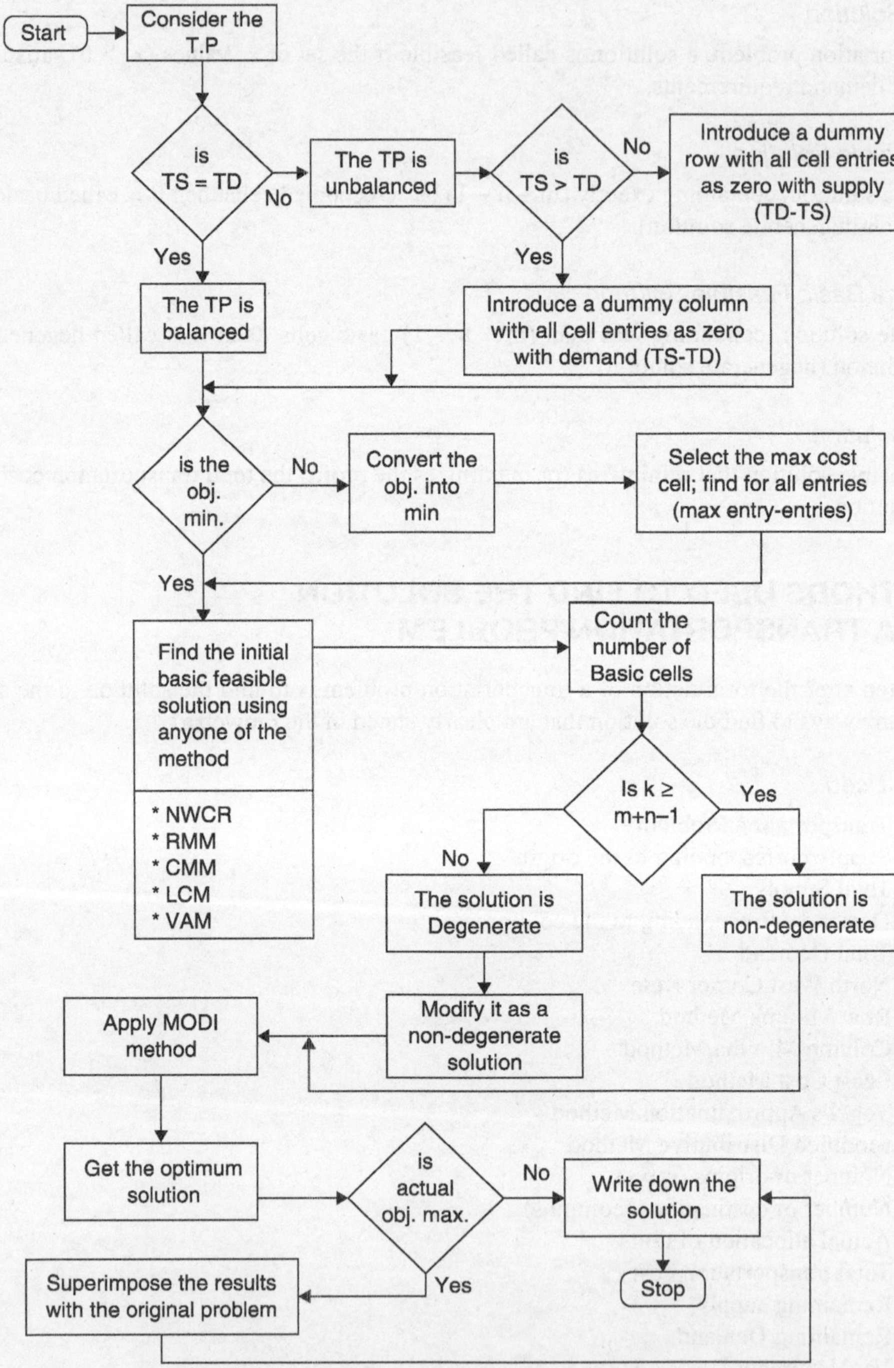

Flowchart 4.1 Methods to find the solution to the TPP

4.6 DESCRIPTION OF VARIOUS METHODS TO FIND THE INITIAL BASIC FEASIBLE SOLUTION

1. North West Corner Rule (NWCR)
2. Row Minima Method (RMM)
3. Column Minima Method (CMM)
4. Least Cost Method or Matrix Minima Method (LCM/MMM)
5. Vogel's Approximation Method (VAM)

The usual procedure of getting the initial solution is, every time the cell to be allocated is selected based on certain principles. By comparing the corresponding supply and demand, the minimum value will be taken and allocated, this process will be repeatedly done until all the supply and demand are satisfied fully.

Remark 1: We name the balanced TP with minimization objective as a **standard transportation problem.**

In this section we will see the methods one by one with examples.

4.6.1 North West Corner Rule

The various steps to be followed to solve a given transportation problem using NWCR are clearly depicted in the following Flowchart 4.2.

Example: 2

Write down the initial basic feasible solution to the following transportation problem by NWCR.

	Destination			Supply
Origin	7	3	4	2
	2	1	3	3
	3	4	6	5
Demand	4	1	5	

Step 1: Find the total supply and the total demand

$$TS = 2 + 3 + 5 = 10 \qquad TD = 4 + 1 + 5 = 10$$

TS = TD; The given problem is balanced. Proceed to Step 2.

Step 2: Consider the problem and select the north west corner cell for allocation;

		Destination			
		D_1	D_2	D_3	S
	O_1	[2] 7	3	4	~~2~~ 0
Origin	O_2	2	1	3	3
	O_3	3	4	6	5
D		~~4~~ 2	1	5	

*These questions have appeared in Indian university examination papers.

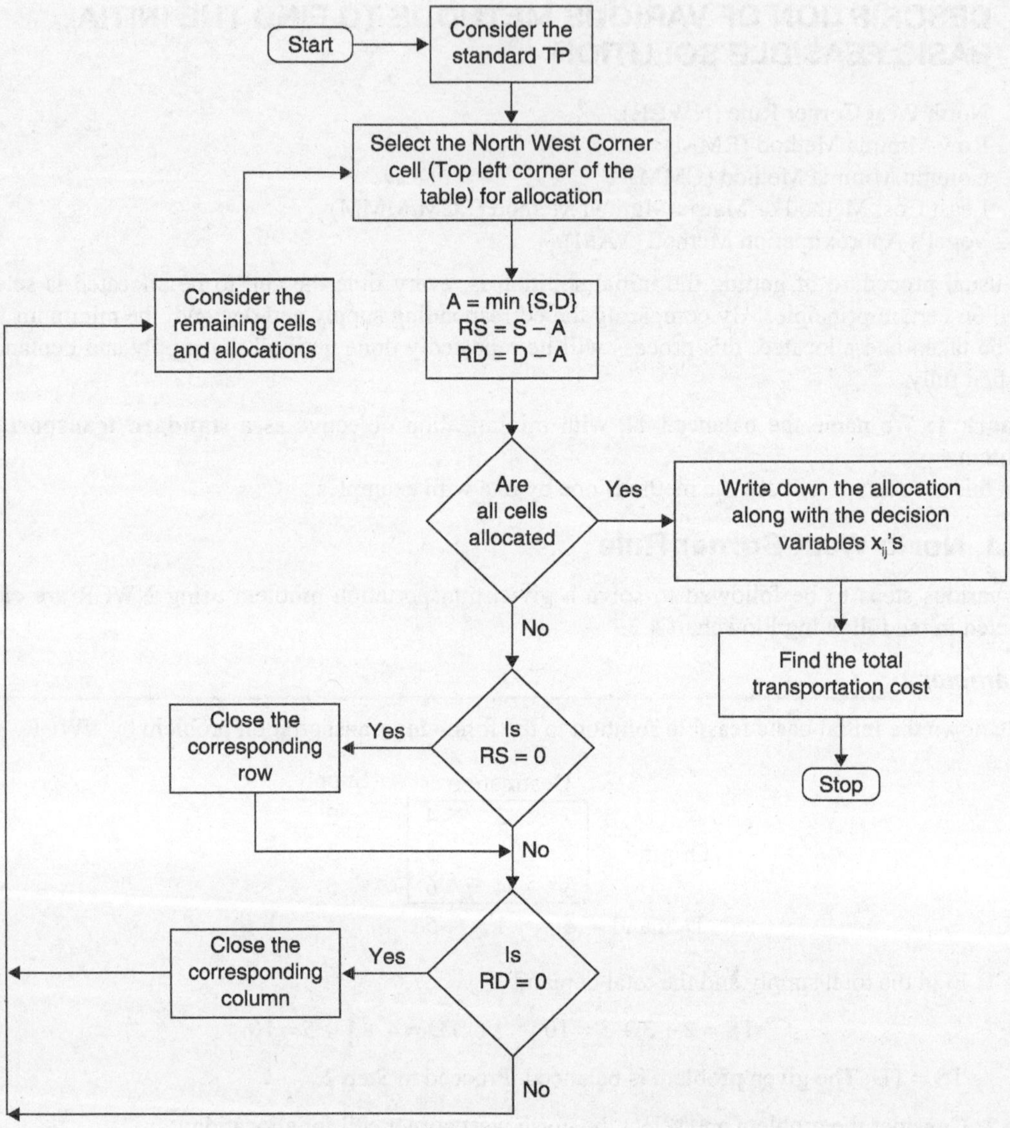

Flowchart 4.2 North West Corner Rule

allocation for that cell (1, 1) = min $\{s_1, d_1\}$ = min {2, 4} = 2
Allocate 2 units to that cell (1, 1)
RS = 2 – 2 = 0; RD = 4 – 2 = 2
Since RS = 0; close down the Row 1

Step 3: Verify whether all the cells are allocated. If allocated go to Step 4; if not repeat Step 2.
* Here, we have to repeat Step 2.

	D_1	D_2	D_3	
O_2	[2] 2	1	3	~~3~~ 1
O_3	3	4	6	5
D	~~2~~ 0	1	5	

Consider the remaining cells.
NW corner cell is (2, 1); so allocate in cell (2, 1),
allocation = min {3, 2} = 2
Allocate 2 units to that cell,

RS = 3 − 2 = 1; RD = 2 − 2 = 0; close down column 1.

	D_1	D_2	S
O_2	[1] 1	3	~~1~~ 0
O_3	4	6	5
D	0 ~~1~~	5	

Since, both RS = RD = 0 close both row 2 and column 2.
See that, for the final allocation, the remaining supply and the demand are equal [This is true for all the problems]

	D_3	S
O_3	[5]	~~5~~ 0
D	6	
	~~5~~ 0	

Hence, the initial basic solution is

[2] 7	3	4
[2] 2	[1] 1	3
3	4	[5] 6

The cost of transportation is $2 \times 7 + 2 \times 2 + 1 \times 1 + 5 \times 6 = 49$.

$x_{11} = 2$; $x_{21} = 1$, $x_{22} = 1$, $x_{33} = 5$ and all other allocations are equal to zero.

Clearly, $m + n - 1 = 3 + 3 - 1 = 5$

number of basic cells = 4

Since $5 \neq 4$, implies that

The initial basic feasible solution is degenerate.

***Example: 3**

$$[\text{Unit cost}] = \begin{bmatrix} 8 & 7 & 3 \\ 3 & 8 & 9 \\ 11 & 3 & 5 \end{bmatrix} \begin{matrix} S \\ 60 \\ 70 \\ 80 \end{matrix}$$

$$D \quad 60 \quad 80 \quad 80$$

Convert the unbalanced transportation problem above into a balanced one.
Consider the given problem,

			a_i
8	7	3	60
3	8	9	70
11	3	5	80
b_j 60	80	80	

TS = 60 + 70 + 80 = 210
TD = 60 + 80 + 80 = 220
TS < TD

⇒ The given problem is not a balanced one. So, add a dummy row with all cost entries as zero and supply as (TD − TS = 220 − 210) 10 units.

				a_i
	8	7	3	60
	3	8	9	70
	11	3	5	80
	0	0	0	10
b_j	60	80	80	220 / 220

*Example: 4

Find an initial basic feasible solution of the following problem using north west corner rule.

	D_1	D_2	D_3	D_4	S
O_1	5	3	6	2	19
O_2	4	7	9	1	37
O_3	3	4	7	5	34
D	16	18	31	25	90

The given problem is balanced.

	D_1	D_2	D_3	D_4	a_i
O_1	16 / 5	3	6	2	~~19~~ 3
O_2	4	7	9	1	37
O_3	3	4	7	5	34
b_j	~~16~~ 0	18	31	25	

	D_2	D_3	D_4	
O_1	3 / 3	6	2	~~3~~ 0
O_2	7	9	1	37
O_3	4	7	5	34
	~~18~~ 15	34	25	

	D_2	D_3	D_4	
O_2	[15] 7	9	1	~~37~~ 22
O_3	4	7	5	34
	~~15~~ 0	31	25	

	D_3	D_4	
O_2	[22] 9	1	~~22~~ 0
O_3	[9] 7	[25] 5	~~34~~ 0
	~~31~~ ~~9~~ 0	~~25~~ 0	

	D_1	D_2	D_3	D_4
O_1	[16] 5	[3] 3	6	2
O_2	4	[15] 7	[22] 9	1
O_3	3	4	[9] 7	[25] 5

The initial solution is,

$x_{11} = 16,\quad x_{12} = 3,\quad x_{22} = 15,$

$x_{23} = 22,\quad x_{33} = 9,\quad x_{34} = 25$

$TC = 16 \times 5 + 3 \times 3 + 15 \times 7 + 22 \times 9 + 9 \times 7 + 25 \times 5 = 580$

The initial solution is a non-degenerate one.

4.6.2 Row Minima Method

The various steps to be followed while applying row minima method are clearly explained in the Flowchart 4.3.

No. of rows = m

Example: 5

Solve the transportation problem using row minima method.

	A	B	C	S
I	6	8	4	14
II	4	9	8	12
III	1	2	6	5
D	6	10	15	31 / 31

Step 1: TS = 31; TD = 31. Hence, the problem is a balanced one.

Step 2: As per the name of the method, we are going to allocate row by row, consider the first row of the given problem; select the min. cost cell. It is cell (1, 3). Do the first allocation in that cell.

	A	B	C	S
I	6	8	[14] 4	~~14~~ 0
II	4	9	8	12
III	1	2	6	5
	6	10	~~15~~ 1	

$A = \min\{14, 15\} = 14$

$RS = 14 - 14 = 0;\quad RD = 15 - 14 = 1$

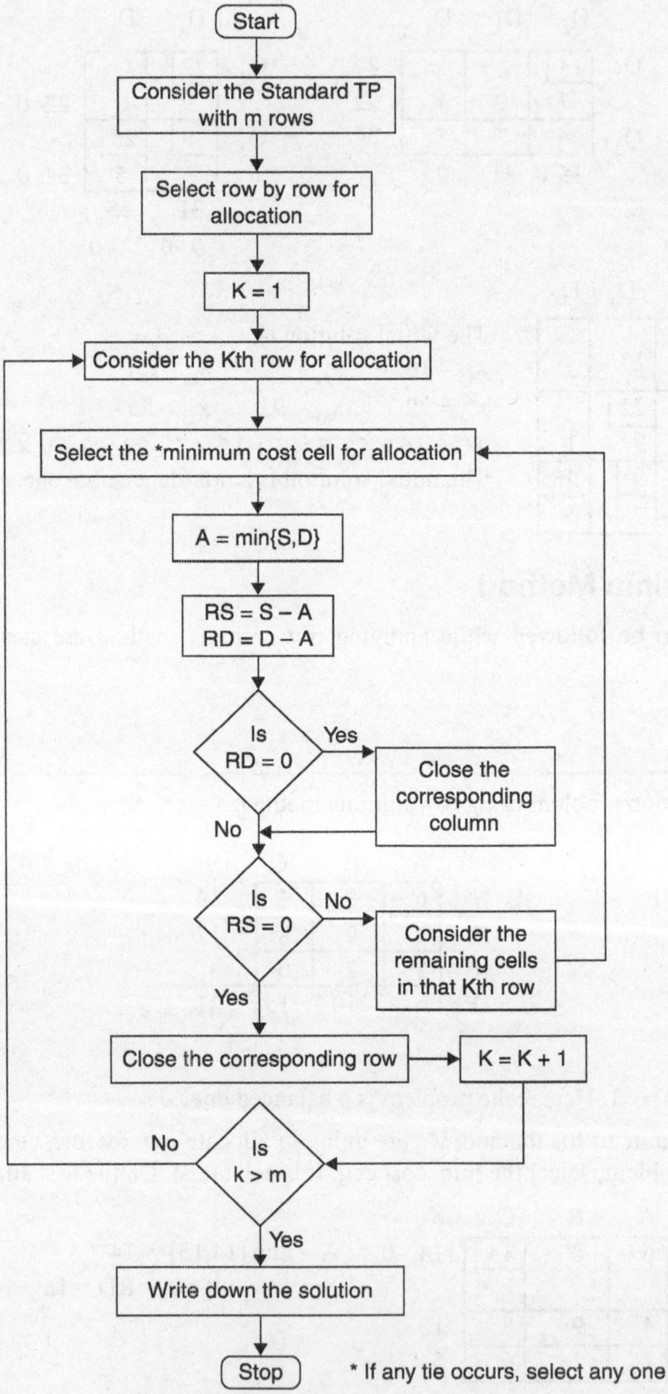

Flowchart 4.3 Row minima method

Since RS = 0; close down Row-1. Since, Row-1, is completely over repeat Step 2. Consider the 2nd Row for allocation.

	A	B	C	S
II	[6] 4	[5] 9	[1] 8	~~12~~ ~~6~~ ~~5~~ 0
III	1	[5] 2	6	5 0
D	0 ~~6~~	~~5~~ ~~10~~ 0	~~1~~ 0	

The min. cost cell is cell (2, 1)

Allocation = min {12, 6} = 6

RD = 0; close column 1.

The row-2 is not completed, again consider the min cell, it is cell (2, 3). Since, RD = 0; close column 3. To get the complete solution repeat the same process.

	A	B	C
I	6	8	[14] 4
II	[6] 4	[5] 9	[1] 8
III	1	[5] 2	6

Here, (m + n − 1) is 5 and the number of basic cells are also 5, hence the solution is non-degenerate.

$x_{13} = 14$; $x_{21} = 6$; $x_{22} = 5$, $x_{23} = 1$, $x_{32} = 5$ and other allocations are 0

Total cost = $14 \times 4 + 6 \times 4 + 5 \times 9 + 1 \times 8 + 5 \times 2 = 143$

4.6.3 Column Minima Method

In Flowchart 4.3, replace wherever the word "row" is there as "column", we get the procedure for column minima method.

Example: 6

Solve the transportation problem using column minima method.

		Destinations			
		1	2	3	Capacities
Sources	1	2	2	3	10
	2	4	1	2	15
	3	1	3	5	40
Demands		20	15	30	

Step 1: TS = 65; TD = 65. Hence the problem is balanced one.

Step 2: As per the name of the method, we are going to allocate column by column.

	1	2	3	
1	2	2	3	10
2	4	1	2	15
3	20 (1)	3	5	~~40~~ 20
	~~20~~ 0	15	30	

Consider the min. cost cell in the column 1; it is cell (3, 1) allocation = min{20, 40} = 20.
RS = 40 – 20 = 20; RD = 0; Since RD = 0; close column 1 and proceed to column 2, and repeat the same process.

	2	3	
1	2	3	10
2	15 (1)	2	~~15~~ 0
3	3	5	20
	~~15~~ 0	30	

Cell (2, 2) is min. Since, RS = 0 and RD = 0, close down column 2 and row 2 and proceed to column 3.

	3	
1	10 (3)	~~10~~ 0
3	20 (5)	~~20~~ 0
	~~30~~ 0	

Since, we left with single column, there is no necessity to follow the rule, we can fill it by looking into the remaining supply and demand.

We have the solution as in the following table.

	1	2	3
1	2	2	10 (3)
2	4	15 (1)	2
3	20 (1)	3	20 (5)

Note that we have (m + n – 1) = 5 and the number of basic cells are 4, hence the solution is degenerate one.

$x_{13} = 10$; $x_{22} = 15$; $x_{31} = 20$; $x_{33} = 20$
and other allocations are zero.

Total cost = 10 × 3 + 15 × 1 + 20 × 1 + 20 × 5 = 165

4.6.4 Least Cost Method/Matrix Minima Method

The sequence of steps to be followed to solve a given transportation problem using LCM/MMM are clearly explained in the Flowchart 4.4.

***Example: 7**

Obtain an initial basic feasible solution to the following transportation problem using matrix minima method.

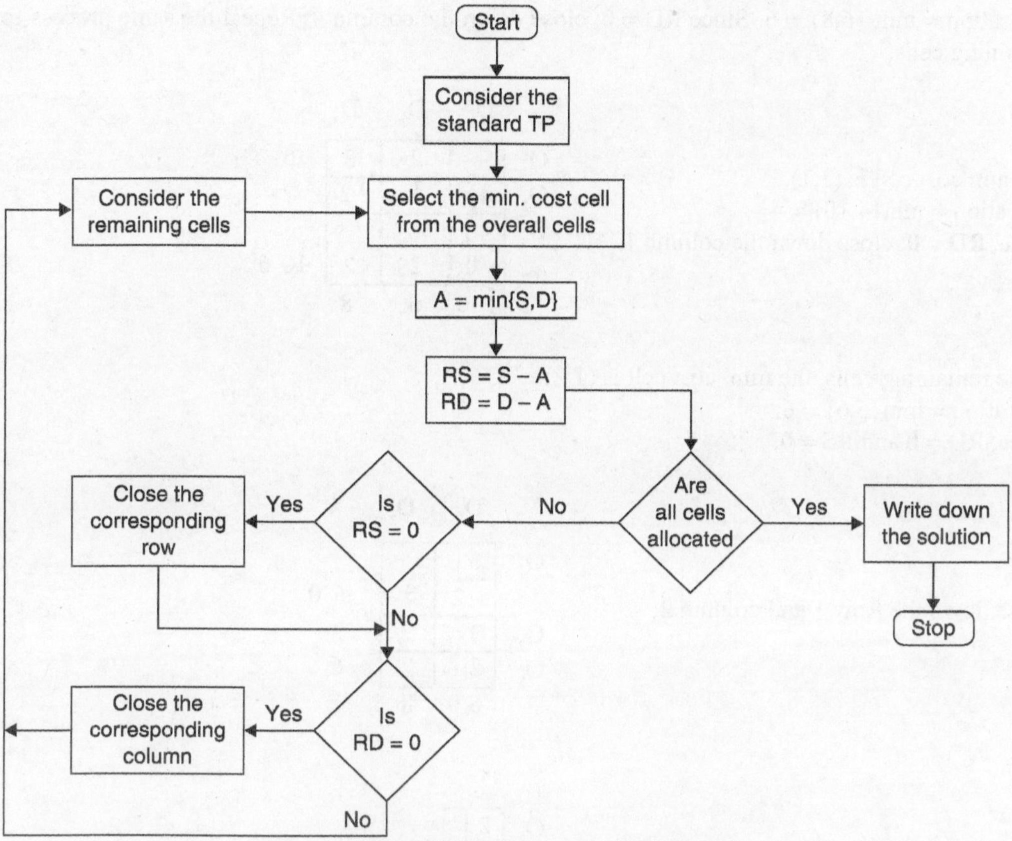

Flowchart 4.4 Least cost method

		Destination				Capacity
		D_1	D_2	D_3	D_4	
Origin	O_1	1	2	3	4	6
	O_2	4	3	2	0	8
	O_3	0	2	2	1	10
Demand		4	6	8	6	24

Step 1: TS = 24; TD = 24; implies the problem is a balanced one.

Step 2: Select a Least Cost (min) cell for allocation, here we have the min. entry as in cells (2, 4) and (3, 1). Select any one cell arbitrarily, take the cell (2, 4).

	D_1	D_2	D_3	D_4	
O_1	1	2	3	4	6
O_2	4	3	2	0 (6)	~~8~~ 2
O_3	0	2	2	1	10
	4	6	8	~~6~~ 0	

Allocation = min {6,8} = 6. Since RD = 0; close down the column 4. Repeat the same process for the remaining cells.

The min cost cell is (3,1).
allocation = min{4,10} = 4
Since, RD = 0; close down the column 1.

	D_1	D_2	D_3	
O_1	1	2	3	6
O_2	4	3	2	2
O_3	[4] 0	2	2	~~10~~ 6
	~~4~~ 0	6	8	

In the remaining cells, the min. cost cell is (1,2).
allocation = min{6,6} = 6
Since, RD = 0 and RS = 0.

Close down the Row 1 and column 2.

	D_2	D_3	
O_1	[6] 2	3	~~6~~ 0
O_2	3	2	2
O_3	2	2	6
	~~6~~ 0	8	

Since, we left with only one column;
we can fill the remaining things directly.

	D_3	
O_2	[2] 2	~~2~~ 0
O_3	[6] 2	~~6~~ 0
	~~8~~ 0	

The initial solution is presented in the following table.

	D_1	D_2	D_3	D_4
O_1	1	[6] 2	3	4
O_2	4	3	[2] 2	[6] 0
O_3	[4] 0	2	[6] 2	1

$x_{12} = 6$; $x_{23} = 2$; $x_{24} = 6$
$x_{13} = 4$; $x_{33} = 6$.
The total cost $= 2 \times 6 + 2 \times 2 + 0 \times 6 + 0 \times 4 + 6 \times 2 = 28$

*Example: 8

Obtain an initial basic feasible solution to the given transportation problem using matrix minima method.

Destination

		A	B	C	D	S
	1	1	5	3	3	34
	2	3	3	1	2	15
Source	3	0	2	3	3	12
	4	2	7	2	4	19
	D	21	25	17	17	

The given problem is a balanced one.

	A	B	C	D	S
1	1	5	3	3	34
2	3	3	1	2	15
3	[12] 0	2	3	3	~~12~~ 0
4	2	7	2	4	19
D	~~21~~ 9	25	17	17	

First allocation in cell (3,1)

	A	B	C	D	
1	1	5	3	3	34
2	3	3	[15] 1	2	~~15~~ 0
4	2	7	2	4	~~19~~
	9	25	~~17~~ 2	17	

Second allocation in cell (2,3)

	A	B	C	D	
1	[9] 1	5	3	3	~~34~~ 25
4	2	7	2	4	19
	~~9~~ 0	25	2	17	

Third allocation in cell (1,1)

	B	C	D	
1	5	3	3	25
4	7	[2] 2	4	~~19~~ 7
	25	~~2~~ 0	17	

Fourth allocation in cell (4,3)

	B	D	
1	5	[17] 3	~~25~~ 8
4	7	4	17
	25	~~17~~ 0	

Fifth allocation in cell (1,4)

	B	
1	[8] 5	~~8~~
4	[17] 7	~~17~~ 0
	~~25~~ 0	

Since, we have one column only allocate the remaining.

The solution is presented in the following table.

9 1	8 5	3	17 3
3	3	15 1	2
12 0	2	3	3
2	17 7	2 2	4

$x_{11} = 9$; $x_{12} = 8$; $x_{14} = 17$;
$x_{23} = 15$; $x_{31} = 12$; $x_{42} = 17$;
$x_{43} = 2$ and

$$TC = 9 \times 1 + 8 \times 5 + 17 \times 3 + 15 \times 1 + 12 \times 0 + 17 \times 7 + 2 \times 2 = 238$$

4.6.5 Vogel's Approximation Method

Vogel (Reinfeld and Vogel 1958) has devised a better method that assures a basic feasible solution close to the optimum called Vogel's Approximation method. This method is based on the concept of cost penalty (Vogel number) in making allocations from origins to destinations. A step-by-step explanation of Vogel's method is depicted in Flowchart 4.5.

Notations Used

N : Number
VN : Vogel Number (penalty cost)
RVN : Row Vogel Number
CVN : Column Vogel Number
R_1 : 1st Row
SC : Selected Cell
CL : Close
Rm : mth Row
Cn : nth Column
UG : Units Given
C_1 : 1st Column

Example: 9

Find the initial basic feasible solution of the following transportation problem by VAM.

	W_1	W_2	W_3	W_4	Available
F_1	19	30	50	10	7
F_2	70	30	40	60	9
F_3	40	8	70	20	18
Demand	5	8	7	14	

Step 1: TS = 34; TD = 34; implies the problem is a balanced one and the objective is minimization.

Step 2: We have to find the Vogel numbers (penalty cost for each row and each column, using the formula difference between smallest from the rest and the smallest transportation cost within rows (columns).

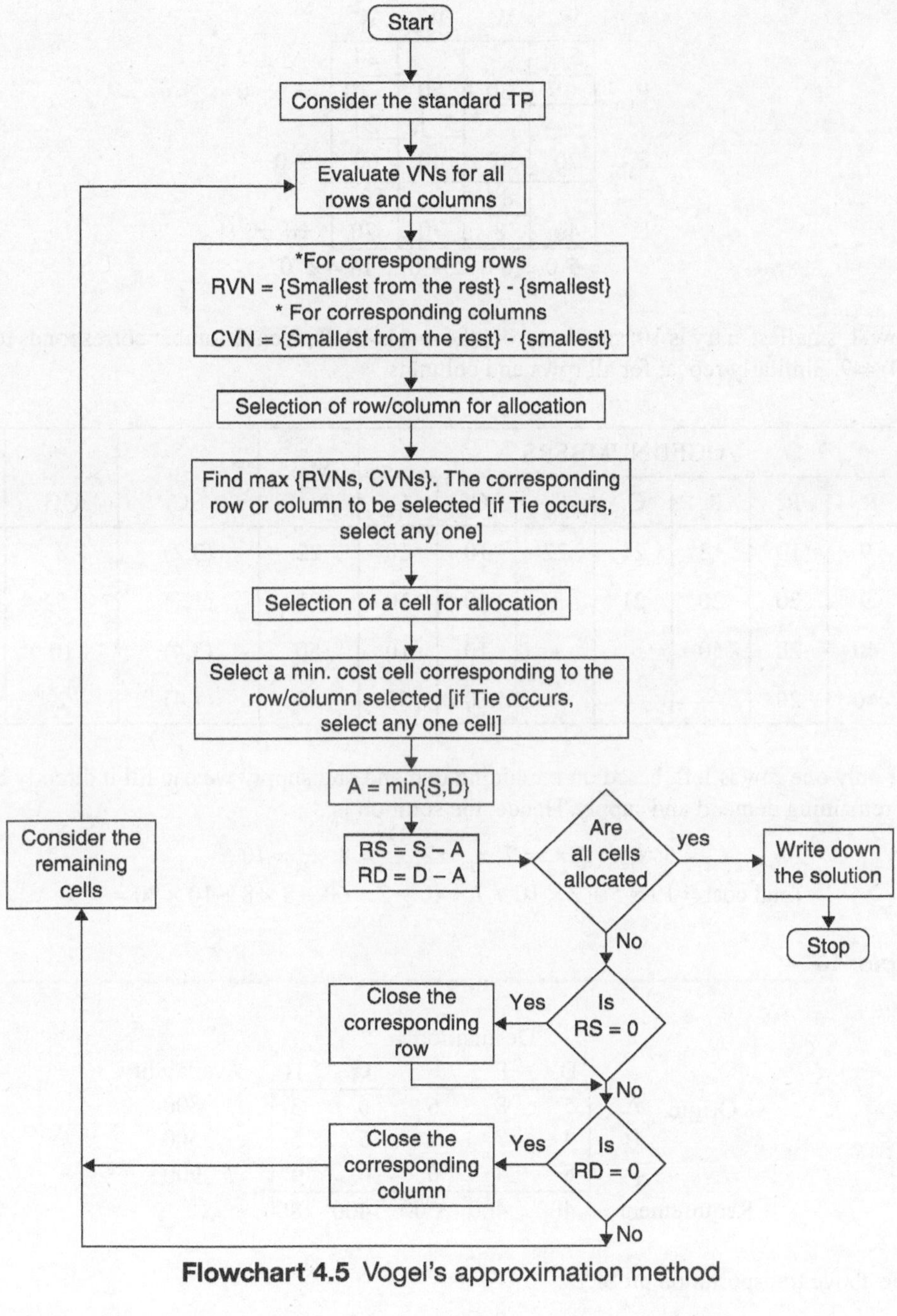

Flowchart 4.5 Vogel's approximation method

	W_1	W_2	W_3	W_4	
F_1	[5] 19	30	50	[2] 10	~~7~~ ~~2~~ 0
F_2	70	30	[7] 40	[2] 60	~~9~~ 0
F_3	40	[8] 8	70	[10] 20	~~18~~ ~~10~~ 0
	~~5~~ 0	~~8~~ 0	~~7~~ 0	~~14~~ ~~4~~ ~~2~~ 0	

In Row-1, smallest entry is 10; smallest from the rest is 19; the vogel number corresponds to Row-1 is (19 – 10) = 9. Similarly repeat for all rows and columns.

	VOGEL NUMBERS							Max			
N	R_1	R_2	R_3	C_1	C_2	C_3	C_4	VN	SC	UG	CL
1	9	10	12	21	22	10	10	22	(3,2)	8	C_2
2	9	20	20	21	-	10	10	21	(1,1)	5	C_1
3	40	20	50	-	-	10	10	50	(3,4)	10	R_3
4	40	20	-	-	-	10	50	50	(1,4)	2	R_1

Since only one row is left, based on remaining demand and supply we can fill it directly by looking into the remaining demand and supply. Hence, the solution is,

$$x_{11} = 5;\ x_{14} = 2;\ x_{23} = 7;\ x_{24} = 2;\ x_{32} = 8;\ x_{34} = 10$$

$$\text{Total cost} = 19 \times 5 + 2 \times 10 + 7 \times 40 + 2 \times 60 + 8 \times 8 + 10 \times 20 = 779$$

***Example:* 10**

Destination

		D	E	F	G	H	Availability
Origin	A	5	8	6	6	3	800
	B	4	7	7	6	5	500
	C	8	4	6	6	4	900
Requirement		400	400	500	400	800	

Solve the above transportation problem.

Step 1: TS = 2200; TD = 2500

Since, TS < TD; the given problem is unbalanced. Add a dummy row (I), with supply (2500–2200) 300 units and all cost entries as "0". Also the objective is minimization.

Step 2: We can solve this using VAM.

	D	E	F	G	H	
A	5	8	6	6	[800] 3	~~800~~ 0
B	[400] 4	7	7	[100] 6	5	~~500~~ ~~100~~ 0
C	8	[400] 4	[200] 6	[300] 6	4	~~900~~ ~~500~~ 0
I	0	0	[300] 0	0	0	~~300~~ 0
	~~400~~ 0	~~400~~ 0	~~500~~ ~~200~~ 0	~~400~~ ~~300~~ 0	~~800~~ 0	

N	VOGEL NUMBER R_1	R_2	R_3	R_4	C_1	C_2	C_3	C_4	C_5	Max VN	SC	UG	CL
1	2	1	0	0	4	4	6	6	3	6	(4,3)	300	R_4
2	2	1	0	-	1	3	0	0	1	3	(3,2)	400	C_2
3	2	1	2	-	1	-	0	0	1	2	(1,5)	800	R_1, C_5
4	-	2	0	-	4	-	1	0	-	4	(2,1)	400	C_1
5	-	1	0	-	-	-	1	0	-	1	(2,4)	100	R_2

Since, we left with only one row R_3, allocate based on the remaining supply and demand. Hence the solution is,

$$x_{15} = 800,\ x_{21} = 400,\ x_{24} = 100,\ x_{32} = 400,\ x_{33} = 200,\ x_{34} = 300,\ x_{43} = 300$$

$$\text{Total Cost} = 3 \times 800 + 4 \times 400 + 6 \times 100 + 4 \times 400 + 6 \times 200 + 6 \times 300 = 9200$$

*Example: 11

The following table shows all the information regarding the available supply at each warehouse, the requirements of each market and the unit transportation cost from each warehouse to each market. Find the initial solution using VAM.

		Market I	II	III	IV	Supply
Warehouse	A	5	2	4	3	22
	B	4	8	1	6	15
	C	4	6	7	5	8
Demand		7	12	17	9	

Step 1: TS = 45; TD = 45

The given problem is balanced. Also the objective is minimization.

Step 2: We can solve this using VAM.

	I	II	III	IV	
A	5	[12] 2	[2] 4	[8] 3	~~22~~ ~~10~~ ~~8~~ 0
B	4	8	[15] 1	6	~~15~~ 0
C	[7] 4	6	7	[1] 5	~~8~~ 0
	~~7~~ 0	~~12~~ 0	~~17~~ ~~2~~ 0	~~9~~ ~~1~~ 0	

N	VOGEL NUMBER							Max VN	SC	UG	CL
	R_1	R_2	R_3	C_1	C_2	C_3	C_4				
1	1	3			4	3	2	4	(1, 2)	12	C_2
2	1	3	1	0	-	3	2	3	(2, 3)	15	R_2
3	1	-	1	1	-	3	2	3	(1, 3)	2	R_3
4	2	-	1	1	-	-	2	2	(1, 4)	8	R_1

Since, we left with only one row i.e., R_3, allocate it based on the remaining supply and demand. Hence, the solution is $x_{12} = 12$, $x_{13} = 2$, $x_{14} = 8$, $x_{23} = 15$, $x_{31} = 7$, $x_{34} = 1$

$$\text{Total cost} = 12 \times 2 + 2 \times 4 + 8 \times 3 + 15 \times 1 + 7 \times 4 + 1 \times 5 = 104.$$

4.6.6 Effectiveness of Various Methods

NWCR

This method provides a simple, systematic approach to find an initial solution without any consideration of the cost factor.

RMM & CMM

These two methods are better than the NWCR, because these two methods take care of the minimum cost cells in Row (Column) wise for allocation.

LCM/MMM

This method is better than the other three methods NWCR, RMM, and CMM because this method considers the over all matrix minimum for allocation.

VAM

This method assures a feasible solution very close to the optimum, because the concept of cost penalty in making allocations. So, it is the best method when we compare with the other methods. When we apply this method to find an initial solution, we can reach the optimum solution using modified distributive method with minimum number of iterations.

Remark: Whenever we are asked to find the optimum solution, better to apply VAM to get the initial basic feasible solution.

4.7 STEPPING STONE METHOD/MODIFIED DISTRIBUTIVE METHOD

The Stepping stone method was presented by COOPER and CHARNES (1954) as a simplification to the actual modified simplex solution to the transportation problem suggested by Datzig (1947) to get the optimum solution for a given problem. It is an iterative method, in each iteration we are going to modify the objective value to get the optimum solution.

An iteration in this method consists of various steps.

1. Condition check for optimality.
2. Selection of the entering variable.
3. Selection of the leaving variable.
4. Evaluation of the new basic feasible solution.
5. Repeat the steps 1 to 4 until the optimum solution is reached.

Also, the clear procedure for getting an optimum solution discussed in the Flowchart 4.6.

Notations Used

NBC : Number of Basic Cells
BCs : Basic Cells
VD : Vertically Down
HL : Horizontally Left
C : Column
V : Vertical
I : Number of Iterations
VU : Vertically Up
HR : Horizontally Right
R : Row
H : Horizontal

Remark: In the optimum iteration, if one or more $(Z_{ij} - C_{ij})$ value equals to zero indicates that the given problem is said to have an alternate solution. To get the alternate solution, select the $(Z_{ij} - C_{ij}) = 0$ valued cell as an entering cell and proceed with the usual process, then we can have the alternate solution.

**Example: 12*

Find the optimal solution for the given transportation problem.

		Destination				
		D_1	D_2	D_3	D_4	Availability
Origin	O_1	1	2	1	4	30
	O_2	3	3	2	1	50
	O_3	4	2	5	9	20
Requirement		20	40	30	10	

Remark: If the method is not mentioned to find initial solution, we can use any one of the method available.

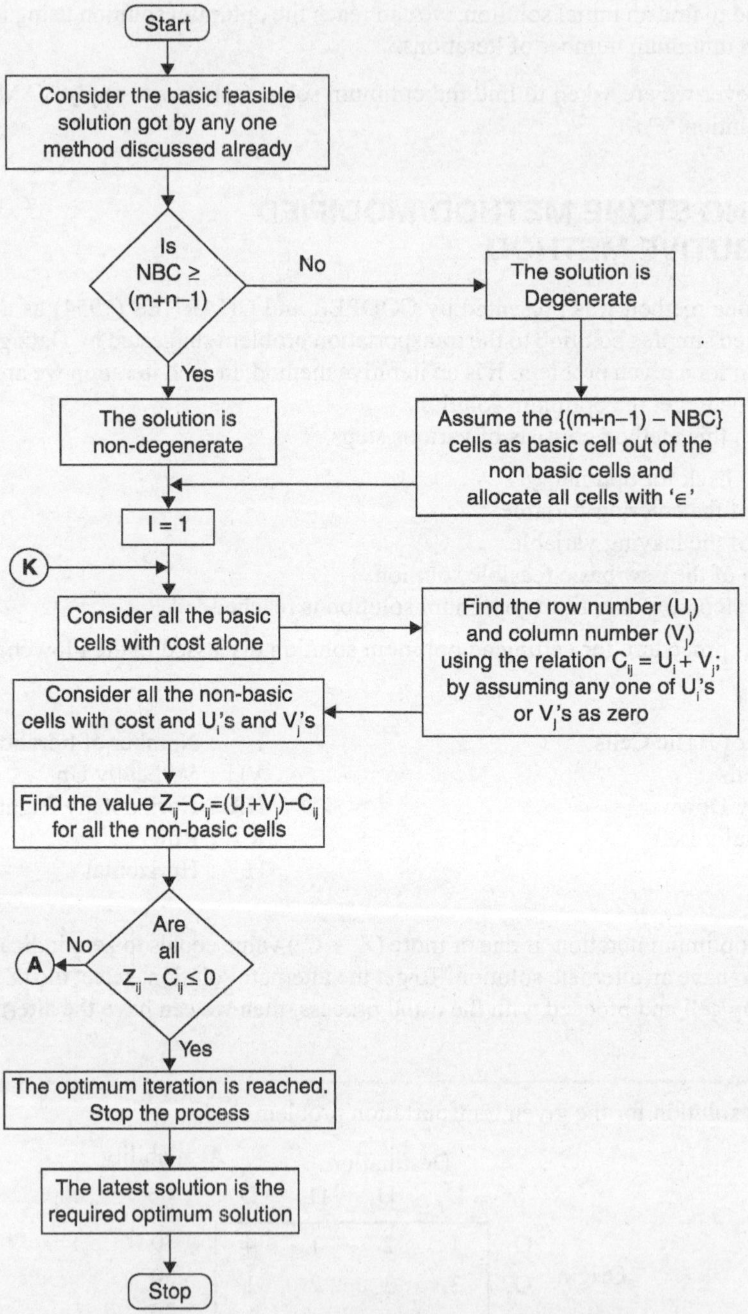

Flowchart 4.6 Modified distributive method

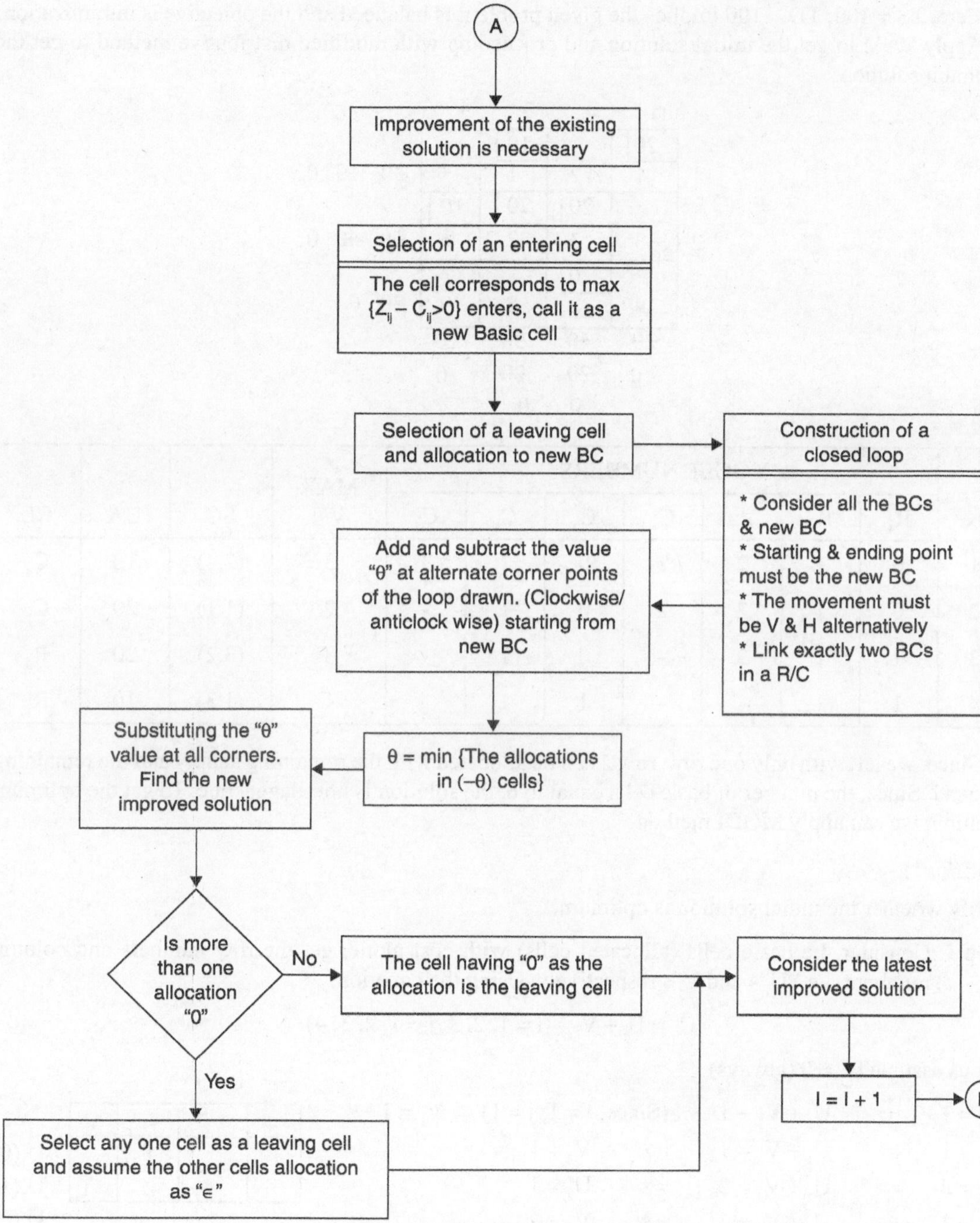

Flowchart 4.6 (Continued)

Here, TS = 100; TD = 100 implies the given problem is balanced and the objective is minimization.

Apply VAM to get the initial solution and proceeding with modified distributive method to get the optimum solution.

	1	2	3	4	Supply
1	[20] 1	2	[10] 1	4	~~30~~ ~~10~~ 0
2	3	[20] 3	[20] 2	[10] 1	~~50~~ ~~40~~ 0
3	4	[20] 2	5	9	~~20~~ 0
Demand	~~20~~	~~40~~	~~30~~	~~10~~	
	0	~~20~~	~~20~~	0	
		0	0		

N	VOGEL NUMBERS							MAX VN	SC	UA	CL
	R_1	R_2	R_3	C_1	C_2	C_3	C_4				
1	0	1	2	2	0	1	3	3	(2,4)	10	C_4
2	0	1	2	2	0	1	-	2	(1.1)	20	C_1
3	1	1	3	-	0	1	-	3	(3,2)	20	R_3
4	1	1	-	-	1	1	-	1	(1.3)	10	R_1

Since, we left with only one row, row 2 can be allocated with the remaining supply and the remaining demand. Since, the number of basic cells equal to 6, the solution is non-degenerate. To get the optimum solution, we can apply MODI method.

Iteration 1

Verify whether the initial solution is optimum.

Step 1: Consider the basic cells (allocated cells) with cost alone, get the row numbers and column numbers (i.e.) U_i's and V_j's respectively. Using the equation,

$$C_{ij} = U_i + V_j \quad (i = 1, 2, 3; j = 1, 2, 3, 4)$$

Let us assume $U_1 = 0$ (always)

$C_{11} = 1 \Rightarrow U_1 + V_1 - 1$ (Since, i = 1; j = 1) $\Rightarrow V_1 = 1$

$C_{13} = 1 \Rightarrow U_1 + V_3 = 1 \Rightarrow V_3 = 1$

$C_{23} = 2 \Rightarrow U_2 + V_3 = 2 \Rightarrow U_2 = 1$

$C_{22} = 3 \Rightarrow U_2 + V_2 = 3 \Rightarrow V_2 = 2$

$C_{24} = 1 \Rightarrow U_2 + V_4 = 1 \Rightarrow V_4 = 0$

$C_{32} = 2 \Rightarrow U_3 + V_2 = 2 \Rightarrow U_3 = 0$

1		1		$U_1(0)$
	3	2	1	$U_2(1)$
	2			$U_3(0)$
V_1 (1)	V_2 (2)	V_3 (1)	V_4 (0)	

Remark: U_i's and V_j's can be (+ve) or (–ve) or "0".

Step 2: Using the cost coefficient of non-basic cells (non-allocated cells) with U_i's and V_j's evaluated in step 1. Find $Z_{ij} - C_{ij} = (U_i + V_j) - C_{ij}$

$Z_{12} - C_{12} = U_1 + V_2 - C_{12} = (2 + 0) - 2 = 0$
$Z_{14} - C_{14} = U_1 + V_4 - C_{14} = (0 + 0) - 4 = -4$
$Z_{21} - C_{21} = U_2 + V_1 - C_{21} = (1 + 1) - 3 = -1$
$Z_{31} - C_{33} = U_3 + V_1 - C_{31} = (0 + 1) - 4 = -3$
$Z_{33} - C_{33} = U_3 + V_3 - C_{33} = (0 + 1) - 5 = -4$
$Z_{34} - C_{34} = U_3 + V_4 - C_{34} = (0 + 0) - 9 = -9$

	(0) 2		(–4) 4	(0)
(–1) 3				(1)
(–3) 4		(–4) 5	(–9) 9	(0)
(1)	(2)	(1)	(0)	

Since, all $Z_{ij} - C_{ij}$ (non-basic cells) are ≤ 0, implies that the iteration 1 is optimum.

Hence, the initial solution got by VAM's method itself is optimum. The optimum solution is, $x_{11} = 20$; $x_{13} = 10$; $x_{22} = 20$; $x_{23} = 20$; $x_{24} = 10$; $x_{32} = 20$ and the total transportation cost, optimum cost = $1 \times 20 + 1 \times 10 + 3 \times 20 + 2 \times 20 + 1 \times 10 + 2 \times 20 = 180$.

***Example: 13**

Solve the following transportation problem.

		To			
		1	2	3	Supply
From	A	5	1	7	10
	B	6	4	6	80
	C	3	2	5	15
Demand		75	20	50	

Here, TS = 105; TD = 145. Since, TS < TD the given problem is unbalanced. Hence, introduce a dummy row with demand (145 – 105) 40 units and all cost as "0". The modified problem is,

		To			
		1	2	3	Supply
From	A	5	1	7	10
	B	6	4	6	80
	C	3	2	5	15
	D	0	0	0	40
Demand		75	20	50	

Applying LCM for the problem, we have

	1	2	3	
A	5	[10] 1	7	~~10~~ 0
B	[30] 6	4	[50] 6	~~80~~ 0
C	[5] 3	[10] 2	5	~~15~~ ~~5~~ 0
D	[40] 0	0	0	~~40~~ 0
	~~75~~ ~~35~~ ~~30~~ 0	~~20~~ ~~10~~ 0	~~50~~ 0	

Clearly, the initial basic solution is a non-degenerate one. So, we can apply MODI method to get the optimum solution.

Iteration 1

Verify whether the initial solution is optimum.

Step 1: Consider the basic cells with cost and find the U_i's and V_j's, by assuming $U_1 = 0$, and using $C_{ij} = U_i + V_j$.

	1		0
6		6	4
3	2		1
0			–2
2	1	2	

Step 2: Consider the non-basic cells with cost and find $Z_{ij} - C_{ij}$. Since the cell (2, 2)'s $Z_{12} - C_{12} = 1 > 0$, implies that the existing solution is not optimum. So, improvement is required.

(–3) 5		(–5) 7	0
	(1) 4		4
		(–2) 5	1
	(–1) 0	(0) 0	–2
2	1	2	

Step 3: Since the $(Z_{22} - C_{22})$ entry is +ve, consider the cell (2, 2) as a basic cell, and draw the loop. Consider the basic cells with allocation and the new basic cell (2, 2).

Starting Basic Cell	Nature of Move	Connecting Cell
(2, 2)	VD	(3, 2)
(3, 2)	HL	(3, 1)
(3, 1)	VU	(2, 1)
(2, 1)	HR	(2, 2)

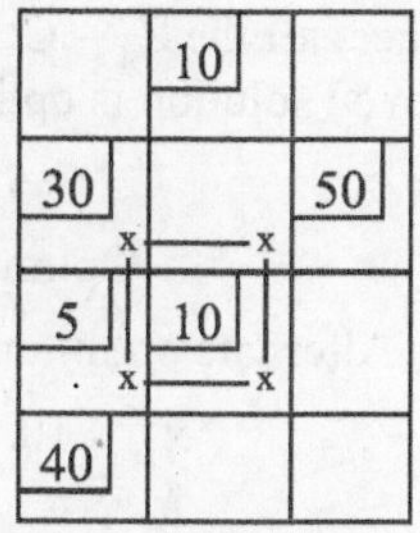

Remark 1: We have to draw the closed loop (unique); by linking possible basic cells. (No necessity to connect all the basic cells).

Remark 2: From the new basic cell, we cannot move vertically up, eventhough the cell (1, 2) is there, because it is a dead end, that is, from that cell we cannot move elsewhere.

Step 4: Add and subtract θ at alternate ends of the loop starting from (Clock wise or anti-clock wise) the new basic cell

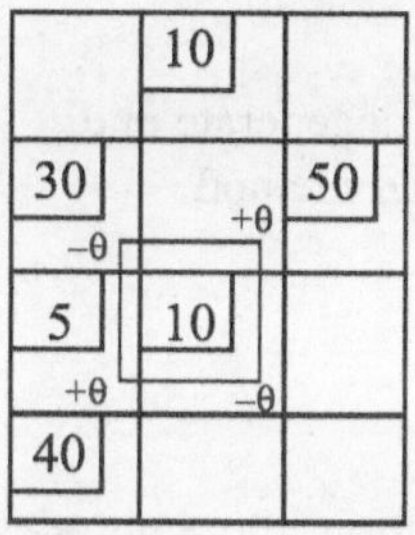

θ = min{(–0) cell allocations}
= min{10, 30}= 10

Substitute θ = 10, wherever θ is there; here the first improved solution is

	10	
20	10	50
15	0	
40		

Since, the allocation of cell (3, 2) is 'Zero', that cell leaves, i.e. becomes non-basic. The solution is non-degenerate. Proceed to iteration 2.

Iteration 2

Verify whether the first improved solution is optimum. Repeat the process.

Step 1:

	1		0
6	4	6	3
3			0
0			–3
3	1	3	

Step 2:

(–2) 5		(–4) 7	0
			3
	(–1) 2	(–2) 5	0
	(–2) 0	(0) 0	–3
3	1	3	

Since, all the $Z_{ij} - C_{ij}$ values are ≤ 0, implies that the optimum iteration is reached. Hence, the first improved solution is optimum.

$$\Rightarrow \quad x_{12} = 10; x_{21} = 20; x_{22} = 10; x_{23} = 50; x_{31} = 15; x_{41} = 40$$

$$\text{Optimum cost} = 1 \times 10 + 6 \times 20 + 4 \times 10 + 6 \times 50 + 15 \times 3 = 515.$$

Note: Alternate optimum solution exists for this problem.

$$x_{21} = 10, x_{21} = 60, x_{22} = 10, x_{23} = 10, x_{31} = 15, x_{43} = 40 \text{ and Min } Z = 515$$

**Example: 14*

Consider the transportation problem find (i) Initial solution using NWCR
(ii) Find the optimum solution.
Is the solution unique? If not find the alternate solution.

	7	3	4	2
2	1	3	3	
	3	4	6	5
	4	1	5	

Clearly, the initial solution is a degenerate one.
As such we cannot apply the MODI method

[2] 7	3	∈ 4
[2] 2	[1] 1	3
3	4	[5] 6

To convert the degenerate solution into a non-degenerate one, we require one more basic cell. So, select any one of the non-basic cells as a basic cell and allocate a (+ve) quantity ∈ to it. Let us consider the cell (1, 3) as basic and allocate with ∈.

Iteration 1

Step 1: Consider the basic cells with cost and find U_i's; V_j's.
Using $C_{ij} = U_i + V_j$

7		4	0
2	1		–5
		6	2
7	6	4	

Step 2: Consider the non-basic cells with cost and U_i's; V_j's. Find $(Z_{ij} - C_{ij})$.

	(3) 3		0
		(–4) 3	–5
(6) 3	(4) 4		2
7	6	4	

In this, some of the $(Z_{ij} - C_{ij})$ values are > 0, implies that the existing solution must be improved.

Step 3: Selection of an entering cell. Consider the Max$\{(Z_{ij} - C_{ij}) > 0\}$ i.e. Max$\{2, 3, 6\} = 6$, corresponds to cell (3, 1). Hence, cell (3, 1) enters.

Step 4: Selection of a leaving cell. Consider the basic cells with allocation and draw the loop.

Starting Cell	Nature of Move	Connecting Cell
(3, 1)	VU	(1, 1)
(1, 1)	HR	(1, 3)
(1, 3)	VD	(3, 3)
(3, 3)	HL	(3, 1)

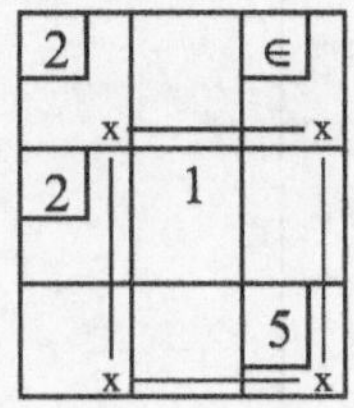

Step 5: Add 6 and Subtract 0 at alternate ends of the loop.

$\theta = \min\{(-\theta) \text{ cell allocations}\}$
$= \min\{2, 5\} = 2$
Substitute $\theta = 2$; wherever θ is there; hence the first improved solution is,

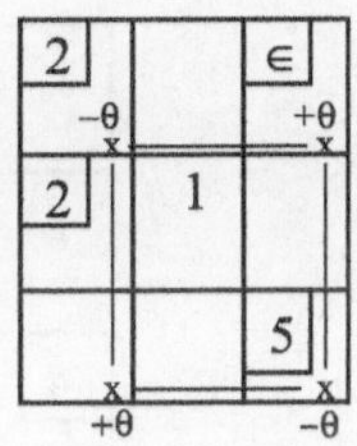

		2+∈
2	1	
2		3

Here, the cell (1,1) leaves. Also the solution is non-degenerate. Proceed to iteration 2.

Iteration 2

Verify whether the first improved solution is optimum. Repeat the process.

Step 1:

		4	0
2	1		1
3		6	2
1	0	4	

Step 2:

(–6) 7	(–3) 3		0
		(2) 3	1
	(–2) 4		2
1	0	4	

Step 3: The new basic cell is (2, 3)

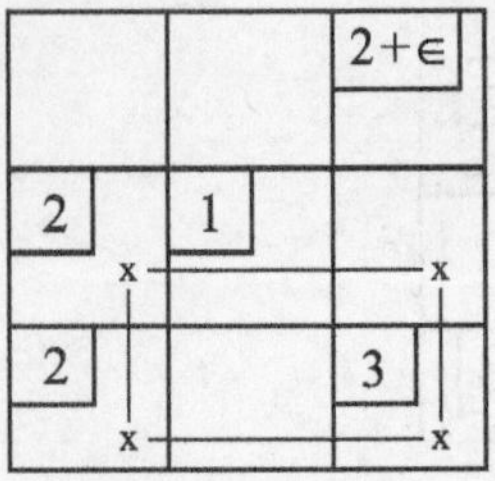

Step 4:

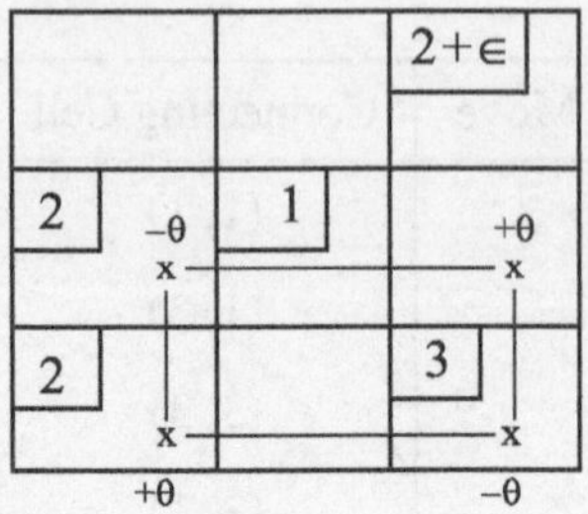

$\theta = \min\{2, 3\} = 2$; the second improved solution is got by substituting $\theta = 2$. Cell (2, 1) leaves. The second improved solution which is non-degenerate. Proceed to iteration 3.

		2+∈
	1	2
4		1

Iteration 3

Verify whether the second improved solution is optimum.

Step 1:

		4	0
	1	3	−1
3		6	2
1	2	4	

Step 2:

(−6) 7	(−1) 3		0
(−2) 2			−1
	(0) 4		2
1	2	4	

Since, all the $(Z_{ij} - C_{ij})$ entries are ≤ 0; implies that the optimum iteration is attained. Hence, the second improved solution is optimum. (Finally, put $\in = 0$)

		2
	1	2
4		1

$x_{13} = 2, x_{22} = 1, x_{23} = 2,$
$x_{31} = 4, x_{33} = 1$ and
Total cost $= 2 \times 4 + 1 \times 1 + 2 \times 3 + 4 \times 3 + 1 \times 6 = 33$

Alternate Optimum Solution

In the optimum iteration, the $Z_{ij} - C_{ij}$ value of the cell (3, 2) is 0. This clearly indicates the given problem having an alternate optimum solution.

Select the cell (3, 2) as a new basic cell and proceed.

		2+∈
	1 (−θ)	2 (+θ)
4	x (+θ)	1 (−θ)

$\theta = \min\{1, 1\}$

$\theta = 1$

Here the cell (3, 2) enters and the cells (2, 2) & (3, 3) leaves. To avoid degeneracy keep the cell (2, 2) as basic and allocate the quantity ∈.

Hence, the 3rd improved solution is

		2+∈
	∈	3
4	1	

Iteration 4

Verify whether the third improved solution is optimum?

Step 1:

			u_i
		4	0
	1	3	−1
3	4		2
1	2	4	

Step 2:

			u_i
(−6) 7	(−1) 3		0
(−2) 2			−1
		(0) 6	2
1	2	4	

Since all $(Z_{ij} - C_{ij})$ values of the non-basic cells are ≤ 0, indicates the optimality is attained.

Hence, the alternate optimum solution is (put ∈ = 0)

		[2] 4
		[3] 3
[4] 3	[1] 4	

$x_{13} = 2 \qquad x_{31} = 4$

$x_{23} = 3 \qquad x_{32} = 1$

Optimum cost $= 2 \times 4 + 3 \times 3 + 4 \times 3 + 1 \times 4 = 33$

Note: Alternate solution is, $x_{13} = 2$; $x_{22} = 1$; $x_{23} = 2$; $x_{31} = 4$ and $x_{33} = 6$

*Example: 15

Determine the optimum basic feasible solution for the following TPP.

		Destination				
		1	2	3	4	Supply
	1	10	0	20	11	15
Source	2	12	7	9	20	25
	3	0	14	16	18	5
Demand		5	15	15	10	

Clearly, the given TPP is a balanced one. Consider the initial solution got by VAM.

		Destination 1	2	3	4
Source	1	[∈] 10	[15] 0	20	[∈] 11
	2	12	7	[15] 9	[10] 20
	3	[5] 0	14	16	18

Since the number of BCs is not equal to (m + n − 1) [4 ≠ 6], indicates that the solution is a degenerate one. To apply the MODI method, first we have to convert the degenerate solution into a non-degenerate one. For this conversion, we need two more basic cells. To overcome from this, consider any two non-basic cells as basic cells and allocate the quantity ∈ (a least +ve quantity). Let us take the cells (1, 1) & (1, 4) as basic. Now, we can apply MODI method.

Iteration 1

Verify whether the initial solution got by VAM is optimum?

Step 1:

				u_i
10	0		11	0
		9	20	9
0				−10
10	0	0	11	

Step 2:

				u_i
		(−20) 20		0
(7) 12	(2) 7			9
	(−24) 14	(−26) 16	(−17) 18	−10
10	0	0	11	

Since some of the $(Z_{ij} - C_{ij})$ entries are > 0, we have to improve the solution.

Step 3: The new basic cell is (2, 1) [Max {2, 7} = 7]

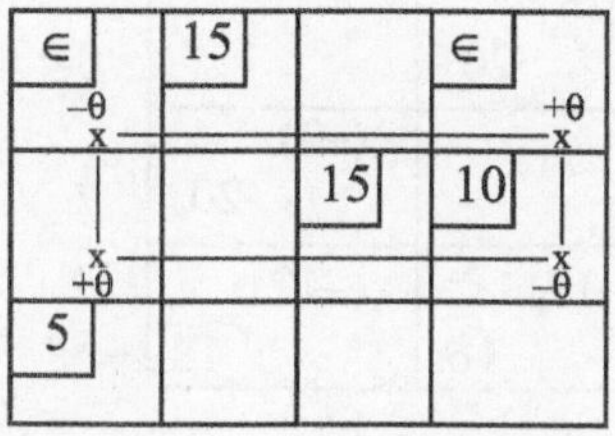

	15		2∈
∈		15	10−∈
5			

$\theta = \min\{\in, 10\} = \in$

The cell (1, 1) leaves the basis. Since one cell enters and one leaves, the first improved solution is non-degenerate.

Iteration 2

Verify whether the first improved solution is optimum?

Step 1:

	0		11	0
12		9	20	9
0				−3
3	0	0	11	

Step 2:

(−1) 10		(−20) 20		0
	(2) 7			9
	(−17) 14	(−19) 16	(−10) 18	−3
3	0	0	11	

Since the cell (2, 2)'s $Z_{ij} - C_{ij}$ entry is > 0. We have to improve the solution.

Step 3: The new basic cell is (2, 2)

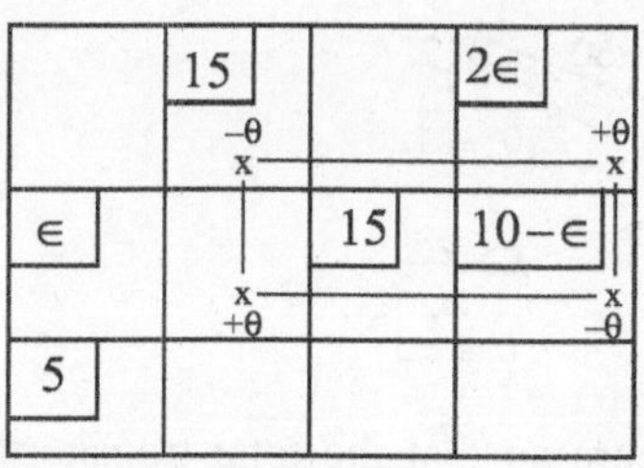

	5+∈		10+∈
∈	10−∈	15	
5			

$\theta = \min\{15, 10 - \in\} = 10 - \in$

The cell (2, 4) leaves the basis. The second improved solution is non-degenerate.

Iteration 3

Verify whether the second improved solution is optimum?

Step 1:

				u_i
	0		11	0
12	7	9		7
0				–5
5	0	2	11	

(–5) 10		(–18) 20		0
			(–2) 20	7
	(–19) 14	(–19) 16	(–12) 18	–5
5	0	2	11	

Since all the $(Z_{ij} - C_{ij})$ entries are ≤ 0, implies that the optimality is attained.
Hence the optimum solution is (put $\in = 0$)

	[5] 0		[10] 11
[0] 12	[10] 7	[15] 9	
[5] 0			

Optimum cost = 0 × 5 + 11 × 10 + 0 × 12 + 10 × 7 + 15 × 9 + 0 × 5 = 315

***Example:* 16**

A company has three refineries R_1, R_2 and R_3 with respective maximum daily capacity of 3, 5 and 10 million gallons of petroleum supply per day. It has three distribution areas D_1, D_2 and D_3 which have a daily demand of 8, 5 and 5 million gallons respectively. The Petroleum is transported through a network of pipelines. The transportation cost is estimated to be ₹2 per thousand gallons per kilometre. The distance between the various refineries and the distribution areas are given in the following figure:

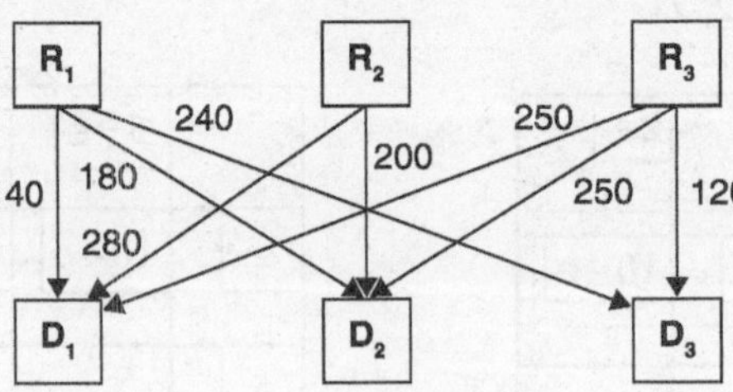

How should the company plan to transport the petroleum so that the total transportation cost is minimum.

First of all we have to formulate the given problem in to a transportation problem structure.

Formulation:

Given, TP cost/thousand gallons per km = ₹2

1 unit of supply or demand = 1 million gallon

TP cost for one million gallons/km = $(10^6/10^3) \times 2 = 2000$

Since the TP cost is proportionate to the distance travelled, we can construct the distance matrix and try to optimize the distance travelled.

	D_1	D_2	D_3	Supply
R_1	140	180	240	3
R_2	280	200	-	5
R_3	250	250	120	10
Demand	8	5	5	

Consider the initial solution got by VAM.

[3] 140	180	240
280	[5] 200	-
[5] 250	[∈] 250	[5] 120

Clearly the initial solution is a degenerate one. To convert it into a non-degenerate one we need one more (5-4) basic cell. Select the non-basic cell (3, 2) as a basic cell and allocate the quantity ∈. Then proceed with MODI method.

Iteration 1

Verify whether the initial solution is non degenerate?

Step 1:

140			0
	200		60
250	250	120	110
140	140	10	

Step 2:

	(–40) 180	(–230) 240	0
(–80) 280			60
			110
140	140	10	

Since all $Z_{ij} - C_{ij} \geq 0$ for all non-basic cells, implies that the optimality is attained. Hence the optimum solution is (put $\in = 0$)

3 140		
	5 200	
5 250		5 120

The optimum distance travelled = (3 × 140) + (5 × 200) + (250 × 5) + (5 × 120) = 3270 kms.
Total TP cost = ₹3270 × 2000 = ₹65,40,000/-

Example: 17

Given x_{13}= 50 units, x_{14} = 20 units, x_{21} = 55 units, x_{31} = 30 units, x_{32} = 35 units and x_{34} = 25 units. Is it an optimal solution of the TPP:

6	1	9	3	70
11	5	2	8	55
10	12	4	7	90
85	35	50	45	

If not, modify it to obtain a better feasible solution.

Since, the number of basic cells equal to (m + n – 1), the given solution is non-degenerate. Verify this solution using MODI.

		50	20
55			
30	35		25

Iteration 1

Verify whether the initial solution is optimum.

Step 1:

		9	3	0
11				5
10	12		7	4
6	8	9	3	

Step 2:

(0) 6	(7) 1			0
	(8) 5	(12) 2	(0) 8	5
		(9) 4		4
6	8	9	3	

Since some of the ($Z_{ij} - C_{ij}$) entries are > 0, implies that the optimality is not attained.

Step 3:

EC: Max {7, 8, 9, 12} = 12, cell (2, 3) enters.

LC: Draw a closed loop.

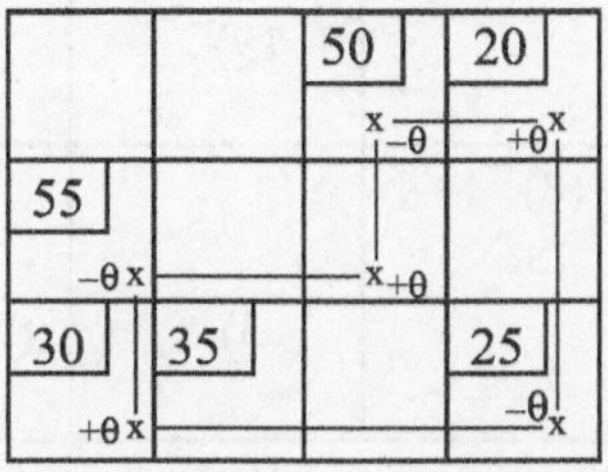

θ = Min {50, 25, 55} = 25

Here, the cell (3, 4) leaves. Clearly, the 1st improved solution is non-degenerate.

		25	45
30		25	
55	35		

Iteration 2

Verify whether the 1st improved solution is optimum.

Step 1:

		9	3	0
11		2		−7
10	12			−8
18	20	9	3	

Step 2:

(12) 6	(19) 1			0
	(8) 5		(−12) 8	−7
		(−3) 4	(−12) 7	−8
18	20	9	3	

Step 3:

Cell (1, 2) enters.

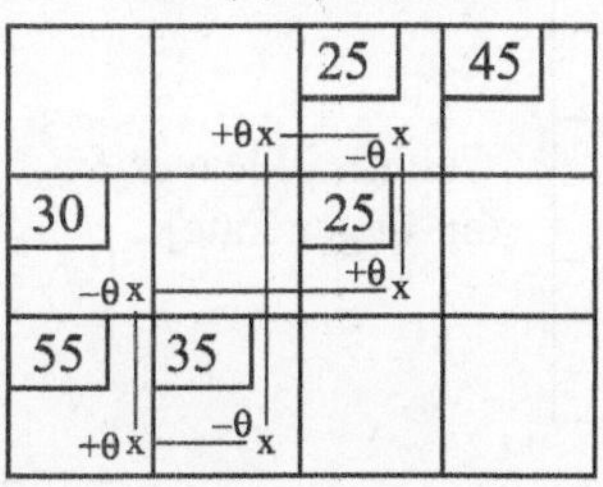

θ = min {25, 35, 30} = 25. The 2nd improved solution is,

	25		45
5		50	
80	10		

[Cell (1, 3) leaves & non-degenerate].

Iteration 3

Verify whether the 2nd improved solution is optimum.

Step 1:

				u_i
	1		3	0
11		2		12
10	12			11
–1	1	–10	3	

Step 2:

				u_i
(–7) 6		(–19) 9		0
	(8) 5		(7) 8	12
		(–3) 4	(7) 7	11
–1	1	–10	3	

Step 3:

Cell (2, 2) enters.

	25		45
5 –θ	+θ	50	
80 +θ	10 –θ		

θ = min{5, 10} = 5. The 3rd improved solution is

	25		45
	5	50	
85	5		

[Cell (2, 1) leaves & non-degenerate]

Iteration 4

Verify whether the 3rd improved solution is optimum.

Step 1:

				u_i
	1		3	0
	5	2		4
10	12			11
–1	1	–2	3	

Step 2:

				u_i
(–7) 6		(–11) 9		0
(–8) 11			(–1) 8	4
		(5) 4	(7) 7	11
–1	1	–2	3	

Step 3:

Cell (3, 4) enters.

	25 +θ		45 –θ
5	50		
85	5 –θ		+θ

θ = min {5, 45} = 5. The 4th improved solution is

	30		40
	5	50	
80			5

[Cell (3, 2) leaves & non-degenerate]

Iteration 5

Verify whether the 4th improved solution is optimum.

Step 1:

	1		3	0
	5	2		4
10			7	4
6	1	−2	3	

Step 2:

(0) 6		(−11) 9		0
(−1) 11			(−1) 8	4
	(−7) 12	(−2) 4		4
6	1	−2	3	

Since all $Z_{ij} - C_{ij}$ are ≤ 0, implies that the optimality is attained. That is the 4th improved solution is the required optimum.

	30 1		40 3
	5 5	50 2	
85	10		5 7

Note: Since the non-basic cell (1,1) having $Z_{11} - C_{11} = 0$. We can take and proceed to get alternate solution. $x_{11} = 40$; $x_{12} = 30$; $x_{22} = 5$; $x_{23} = 50$; $x_{31} = 45$; $x_{44} = 45$ and optimum cost = ₹1160.

Example: 18

Krishna enterprise has three factories at locations x, y and z which supplies three warehouses located at A, B & C. Monthly factory capacities are 10, 80 and 15 units respectively. Monthly warehouse requirements are 75, 20 and 50 units respectively. Unit shipping costs (in ₹) are given below.

Factory	Warehouses A	B	C	Capacity
X	5	1	7	10
Y	6	4	6	80
Z	3	2	5	15
Requirement	75	20	50	

The penalty cost for not satisfying demand at the warehouse A, B and C are ₹5, 3 and 2 per unit respectively. Determine optimal distribution for the company.

The given problem is an unbalanced one. That is, TS = 105 and TD = 145. The insufficient demand is 40 units. We have to introduce a dummy factor for the corresponding insufficient demand.

Use the given penalty cost for the insufficient demand based on the individual warehouses in the dummy factory (W) created. Consider the solution got by VAM.

	A	B	C	
X	5	[10] 1	7	10
Y	[60] 6	[10] 4	[10] 6	80
Z	[15] 3	2	5	15
W	5	3	[40] 2	40
	75	20	50	145 / 145

No. of basic cells = 6

$m + n - 1 = 4 + 3 - 1 = 6$

The initial solution got by VAM is non-degenerate.

Apply MODI, to verify the optimality.

Iteration 1

Verify whether the initial solution is optimum.

Step 1:

			u_i
	1		0
6	4	6	3
3			0
		2	–1
3	1	3	

Step 2:

			u_i
(–2) 5		(–4) 7	0
			3
	(–1) 2	(–2) 5	0
(–3) 5	(–3) 3		–1
3	1	3	

Since all $(Z_{ij} - C_{ij})$ are ≤ 0, implies that the optimality is attained. That is the initial solution got by VAM itself is optimum. Hence the optimum solution is,

	A	B	C
X		[10] 1	
Y	[60] 6	[10] 4	[10] 6
Z	[15] 3		
W			[40] 2

The optimum cost = ₹595. Out of this the Krishna enterprise payout ₹80 in the form of penalty. Because the demand of 40 units to the warehouse-C is not possible to meet out.

***Example:* 19**

Abinaya Company has three production centres P_1, P_2 and P_3 producing the same product and sells in three market places M_1, M_2 and M_3. At all production centres the production cost and raw material costs differ. Also the selling price differs in the three markets. They are clearly stated in the following tables:

Item	Production Centres		
	P_1	P_2	P_3
Raw materials cost (in ₹)	16	19	15
Production cost (in ₹)	11	10	13

Item	Market		
	M_1	M_2	M_3
Selling price	38	33	32

The unit transportation cost from the production centres to markets are given below with capacity and demands.

	M_1	M_2	M_3	Capacity
P_1	10	13	5	40
P_2	8	5	5	30
P_3	9	2	4	30
Demand	40	20	40	100

Find the initial feasible solution to this problem and find the optimum solution.

Step 1: Using the production cost, raw material cost, transportation cost and selling price, find the profit matrix. That evaluates the profit for each cell.

Total cost for the cell (i, j) = Production cost in ith centre + Raw material cost in ith centre + Transportation cost for cell (i, j)

Profit for the cell (i, j) = Selling price in jth market – Total cost in (i, j) cell

The profit matrix, we want to maximize is

	M_1	M_2	M_3
P_1	1	–7	0
P_2	1	–1	–2
P_3	1	3	0

The problem is balanced.

Convert the max objective into minimization. For which select the maximum entry and subtract all entries from it. The new matrix with the minimization objective is.

(10) 2	10	(30) 3	~~40~~ ~~30~~ 0
(30) 2	4	5	~~30~~ 0
2	(20) 0	(10) 3	~~30~~ ~~10~~ 0
~~40~~ ~~10~~ 0	~~20~~ 0	~~40~~ 0	

N	VOGEL NUMBER						MAX VN	SC	UA	CL
	R_1	R_2	R_3	C_1	C_2	C_3				
1	1	2	2	0	4	0	4	(3,2)	20	C_2
2	1	3	1	0	-	0	3	(2,1)	30	R_2
3	1	-	1	0	-	0	1	(1,1)	10	C_1

Since, we left with only one column, fill with the rest of the demand and supply. Superimposing the results in the original matrix.

(10) 1	−7	(30) 0
(30) 1	−1	−2
1	(20) 3	(10) 0

The total profit = 10 × 1 + 0 × 30 + 30 × 1 + 20 × 3 + 10 × 0 = ₹100

(10) 2	10	(30) 3
2	4	5
2	(20) 0	(10) 3

Number of basic cells = 5
m + n − 1 = 3 + 3 − 1 = 5. Clearly the initial solution is non-degenerate. Apply MODI method to find the optimum solution.

Iteration 1

Verify whether the initial solution got by VAM is optimum.

Step 1:

2		3	0
2			0
	0	3	0
2	0	3	

Step 2:

	(−10) 10		0
	(−4) 4	(−2) 5	0
(0) 2			0
2	0	3	

Since all $Z_{ij} - C_{ij}$ are ≤ 0, implies that the optimality is attained. That is, the initial solution itself is optimum. Superimposing the existing solution in the original matrix, we have,

	M_1	M_2	M_3
P_1	[10] 1	−7	[30] 0
P_2	[30] 1	−1	−2
P_3	1	[20] 3	[10] 0

The optimum profit is ₹100

4.8 TRANSSHIPMENT PROBLEMS

A transportation problem allows only shipments that 90 directly from a supply point to a demand point. In many situations, shipments are allowed between supply points or between demand points. Occasionally there may also be points through which goods can be transshipped on their journey from a supply point to a demand point called transshipment points. Shipping problems of the above stated characteristics (fully or partially) are called transshipment problems. Also, the optimal solution to a transshipment problem can be found by solving a transportation problem.

The extended version of transportation stated above was developed and solved by ORDER (1956). The basic idea underlying the solution method is, first, to convert the transshipment problem into a transportation by presenting the data in a transshipment table.

***Example:* 20**

Consider the following TPP along with the optimum solution.

		Destinations			
		D	E	F	Availability
Origin	1	[800] 5	10	12	800
	2	[200] 8	[600] 4	[400] 2	1200
Requirement		1000	600	400	2000

The optimum TP cost is ₹8800. The management would like to know whether the present optimum TP cost can be further reduced by utilizing the transshipment possibilities between all destinations and origin of the problem. The following table shows that origins 1 and 2 of the actual TP became potential points of transshipment that can act simultaneously as origins and destinations. Similarly, the destinations D, E and F of the actual TP also became points of shipment.

Note that in Table 4.1 the unit costs of transportation between the origins and destinations of the actual problem have also been known. The diagonal cells of the table represent shipments within the same point. Thus, the transshipment cost C_{ij} where i = j will reduce to zero. The unit costs of transshipment for the other cells can be obtained from appropriate sources. Assume that, for the present example, the transshipment costs have been substituted into the proper cells of the table below:

	Destinations					
Origin	**1**	**2**	**D**	**E**	**F**	**Available**
1	0	6	5	10	12	2000 + 800
2	6	0	8	4	2	2000 + 1200
D	5	8	0	2	5	2000
E	10	4	2	0	3	2000
F	12	2	5	3	0	2000
Requirement	2000	2000	2000 +1000	2000 +600	2000 +400	12000

Regarding the requirement or availability of the transshipment points created, we can take it as buffer stock of size B and for the others {actual + B}. Clearly, at any one of these points the availability or requirement cannot exceed the total units available for the transshipment as a whole. Thus, an upper limit for our buffer stock (B) would be,

$$B = \sum_{i=1}^{m} s_i = \sum_{j=1}^{n} d_j = 2000 \text{ units}$$

Here B = 800 + 1200 = 2000 (1000 + 600 + 400)

Since, the above transshipment table satisfied, all characteristics of a TPP, the problem can be solved by the application of TP method used for the solution of the standard transportation problem. Hence, the optimum solution of the above transshipment problem is

	1	2	D	E	F
1	2000 (0)		800 (5)		
2		2000 (0)		800 (4)	400 (2)
D			2000 (0)		
E			200 (2)	1800 (0)	
F					2000 (0)

The optimum solution and the computation of the optimal transshipment cost can be given in the following table.

x_{ij}	$C_{ij} x_{ij}$ (₹)
$x_{1D} = 800$ $x_{2E} = 800$ $x_{2F} = 400$ $x_{ED} = (-200)$	$5 \times 800 = 4000$ $4 \times 800 = 3200$ $2 \times 400 = 800$ $2 \times 200 = 400$
$\sum_{i=1}^{5} s_i = \sum_{j=1}^{5} d_j = 2000$	₹ 8400

It should be noted that $x_{ED} = 200$ units transshiped from B to A is already included in $X_{2E} = 800$ units shipped from 2 to E. Therefore, in the calculation of the total availability $\sum_{i=1}^{5} s_i$ or total requirement $\sum_{j=1}^{5} d_j$, x_{ED} must appear with a coefficient of –1. In general, for the calculation of total availability and requirements in the solution of the transshipment problem, the following rules must be taken care of.

x_{ij} appears with a coefficient of +1 if i = 1, 2, ... m.

x_{ij} appears with a coefficient of –1 if i = m + 1, ... m + n

4.9 SENSITIVITY ANALYSIS FOR TRANSPORTATION PROBLEM

In this section, we discuss the following three aspects of sensitivity analysis for the transportation problem.

1. Changing the objective function coefficient of a non-basic variable.
2. Changing the objective function coefficient of a basic variable.
3. Increasing a single supply by Δ and a single demand by Δ.

Consider a TPP along with the optimum solution.

	City 1	City 2	City 3	City 4	Supply	U_i
Plant 1	8	[10] 6	[25] 10	9	35	0
Plant 2	[45] 9	12	[5] 13	7	50	3
Plant 3	14	[10] 9	16	[30] 5	40	3
Demand	45	20	30	30		
V_j	6	6	10	2		

4.9.1 Change in the Objective Function Coefficient by a Non-basic Variable

Let us consider the non-basic cell (1, 1) for our study. Suppose we change C_{11} from 8 to $8 + \Delta$. For what values of Δ will the current basis remain optimal? Consider the non-basic cells with cost alone and the U_i's & V_j's

				U_i
$-(2+\Delta)$ $8+\Delta$			(-7) 9	0
	(-3) 12		(-2) 7	3
(-5) 14		(-3) 16		3
6	6	10	2	

i.e., $-(2 + \Delta) \leq 0 \Rightarrow 2 + \Delta \geq 0; \quad \Delta \geq -2$

Hence the current basis remains optimal for $6 \leq C_{11} \leq 8$.

4.9.2 Change in the Objective Function Coefficient of a Basic Variable

Suppose we change C_{13} from 10 to $10 + \Delta$.

Consider the basic cells with cost alone and find the U_i's & V_j's.

				U_i
	6	$10+\Delta$		0
9		13		$3-\Delta$
	9		5	3
$6+\Delta$	6	$10+\Delta$	2	

Consider the non-basic cells with cost alone and with U_i's & V_j's.

				U_i
$(\Delta-2)$ 8			(-7) 9	0
	$(-3-\Delta)$ 12		$(-2-\Delta)$ 7	$3-\Delta$
$(\Delta-5)$ 14		$(\Delta-3)$ 16		3
$6+\Delta$	6	$10+\Delta$	2	

The current basis remains optimal for

$\Delta - 2 \leq 0; \quad -3 - \Delta \leq 0; \quad \Delta - 5 \leq 0; \quad \Delta - 3 \leq 0 \quad$ and $\quad -2 - \Delta \leq 0$

i.e., $\Delta \leq 2; \quad \Delta \geq -3; \quad \Delta \geq -2; \quad \Delta \leq 5 \quad \& \quad \Delta \leq 3$

$\Rightarrow \quad \Delta \geq -2 \quad \& \quad \Delta \leq 2$

$\Rightarrow \quad -2 \leq \Delta \leq 2$

$\Rightarrow \quad 10 - 2 \leq C_{13} \leq 10+2$

$\Rightarrow \quad 8 \leq C_{13} \leq 12$

Hence the current basis remains optimal for $8 \leq C_{13} \leq 12$.

4.9.3 Increasing Both Supply S_i and Demand d_j by Δ

Observe that this change maintains a balanced transportation problem.

Case (i): If x_{ij} is a basic variable in the optimal solution, increase x_{ij} by Δ. The optimal value can be given by, New optimal value = Old optimal value + $\Delta.U_i + \Delta.V_j$

Suppose we increase S_1 by 3 ($S_1 = 35 + 3$) and d_3 by 3 ($d_3 = 30 + 3$). Since x_{13} is a basic variable in the optimum solution, the new optimum solution can be given as

					U_i's
	[10] 6	[28] 10		38	(0)
[45] 9		[5] 13		50	(3)
	[10] 9		[30] 5	40	(3)
45	20	33	30	128	
(6)	(6)	(10)	(2)		

V_j's (row above: (6), (6), (10), (2))

$$\text{New optimum value} = 1020 + 3.U1 + 3.V2$$
$$= 1020 + 3 \times 0 + 3 \times 10 = 1050$$

Case (ii): If x_{ij} in a non-basic variable in the optimum solution, find the loop involving x_{ij} and some of the basic cells, select an odd cell in the loop that is in row i. Give the increment as Δ for the odd cell and go around the loop, alternatively increasing and decreasing current basic variables in the loop by Δ. Suppose we increase s_1 by 1 ($s_1 = 36$) and d_1 by 1 ($d_1 = 46$).

Consider the basic cells with allocations. Since x_{11} is a nonbasic variable in the current optimal solution, we must find the loop involving x_{11} and linking some of the basic cells.

	[10] x	[25] x		36	(0)
[45]	x	[5] x		50	(3)
	[10]		[30]	40	(3)
46	20	30	30		
(6)	(6)	(10)	(2)		

The odd cell [The basic cell corresponds to the increased supply is cell (1,3)] among $\{x_{11}\ x_{13}, x_{21}, x_{23}\}$ is x_{23}. Hence the new optimum solution will be obtained by increasing both x_{13} and x_{21} by 1 and decreasing x_{23} by 1. The new optimal solution is,

$$\text{New optimal value} = \text{Old optimal value} + U_1 + V_1$$
$$= 1020 + 0 + 6 = 1026$$

EXERCISE PROBLEMS

*1. Write down the initial basic solution to the given transportation problem using MMM.

				s_i
	8	4	5	2
	3	2	4	3
	4	5	7	5
d_j	4	1	5	

*2. Write down the initial basic feasible solution to the TP by NWCR.

				s_i
	7	3	4	2
	2	1	3	3
	3	4	6	5
d_j	4	2	4	

*3. Solve the following TP using VAM.

	Consumers			
	A	B	C	Available
Suppliers I	6	8	4	14
II	4	9	8	12
III	1	2	6	5
Required	6	15	15	

*4. Solve the following transportation problem.

	a	b	c	d	Supply
A	8	9	6	3	18
B	6	11	5	10	20
C	3	8	7	9	18
Demand	15	16	12	13	

*5. Describe any one method for finding an initial solution of a TP.

*6. What are the methods available for finding an initial basic feasible solution to a TP? Explain the distinctions among them.

*7. Explain VAM's method to find the initial solution to a TP?

*8. Find the optimum solution of the following transportation problem.

	D_1	D_2	D_3	Available
O_1	7	4	0	5
O_2	6	8	0	15
O_3	3	9	0	9
Demand	15	6	8	

*9. Find the optimum solution to the TP.

		To			
		A	B	C	Supply
From	W	14	8	8	76
	X	16	24	16	82
	Y	8	16	24	77
Demand		72	102	41	

*10.

		Destination				
		D_1	D_2	D_3	D_4	Supply
	O_1	6	4	1	5	14
Origin	O_2	8	9	2	7	16
	O_3	4	3	6	2	5
Demand		6	10	15	4	35

Find the optimum solution.

*11.

		To				
		I	II	III	IV	Supply
	O_1	23	27	16	18	30
Sources	O_2	12	17	20	51	40
	O_3	22	28	12	32	53
Demand		22	35	25	41	

Solve the above TP.

12.

	I	II	III	IV	Supply
A	3	4	5	6	6
B	2	4	4	3	10
D	4	6	8	6	

*13. Evaluate the optimum solution to the following TP.

	Warehouses D_1	D_2	D_3	D_4	Supply
Origin O_1	42	48	38	37	160
O_2	40	49	52	51	150
O_3	39	38	40	43	190
Demand	80	90	110	160	

*14. M/s Mahindra and Mahindra (M & M Ltd) Ltd manufacturers jeeps in two different plants, P_1 and P_2. It supplies three major markets M_1, M_2 and M_3 with its products. The plants (yearly capacities), markets (projected yearly demands) and unit transportation cost (₹ per unit) are given below. Determine the minimum-cost distribution plan for meeting market demand.

Plants \ Markets	M_1	M_2	M_3	Capacity
P_1	80	50	60	100
P_2	85	55	40	150
Demand	50	125	75	

Suppose a marketing consultant is forecasting that the demand in market's M_1 and M_2 likely to raise to 75 and 225 units per year within next three years and advises the company to go for building a new plant P_3 with a capacity of 300 units (Transportation costs to different markets from plant P_3 is ₹75, ₹50 and ₹50 respectively) determine the changes in minimum-cost distribution. [Hint: Solve the first case as it is and find the optimum solution. For the second case, remodify the given problem in the following fashion and find the optimum solution. Then compare the output of first case with the second case.

Plants \ Markets	M_1	M_2	M_3	Capacity
P_1	80	50	60	100
P_2	85	55	40	150
P_3	75	50	50	300
Demand	125	350	75	

*15. A company produces a small component for an industrial product and distributes it to five wholesalers at a fixed delivered price of ₹250 per unit. Sales forecast indicates that monthly deliveries will be 300, 300, 100, 500 and 400 units to wholesalers 1, 2, 3, 4 and 5 respectively. The direct costs of production of each unit are ₹100, ₹90 and ₹80 at plants 1, 2 and 3 respectively. The transportation cost of shipping a unit from plants to wholesalers are given in the following page.

		Wholesalers				
		1	2	3	4	5
	1	5	7	10	15	15
Plants	2	8	6	9	12	14
	3	10	9	8	10	15

Find how many components of each plant must supply to each wholesaler to maximize the profit? Take the monthly production capacities of plants 1, 2 and 3 as 500, 100 and 1250 units respectively. [Hint: By considering the given data, we have to arrive the profit matrix, using the relation. Profit = delivered price – (direct cost + TP cost)]

*16. Using Vogel's approximation method, find the initial feasible solution to the transportation problem given below.

		Godown			
		A	B	C	Supply
	P	8	7	3	60
Factory	Q	3	8	9	70
	R	11	3	5	80
Demand		50	80	80	

*17. Solve

		To				
		1	2	3	4	Supply
	I	3	6	8	5	20
From	II	6	1	2	5	28
	III	7	8	3	9	17
Demand		15	19	13	18	

Using VAM obtain an initial basic feasible solution and hence solve the above TPP.

*18. Solve the given TPP:

	A	B	C	D	Supply
X	8	10	9	6	100
Y	8	9	12	7	430
Z	12	16	16	9	200
Demand	150	180	190	210	730

*19. Solve the following TPP using LCM.

	D_1	D_2	D_3	D_4	Supply
O_1	1	2	3	4	6
O_2	4	3	2	0	8
O_3	0	2	2	1	10
Demand	4	6	8	6	24

*20.

		Retailer					
		A	B	C	D	E	Supply
Wholesaler	P	10	8	6	9	12	50
	Q	5	3	8	4	10	90
	R	7	9	6	10	4	60
Demand		100	80	70	40	20	

Using VAM find the initial basic feasible solution to the TPP given above.

*21.

	A	B	C	D	E	Supply
X	55	30	40	50	50	40
Y	35	30	100	45	60	20
Z	40	60	95	35	30	40
Demand	25	10	20	30	15	

Using Vogel's approximation method obtain an initial basic feasible solution and hence solve the above transportation problem.

*22. A company has received a contract to supply gravel for three new construction projects located in towns A, B and C. Construction engineers have estimated the required amounts of gravel which will be needed at these construction projects.

Project Location	Weekly Requirements (truck loads)
A	72
B	102
C	41

The company has 3 gravel and pits located in towns W, X and Y. The gravel required by the construction projects can be supplied by three pits. The amount of gravel which can be supplied by each pit is as follows:

Plant	W	X	Y
Amount available (truck loads)	76	82	77

The company has computed the delivery cost from each pit to each project site. These costs are shown in the following table:

	Project Location		
Pit	A ₹	B ₹	C ₹
W	4	8	8
X	16	24	16
Y	8	16	24

Schedule the shipment from each pit to each project so as to minimize the total transportation cost within the constraints by pit capacities and project requirements.

Find the minimum cost. Is the solution unique? If it is not, find alternate schedule with the same minimum cost.

*23. Find the optimum solution to the following transportation problem.

	W_1	W_2	W_3	W_4	W_5	Available
F_1	7	6	4	5	9	40
F_2	8	5	6	7	8	30
F_3	6	8	9	6	5	20
F_4	5	7	7	8	6	10
Required	30	30	15	20	5	100

*24. A manufacturer of jeans is interested in developing an advertising campaign that will reach four different age groups. Advertising campaigns can be conducted through TV, Radio and Magazines. The following table gives the estimated cost in paise per exposure for each age group according to the medium employed. In addition, maximum exposure levels possible in each of the media, namely TV, Radio and Magazines are 40, 30 and 20 millions respectively. Also minimum desired exposures within each age group, namely 13–18, 19–25, 26–35, 36 and above, are 30, 25, 15 and 10 millions. The objective is to minimize the cost of attaining the minimum exposure level in each age group.

	Age Groups			
Media	13–18	19–25	26–35	36 & above
TV	12	7	10	10
Radio	10	9	12	10
Magazines	14	12	9	12

Formulate the above as a transportation problem, and find the optimal solution.

*25. A company wishes to determine an investment strategy for each of the next four years. Five investment types have been selected, investment levels have been established for each investment type. An assumption is that amount invested in any year will remain invested until the end of the planning horizon of four years. The following table summaries the data for this problem. The values in the body of the table represent net return on investment of one rupee upto the end of the planning horizon. For example, a rupee invested in investment type B at the beginning of the year 1 will grow to ₹1.90 by the end of the fourth year, yielding a net return of ₹0.90.

Investment Made at the Beginning of Year	Investment Type NET RETURN DATA A	B	C	D	E	Rupees Available (in 000's)
1	0.8	0.9	0.6	0.75	1.00	500
2	0.55	0.65	0.4	0.6	0.5	600
3	0.3	0.25	0.3	0.5	0.2	750
4	0.15	0.12	0.25	0.35	0.1	800
Maximum Rupees Investment (in 000's)	750	600	500	800	1000	

The objective in this problem is to determine the amount to be invested at the beginning of each year in an investment type so as to maximize the net rupee return for the four-year period. Solve the above TP and get an optimal solution. Also calculate the net return on investment for the planning horizon of four-year period.

ANSWERS TO THE EXERCISE PROBLEMS

1. $x_{13} = 2; x_{21} = 2; x_{22} = 1; x_{31} = 2; x_{33} = 3$ and the total cost is ₹47.
2. $x_{11} = 2; x_{21} = 2; x_{22} = 1; x_{32} = 1; x_{33} = 4$ and the total cost is ₹47.
3. $x_{12} = 4; x_{13} = 10; x_{21} = 6; x_{22} = 6; x_{32} = 5; x_{43} = 5$ and the total cost is ₹160.
4. $x_{12} = 5; x_{14} = 13; x_{22} = 8; x_{23} = 12; x_{31} = 15; x_{32} = 3$ and the minimum cost is ₹301.
8. $x_{12} = 5; x_{21} = 6; x_{22} = 1; x_{23} = 8; x_{31} = 9$ and the minimum cost is ₹91.
9. $x_{12} = 76; x_{22} = 21; x_{23} = 41; x_{24} = 20; x_{31} = 72; x_{32} = 5$ and the minimum cost is ₹2424.

   Alternate optimum:
   $x_{12} = 76; x_{21} = 21; x_{23} = 41; x_{24} = 20; x_{31} = 51, x_{32} = 26$ and the minimum cost is ₹2424.
10. $x_{11} = 4; x_{12} = 10; x_{21} = 1; x_{23} = 15; x_{31} = 1; x_{34} = 4$ and the optimum cost is ₹114.
11. $x_{14} = 30; x_{21} = 5; x_{22} = 35; x_{31} = 17; x_{33} = 25; x_{34} = 11$ and the optimum cost is ₹2221.
12. $x_{12} = 6; x_{21} = 4; x_{24} = 6; x_{33} = 8$ and the optimum cost is ₹50.
13. $x_{14} = 160; x_{21} = 80; x_{22} = 10; x_{25} = 60; x_{32} = 80; x_{33} = 110$ and the optimum cost is ₹17050.
14. $x_{12} = 100; x_{21} = 50, x_{22} = 25; x_{23} = 75$ and the optimum cost is ₹13625.

   Extension:
   $x_{12} = 100; x_{22} = 75; x_{23} = 75; x_{31} = 125; x_{32} = 175$ and the optimum cost is ₹30250.
15. $x_{11} = 300; x_{12} = 200; x_{22} = 100; x_{33} = 100; x_{34} = 500; x_{35} = 400; x_{35} = 250$ and the optimum cost is ₹2,88,200.

16. $x_{13} = 60$; $x_{21} = 50$; $x_{23} = 20$; $x_{32} = 80$ and the total cost is ₹750.
17. $x_{11} = 15$; $x_{14} = 5$; $x_{22} = 19$; $x_{24} = 9$; $x_{33} = 13$; $x_{34} = 4$ and the optimum cost is ₹209.
18. $x_{13} = 100$; $x_{21} = 150$; $x_{22} = 180$; $x_{23} = 90$; $x_{24} = 10$; $x_{34} = 200$ and the optimum cost is ₹6670.
19. $x_{13} = 6$; $x_{23} = 2$; $x_{24} = 6$; $x_{31} = 4$; $x_{32} = 6$ and the total cost is ₹34.
20. $x_{11} = 20$; $x_{14} = 30$; $x_{22} = 80$; $x_{24} = 10$; $x_{31} = 40$; $x_{35} = 20$; $x_{41} = 40$; $x_{43} = 70$ and the total cost is ₹1110.
21. $x_{12} = 10$; $x_{13} = 20$; $x_{14} = 10$; $x_{21} = 20$; $x_{31} = 5$; $x_{34} = 20$; $x_{35} = 15$ and the optimum cost is ₹3650.
22 $x_{12} = 76$; $x_{22} = 21$; $x_{23} = 41$; $x_{24} = 20$; $x_{31} = 72$; $x_{32} = 5$ and the optimum cost is ₹2424.
Alternate optimum solution exists. That is the solution is not unique. $x_{12} = 76$; $x_{21} = 21$; $x_{23} = 41$; $x_{24} = 20$; $x_{31} = 51$; $x_{32} = 26$ and the optimum cost is ₹2424.
23. $x_{11} = 5$; $x_{13} = 15$; $x_{14} = 20$; $x_{22} = 30$; $x_{31} = 15$; $x_{35} = 5$; $x_{41} = 10$ and the optimum cost is ₹510.
24. $x_{12} = 25$ millions; $x_{14} = 10$ millions; $x_{15} = 5$ millions; $x_{21} = 30$ millions; $x_{33} = 15$ millions; $x_{35} = 5$ millions and the minimum cost is ₹71 lakhs ($710/10^5/10^2 = 71 \times 10^5$ = ₹71 lakhs).
25. The given problem is of maximization type.
$x_{15} = 500$; $x_{22} = 600$; $x_{34} = 750$; $x_{41} = 250$; $x_{43} = 500$; $x_{44} = 50$; $x_{51} = 500$; $x_{55} = 500$ and the maximum net return is ₹14,45,000.

REVIEW QUESTIONS

Short Questions

1. Define a transportation problem. (Section 4.1.1)
2. State the methods used to find the initial solution to a TPP. (Section 4.6)
3. Define an unbalanced TPP. (Section 4.2)
4. Define a balanced TPP. (Section 4.2)
5. Formulate the general TP as an LPP. (Section 4.2)
6. When a constant value is added to every cost element (C_{ij}) in the TPP, will there be a change in the optimum values of the variables x_{ij} and the optimal cost? (Regarding x_{ij}'s there won't be any change but the optimality cost will differ).
7. Define an optimum solution to a TPP. (Section 4.4)
8. Give an example of a TPP. (Section 4.3)
9 State a necessary and sufficient condition for the existence of a feasible solution to the TPP.
   i. All costs, supplies and demand must be known and fixed.
   ii. Number of destinations and origins must be finite.
   iii. The total demand and the total supply must be equal.

10. What is the purpose of a TPP?
To give the optimum schedule regarding the shipment of items from various origins to various destinations.
11. How do you resolve the degenerate solution in a TPP? (Refer Example 14)
12. Write the standard form of a TPP. (Section 4.6)
13. State the rules to draw the loop. (Refer the Flowchart 4.6)
14. State the indication regarding the existence of an alternate optimum solution. (Refer Example 14)

State True or False

15. The transportation problem can be formulated as an LPP. (T)
16. We can solve the given transportation problem directly. (F)
17. For a given TPP, the initial solution will differ based on the method used. (T)
18. Whenever Tie occurs during selection of the cell for allocation, we can have different initial solutions. (T)
19. VAM method will give a better starting solution. (T)
20. In a TPP, the number of origins must be equal to the number of destinations. (F)
21. While selecting the leaving variable, there is a chance for more than one cell to leave. (T)
22. Maximization problem can not be solved directly by any method. (T)
23. Unbalanced TPP can be solved directly. (F)
24. Solution got by VAM will be always optimum. (F)
25. VAM gives the better optimum solution. (T)
26. LCM is better than VAM. (F)
27. LCM is better than RMM and NWCR. (T)

Fill in the Blanks

28. The aim of the TPP is _________. (to find the optimum TP cost)
29. The condition to be satisfied for a non-degenerate solution is _________. (no. of basic cells = m + n – 1)
30. For optimality test we should consider only a _________ solution. (non-degenerate)
31. The NWCR, RMM, CMM, LCM and VAM are used to find the _________ solution. (initial)
32. The MODI method is used to find the _________ solution. (optimum)
33. The cell with allocation can be called _________ cell. (an allocated or basic)
34. The cell without allocation, (i.e.) Zero allocation is termed as a _________ cell. (non-basic)

35. The optimality condition is ________. (for all non-basic cells the value of $Z_{ij} - C_{ij}$ must be ≤ 0).
36. MODI method can otherwise be called ________. (Stepping stone)
37. The stepping stone method was presented by ________ and ________ in ________. (Cooper, Charnes, 1959)
38. TPP was first introduced by ________ and ________ and solved by ________ using the ________ method. (Hitchcock, Koopmns, Dantzig, RSM)
39. Least cost method is otherwise called ________. (Matrix minima method)
40. We name the balanced TPP with minimization objective as a TPP in ________. (standard form)

5

Assignment Problem

5.1 INTRODUCTION

The assignment problem is one of the special type of transportation problem for which more efficient (less-time consuming) solution method has been devised by KUHN (1956) and FLOOD (1956). The justification of the steps leading to the solution is based on theorems proved by Hungarian mathematicians KONEIG (1950) and EGERVARY (1953), hence the method is named Hungarian.

5.2 GENERAL MODEL OF THE ASSIGNMENT PROBLEM

Consider n jobs and n persons. Assume that each job can be done only by one person and the time a person required for completing the ith job (i = 1,2,...n) by the jth person (j = 1,2,...n) is denoted by a real number C_{ij}. On the whole this model deals with the assignment of n candidates to n jobs in such a way that to minimize the time (cost) for over all completion of the jobs.

$$
\begin{array}{cc|cccc|}
 & & \multicolumn{4}{c}{\text{Person}} \\
 & & 1 & 2 & 3 \ldots\ldots & n \\
\hline
 & 1 & C_{11} & C_{12} & \cdots\cdots & C_{1n} \\
\text{J} & 2 & C_{21} & C_{22} & \cdots\cdots & C_{2n} \\
\text{o} & \vdots & \vdots & \vdots & & \vdots \\
\text{b} & \vdots & \vdots & \vdots & & \vdots \\
\text{s} & \vdots & \vdots & \vdots & & \vdots \\
 & n & C_{n1} & C_{n2} & \cdots\cdots & C_{nn} \\
\hline
\end{array}
$$

5.3 CONVERSION INTO AN EQUIVALENT LPP

We know that, C_{ij} be the time/cost required to complete the ith job by the jth person. Let us define x_{ij}, such that

$$
x_{ij} = \begin{cases} 1 & \text{if } i^{th} \text{ job is assigned to } j^{th} \text{ person} \\ 0 & \text{if } i^{th} \text{ job is not assigned to } j^{th} \text{ person} \end{cases}
$$

Here, the objective is to minimize the over all time (cost) required to complete the n jobs.

$$\text{Minimize } Z = \sum_{i=1}^{n} \sum_{j=1}^{n} c_{ij}\, x_{ij}$$

$$\text{Subject to } \sum_{i=1}^{n} x_{ij} = 1 \quad (j = 1, 2, \ldots n)$$

$$\sum_{j=1}^{n} x_{ij} = 1 \quad (i = 1, 2, \ldots n)$$

$$x_{ij} = 0 \text{ or } 1$$

Clearly, the first set of constraints indicates that each person must be assigned to exactly one job and the second set of constraints assures that each job can be performed by only one person.

Obviously, the assignment problem is a special case of the standard transportation problem with the following properties:

(i) The total number of origins is equal to the number of total destinations.
(ii) Each origin is represented by one job that is $s_i = 1$, where $i = 1, 2, \ldots, n$.
(iii) Each destination is represented by one person, that is $d_j = 1$, where $j = 1, 2, \ldots, n$.

5.4 SOLUTION TO THE ASSIGNMENT PROBLEM

The various steps to be followed to solve the given assignment problem using Hungarian method are clearly explained in the Flowchart 5.1.

***Example:* 1**

	1	2	3	4
A	10	12	19	11
B	5	10	7	8
C	12	14	13	11
D	8	15	11	9

Solve the above assignment problem.

Step 1: Here the matrix is a square matrix and the objective is minimization. So, proceed to Step 2.

Step 2: Select a minimum in each row and subtract the same with all the elements in the respective rows.

0	2	9	1
0	5	2	3
1	3	2	0
0	7	3	1

*These questions have appeared in Indian university examination papers.

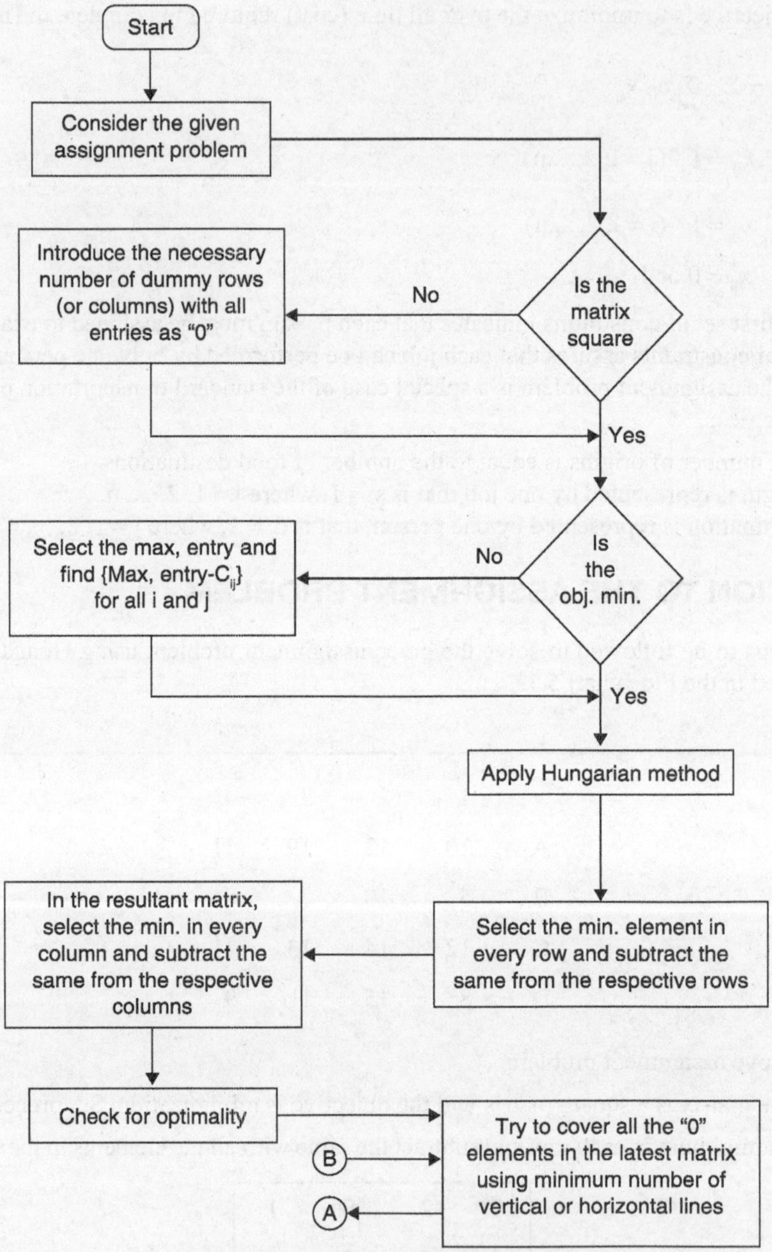

Flowchart 5.1 Hungarian method

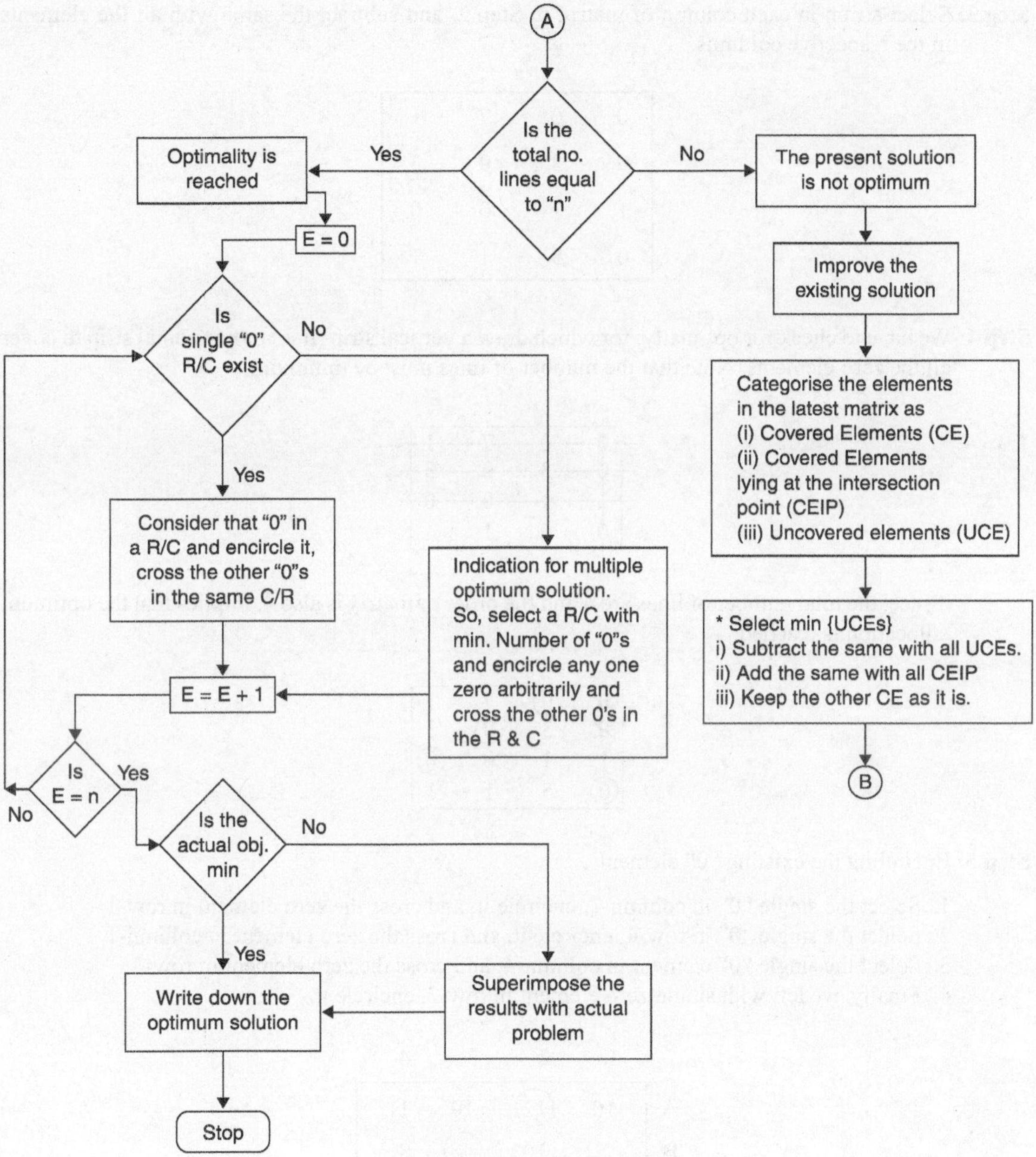

Flowchart 5.1 Continued

Step 3: Select a min in each column of matrix in Step 2, and subtract the same with all the elements in the respective columns.

0	0	7	1
0	3	0	3
1	1	0	0
0	5	1	1

Step 4: We have to check for optimally; for which draw a vertical strip (line) or horizontal strip to cover all the zero elements. Note that the number of lines must be minimum.

0	0	7	1
0	3	0	3
1	1	0	0
0	5	1	1

Since, the total number of lines are 4 and the order of matrix is also 4, implies that the optimum allocation is reached.

~~0~~	(0)	7	1
~~0~~	3	(0)	3
1	1	~~0~~	(0)
(0)	5	1	1

Step 5: Encircling the existing "0" elements.

1. Select the single "0" in column-2, encircle it, and cross the zero element in row-1.
2. Select the single "0" in row-4, encircle it, and cross the zero element in column-1.
3. Select the single "0" element in column-4, and cross the zero element in row-3.
4. Finally, we left with single zero element in row-2, encircle it.

	1	2	3	4
A	10	(12)	19	11
B	5	10	(7)	8
C	12	14	13	(11)
D	(8)	15	11	9

Step 6: Hence, the allocation is, A→2; B→3; C→4; D→1

=> Minimum value = 12 + 7 + 11 + 8 = 38.

*Example: 2

Solve the following Assignment problem.

	I	II	III	IV	V
A	6	5	8	11	16
B	1	13	16	1	10
C	16	11	8	8	8
D	9	14	12	10	10
E	10	13	11	8	16

Step 1: Since, the matrix is square and the objective is minimization, we can proceed to Step 2.

Step 2:

1	0	3	6	11
0	12	15	0	9
8	3	0	0	0
0	5	3	1	1
2	5	3	0	8

Step 3:

1	0	3	6	11
0	12	15	0	9
8	3	0	0	0
0	5	3	1	1
2	5	3	0	8

[Note that in Step 2 all columns containing "0" elements. So, we have the same matrix of Step 2]

Step 4:

1	0	3	6	11
0	12	15	0	9
8	3	0	0	0
0	5	3	1	1
2	5	3	0	8

In Step 4, we have only 4 lines, but the order is 5. So, the existing case is not optimum.

Step 5: Select the minimum out of the elements which are not covered (Uncovered; by lines.
i.e., min{12, 15, 9, 5, 3, 1, 5, 3, 8} = 1
Subtract the value 1, from all the uncovered elements, add the same with elements; lying at the intersection points of the lines {1, 6, 8, 0}, and keep the other covered elements as it is.

Again draw the lines. Since, we have five lines which equals to the order, implies the optimum allocation is attained.

2	0	3	7	11
0	11	14	0	8
9	3	0	1	0
0	4	2	1	0
2	4	2	0	7

Step 6: Encircle the existing "0" elements.

2	(0)	3	7	11
(0)	11	14	~~0~~	8
9	3	(0)	1	~~0~~
~~0~~	4	2	1	(0)
2	4	2	(0)	7

No.	Encircled Element in a Single "0" Row/Column	Crossed Elements
1	(5, 4)	(2, 4)
2	(2, 1)	(4, 1)
3	(4, 5)	(3, 5)
4	(3, 3)	–
5	(1, 2)	–

Hence, the solution is,

		I	II	III	IV	V
A -> II	A	6	(5)	8	11	16
B -> I	B	(1)	13	16	1	10
C -> III	C	16	11	(8)	8	8
D -> V	D	9	14	12	10	(10)
E -> IV	E	10	13	11	(8)	16

Minimum value = 5 + 1 + 8 + 10 + 8 = 32.

*Example: 3

A manager has four salesmen and four sales districts. He estimates that the profit per day from each salesman in each district would be as follows (in ₹). Find the assignment of salesmen to districts that will result in maximum profit.

		District			
		A	B	C	D
Salesman	1	33	21	35	23
	2	29	23	37	31
	3	31	31	33	29
	4	27	25	35	37

The given matrix is square.

Step 1: Since, the objective is maximizing the profit, the problem must be converted into a minimization problem. For which select a maximum entry and subtract all the entries from it. Here, the Max entry = 37, Find the values $\{37 - C_{ij}\}$; (j = 1, 2, 3, 4, j =1.2,3,4)

	A	B	C	D
1	4	16	2	14
2	8	14	0	6
3	6	6	4	8
4	10	12	2	0

The objective of the modified problem is of minimization.

Step 2: Now, we can apply the Hungarian method.

2	14	0	12
8	14	0	6
2	2	0	4
10	12	2	0

Step 3:

0	12	0	12
6	12	0	6
0	0	0	4
8	10	2	0

Step 4:

0	12	0	12
6	12	0	6
0	0	0	4
8	10	2	0

Since, the number of lines equals to the order of the matrix, the optimum table is reached.

Step 5: Encircling the "0"s.

⓪	12	~~0~~	12
6	12	⓪	6
~~0~~	⓪	~~0~~	4
8	10	2	⓪

Encircled Element	Crossed Element
(4, 4)	–
(3, 2)	(3, 1), (3, 3)
(1, 1)	(1, 3)
(2, 3)	–

Step 6: Since, the actual objective is maximization, we have to superimpose the allocations with the actual problem.

	A	B	C	D
1	(33)	21	35	23
2	29	23	(37)	31
3	31	(31)	33	29
4	27	25	35	(37)

Hence, the solution is,

1→A
2→C
3→B
4→D

Maximum profit = 33 + 37 + 31 + 37 = 138.

*Example: 4

Solve the following problem to maximize the profit. Which job is to be left?

		Job				
		A	B	C	D	E
Mechanic	1	62	78	50	101	82
	2	71	84	55	73	59
	3	87	92	111	71	81
	4	48	64	87	77	80

Step 1: The given matrix is not a square, otherwise called Unbalanced Assignment problem. To make it as a balanced one, we have to add a dummy row, with all entries as "0". The restructured problem is,

		Job				
		A	B	C	D	E
Mechanic	1	62	78	50	101	82
	2	71	84	55	73	59
	3	87	92	111	71	81
	4	48	64	87	77	80
	5	0	0	0	0	0

Also, the objective is maximization, this is to be converted into minimization problem. Maximum entry = 111; subtract all entries from this maximum entry.

	A	B	C	D	E
1	49	33	61	10	29
2	40	27	56	38	52
3	24	19	0	40	30
4	63	47	24	34	31
5	111	111	111	111	111

Step 2:

39	23	51	0	19
13	0	29	11	25
24	19	0	40	30
39	23	0	10	7
0	0	0	0	0

Step 3: Since, all the columns of the matrix in Step 2, have "0" entries, we have the same matrix in Step 3 also.

Step 4:

39	23	51	0	19
13	0	29	11	25
24	19	0	40	30
39	23	0	10	7
0	0	0	0	0

Since, the number of lines are not equal to the order of the matrix, modify.

Step 5:

39	23	58	0	19
13	0	36	11	25
17	12	0	33	23
32	16	0	3	0
0	0	7	0	0

Step 6:

39	23	58	(0)	19
13	(0)	36	11	25
17	12	(0)	33	23
32	16	~~0~~	3	(0)
(0)	~~0~~	7	~~0~~	~~0~~

The solution is,

	A	B	C	D	E
1	62	78	50	(101)	82
2	71	(84)	61	73	59
3	87	92	(111)	71	81
4	48	64	87	77	(80)
5	(0)	0	0	0	0

1→D
2→B
3→C
4→E
5→A

Maximum profit = 101 + 84 + 111 + 80 + 0 = 376.
Since, the dummy mechanic 5, is allocated to Job A; Job A is not to be considered. In order to maximize the profit, we have to consider the Jobs B, C, D and E. Hence, the maximum profit is 376.

Example: 5

Impossible Assignment

Consider the problem of assigning four persons to four machines. The assignment costs in rupees are given below. Person 1 cannot be assigned to machine 2. Also, person 3 cannot be assigned to machine 3. Find the optimal assignment. If possible find an alternate optimum.

		Machine			
		1	2	3	4
Person	1	5	-	5	2
	2	7	4	2	3
	3	9	3	-	5
	4	7	2	6	7

The given model consists of an impossible assignment i.e. Person-1 cannot be assigned to machine-2 likewise Person-2 cannot be with machine-4.

Step 1: Clearly the matrix is square and objective is minimization, then proceed to Step 2.

Step 2:

3	-	3	0
5	2	0	1
6	0	-	2
5	0	4	5

Step 3:

0		3	0
2	2	0	1
3	0	-	2
2	0	4	5

Step 4:

0	-	3	0
2	2	0	1
3	0	-	2
2	0	4	5

Step 5:

0	-	3	0
2	4	0	1
1	0	-	0
0	0	2	3

Step 6:

(0)	-	3	~~0~~
2	4	(0)	1
1	~~0~~	-	(0)
~~0~~	(0)	2	3

Note that, after selecting the element (2, 3) as a single zero element; for further selection, we have all the rows containing more than one zero. [This implies that the given problem is having a multiple optimum solution.] Select any one zero arbitrarily and proceed with the method. Encircling the element (1, 1) and proceeding, the optimum solution is,

1→1; 2→3; 3→4 and 4→2.

The optimum cost is (5 + 2 + 5 + 2) = ₹14.

Alternate solution: In Step 6, instead of selecting the element (1.1), if we select the element (3, 2).

~~0~~	-	3	(0)
2	4	(0)	1
1	(0)	-	~~0~~
(0)	~~0~~	2	3

The alternate allocation is, 1→4; 2→3; 3→2 and 4→ 1.

The optimum cost is (7 + 3 + 2 + 2) = ₹14

*Example: 6

Airline Crew Assignment Parchi Airlines that operate seven days a week has a time table shown below. Crews must have a minimum lay over of 6 hrs between flights. Obtain the pairing of flights that minimizes layover time away from home. For any given pairing the crew will be based at the city results in the smaller layover:

Flight No.	DELHI-CALCUTTA		Flight No.	CALCUTTA-DELHI	
	Depart	Arrive		Depart	Arrive
1	7.00 am	9.00 am	101	9.00 am	11.00 am
2	9.00 am	11.00 am	102	10.00 am	12.00 noon
3	1.30 pm	3.30 pm	103	3.30 pm	5.30 pm
4	7.30 pm	9.30 pm	104	8.00 pm	10.00 pm

For each pair also mention the town where the crew should be based.

For the above problem first we have to get a matrix based on the layover time (in hrs). The layover time is the time gab (rest time) between arrival time to Calcutta and departing time from Calcutta [arrival time to Delhi and departing time from Delhi]

Flight No.	Delhi Based				Calcutta Based			
	101	102	103	104	101	102	103	104
1	0	1	6.5	11	20	19	13.5	9
2	22	23	4.5	9	22	21	15.5	11
3	17.5	18.5	0	4.5	2.5	1.5	20	15.5
4	11.5	14.5	18	22.5	8.5	7.5	2	21.5

Given a condition that the minimum layover time must be 6 hrs. If any element in the above matrix is less than 6, add 24 along with this. Hence the modified layover matrix is

	Delhi Based				Calcutta Based			
Flight No.	101	102	103	104	101	102	103	104
1	24	25	6.5	11	20	19	13.5	9
2	22	23	28.5	9	22	21	15.5	11
3	17.5	18.5	24	28 5	26.5	25.5	20	15.5
4	11.5	14.5	18	22.5	8.5	7.5	26	21.5

The composite layover time matrix can be written by selecting the minimum layover time corresponding positional elements of the above matrix.

	101	102	103	104
1	20^C	19^C	6.5	9^C
2	22	21^C	15.5^C	9
3	17.5	18.5	20^C	15.5^C
4	8.5^C	7.5^C	18	21.5^C

Since the objective is to minimize the overall layover time, we can apply the Hungarian Method for the latest matrix

[Note: C – The minimum value selected from the Calcutta based crew's lay over time.]

13.5	12.5	0	2.5
13	12	6.5	0
2	3	4.5	0
1	0	10.5	14

12.5	12.5	0	2.5
12	12	6.5	0
1	3	4.5	0
0	0	10.5	14

12.5	12.5	0	2.5
12	12	6.5	0
1	3	4.5	0
0	0	10.5	14

Clearly the existing solution is not optimum. Improve the solution

12.5	12.5	0	3.5
11	11	5.5	0
0	2	3.5	0
0	0	10.5	15

The existing solution is optimum.

12.5	12.5	(0)	3.5
11	11	5.5	(0)
(0)	2	3.5	~~0~~
~~0~~	(0)	10.5	15

Hence the optimum solution is 1→103, 2→104; 3→101; 4→102.

The required optimum flight schedule is

Flights	Base of the Crew	Layover Time in hrs
1→103	Delhi	6.5
2→104	Delhi	9.0
3→101	Delhi	17.5
4→102	Calcutta	7.5

Total lay over time is 40.5 hrs

5.5 TRAVELLING SALESMAN PROBLEM

Travelling Salesman Problem is very similar to the Assignment Model except that in the former, there is an additional condition that a salesman who starts from his home city, visits each city only once and returns to his home city with the objective that the distance travelled or time required to cover all the n-different cities must be minimum.

Regarding the algorithm for this model, we have to apply the Hungarian Method and get the optimum solution. If the solution satisfies the extra restriction that one city should be visited only once we can stop as it is. If the optimum solution not satisfies the extra restriction then we have to go for the Modification process.

In the modification process, for encircling along with 0's we can consider the minimum positive element available in the latest matrix. If once again the condition is failed, take a next minimum positive value along with the element already selected. Proceed till the additional condition is satisfied.

*Example: 7

A machine operator process five types of items on his machine each week, and must choose a sequence for them. The setup cost percentage depends on the item presently on the machine and the set-up cost to be made according to the following table.

		To item				
		A	B	C	D	E
From item	A	∞	4	7	3	4
	B	4	∞	6	3	4
	C	7	6	∞	7	5
	D	3	3	7	∞	7
	E	4	4	5	7	∞

If he processes each type of item only once each week, how should he sequence the items on his machine in order to minimize the total setup cost.

Clearly the given problem satisfies all the conditions related to an Assignment model. So apply the Hungarian method to get the optimum solution and further we can verify the additional restriction of processing each type of item only once.

Step 1: Application of Hungarian Method

Note: Keep the diagonal entries as they are.

∞	1	4	0	1
1	∞	3	0	1
2	1	∞	2	0
0	0	4	∞	4
0	0	1	3	∞

∞	1	3	0	1
1	∞	2	0	1
2	1	∞	2	0
0	0	3	∞	4
0	0	1	3	∞

∞	1	4	0	1
1	∞	3	0	1
2	1	∞	2	0
0	0	3	∞	4
0	0	1	3	∞

Since the number of lines drawn is not equal to the order, it implies that the solution must be modified.

∞	0	2	0	0
0	∞	1	0	0
2	1	∞	3	0
0	0	3	∞	4
0	0	1	4	∞

Clearly the optimality is attained. Then find the allocations.

∞	~~0~~	2	(0)	0
(0)	∞	1	0	0
2	1	∞	3	(0)
~~0~~	(0)	3	∞	4
~~0~~	~~0~~	(0)	4	∞

Hence, the optimum solution is,

A→D; B→A; C→E; D→B; E→C

The above optimum solution does not satisfy the extra restriction. So, we have to modify the existing solution. For encircling consider the minimum positive entry 1. The element 1 occurs in two positions, (i) (2,3) (ii) (3,2).

Now this problem is branched to two cases. In case-1 we can consider the assignment "1" in the position (2, 3) instead of "0" in (5, 3) and in case-2 we can consider the assignment "1" in the position (3, 2) instead 0 in (4, 2) and find the solutions.

Case 1:

∞	~~0~~	2	(0)	~~0~~
~~0~~	∞	(1)	~~0~~	~~0~~
2	~~1~~	∞	3	(0)
~~0~~	(0)	3	∞	4
(0)	~~0~~	~~0~~	4	∞

The modified solution is
A→D; D→C; C→E; E→B; B→A

Clearly the extra restriction is satisfied. The total setup cost is ₹21.

Case 2:

∞	~~0~~	2	~~0~~	(0)
~~0~~	∞	~~1~~	(0)	~~0~~
2	(1)	∞	3	~~0~~
(0)	~~0~~	3	∞	4
~~0~~	~~0~~	(0)	4	∞

The modified solution is
A→E; E→C; C→B; B→D; D→A

Hence the total setup cost is 21. In both case-1 and case-2 the extra condition is satisfied. Hence the optimum solution is

(i) A→D→C→E→B→A
(ii) A→E→C→B→D→A and the optimum cost is ₹21.

Example: 8

Solve the travelling salesman problem given the distance between the cities. Find the optimum route.

	A	B	C	D	E
A	∞	20	4	10	∞
B	20	∞	5	∞	10
C	4	5	∞	6	6
D	10	∞	6	∞	20
E	∞	10	6	20	∞

First we solve the given problem using Hungarian method.

∞	16	0	6	∞
15	∞	0	∞	5
0	1	∞	2	2
4	∞	0	∞	14
∞	4	0	14	∞

∞	15	0	4	∞
15	∞	0	∞	3
0	0	∞	0	0
4	∞	0	∞	12
∞	3	0	12	∞

∞	15	0	4	∞
15	∞	0	∞	3
0	1	∞	0	0
4	∞	0	∞	12
∞	4	0	12	∞

∞	12	0	1	∞
12	∞	0	∞	0
0	0	∞	0	0
1	∞	8	∞	9
∞	0	0	9	∞

∞	11	0	0	∞
12	∞	1	∞	0
0	0	∞	0	0
0	∞	3	∞	8
∞	0	1	9	∞

Hence the optimum solution is attained.

∞	11	(0)	~~0~~	∞
12	∞	1	∞	(0)
~~0~~	~~0~~	∞	(0)	~~0~~
(0)	∞	3	∞	8
∞	(0)	1	9	∞

A→C→D→A; B→E→B

Clearly the extra condition is not satisfied

Select the next minimum positive element and consider it for allocation. Here we have the element 1. 1 occurring in two places (i) (2, 3) (ii) (5, 3)

Case 1: Consider the element 1 in (2, 3)

∞	11	~~0~~	(0)	∞
12	∞	(1)	∞	~~0~~
~~0~~	~~0~~	∞	~~0~~	(0)
(0)	∞	3	∞	8
∞	(0)	~~1~~	9	∞

A→D→A; B→C→E→B. This solution is not the required optimum.

Case 2: Consider the element 1 in (5, 3)

∞	11	~~0~~	(0)	∞
12	∞	~~1~~	∞	(0)
~~0~~	(0)	∞	~~0~~	~~0~~
(0)	∞	3	∞	8
∞	~~0~~	(1)	9	∞

A→D→A; B →E →C→B.
This solution is not the required optimum.

Since both cases are failed, consider the next minimum positive element 3 in (4, 3). This case is also failed. Select the next minimum positive element 8 in (4, 5).

∞	11	~~0~~	(0)	∞
12	∞	(1)	∞	~~0~~
(0)	~~0~~	∞	~~0~~	~~0~~
~~0~~	∞	~~3~~	∞	(8)
∞	(0)	~~1~~	9	∞

A→D→E→B→C→A
Hence, the required optimum solution is attained. The optimum distance is 49.

EXERCISE PROBLEMS

*1. Write down the mathematical formulation of the general assignment problem.

*2.

		Jobs			
		I	II	III	IV
	A	18	26	17	11
Men	B	13	28	41	26
	C	38	19	18	15
	D	19	26	24	10

Solve the above assignment problem. (Cell entries gives the time taken in hours by each men to complete a task).

*3. Describe the Hungarian Assignment method for solving an assignment problem.

*4. Solve the following assignment problem.

		Jobs				
		1	2	3	4	5
Persons	A	8	4	2	6	1
	B	0	9	5	5	4
	C	3	8	9	2	6
	D	4	3	1	0	3
	E	9	5	8	9	5

*5. The following table presents the cost of assigning 3 jobs to 3 workers. Work-out the optimal assignment.

		Job		
		A	B	C
Worker	1	4	7	8
	2	5	3	2
	3	6	5	4

*6. A company has 5 jobs to be done. The following matrix shows the return of ₹ of assigning i^{th}($i = 1, 2, 3, 4, 5$) machine to the jth job ($j = 1, 2, 3, 4, 5$). Assign the five jobs to the five machines so as to maximize the total return.

Job \ Machine	1	2	3	4	5
1	5	11	10	12	4
2	2	4	6	3	5
3	3	12	5	14	6
4	6	14	4	11	7
5	7	9	8	12	5

*7. Solve the following minimal assignment problem

		Man			
		1	2	3	4
	I	12	30	21	15
Job	II	18	33	9	31
	III	44	25	24	21
	IV	23	30	28	14

*8. With the given set-up cost table, show how to sequence the production so as to minimize the set-up cost per cycle

		To				
		A	B	C	D	E
From	A	-	2	5	7	1
	B	6	-	3	8	2
	C	8	7	-	4	7
	D	12	4	6	-	5
	E	1	3	2	8	-

[Hint: Travelling Salesman Model]

*9. A Head of the department has 5 jobs A, B, C, D, E and five subordinates H_1, H_2, H_3, H_4, H_5. The number of hours each man would take to perform each job is given below. How should the jobs be allocated to minimize the total time.

	H_1	H_2	H_3	H_4	H_5
A	3	5	10	15	8
B	4	7	15	18	8
C	8	12	20	20	12
D	5	5	8	10	6
E	10	10	15	25	10

*10.

		Job				
		1	2	3	4	5
Machine	1	2.5	5	1	5	1
	2	2	5	1.5	7	3
	3	3	6.5	2	8	3
	4	3.5	7	2	9	4.5
	5	4	7	3	9	6
	6	6	9	5	10	6

A company faces the problem of assigning six different machines to five different jobs. The costs are given in the above table (in hundreds of rupees). Solve the problem to minimize the total cost.

*11. The Marketing Manager of a company decides to assign five sales representatives of his company to five different regions in order to maximize the total sales. The representatives A and C expressed their inability to go to Zone III and IV respectively. The following table gives the expected sales by each representative for different zones in thousands of rupees. Find the optimum assignment schedule.

		Zones				
		I	II	III	IV	V
Representatives	A	9	9	-	6	10
	B	11	8	6	7	8
	C	13	7	9	-	7
	D	11	6	10	11	6
	E	10	9	11	13	5

*12. Five jobs are to be processed and five machines are available. Any machine can process any job with the resulting profit (in rupees) as follows:

		Machine				
		A	B	C	D	E
Job	1	32	38	40	28	40
	2	40	24	28	21	36
	3	41	27	33	30	37
	4	22	38	41	36	36
	5	29	33	40	35	39

What is the Maximum profit that may be expected if an optimum assignment is made?

*13. Six salesmen are to be allocated to six sales regions. The earnings of each salesman at each region is given below. How can you find an allocation, so that the earnings will be maximum?

	1	2	3	4	5	6
A	15	35	0	25	10	45
B	40	5	45	20	15	20
C	25	60	10	65	25	10
D	25	20	35	10	25	60
E	30	70	40	5	40	50
F	10	25	30	40	50	15

*14. Find a solution to the assignment problem given below.

		Salesman			
		I	II	III	IV
Sales territory	A	50	40	30	20
	B	50	60	40	20
	C	50	40	70	30
	D	50	40	60	80

(Hint: Solve the problem to maximize the sales)

*15. A car hire company has one car at each of five depots a, b, c, d and e. A customer in each of the five towns A, B, C, D & E requires a car. The distance (in miles) between the depots (origins) and the towns (designations) where the customers are, in the following distance matrix.

	a	b	c	d	e
A	160	130	175	190	200
B	135	120	130	160	175
C	140	110	155	170	185
D	50	50	80	80	110
E	55	35	70	80	105

How should the cars be assigned to the customers so as to minimize the distance travelled?

*16. For the following travelling salesman problem find the best route to come back to the starting point so as to minimize the distance travelled:

	A	B	C	D	E
A	∞	4	10	14	2
B	12	∞	6	16	14
C	16	14	∞	8	14
D	24	8	12	∞	10
E	2	6	4	16	∞

*17. Assign four trucks 1, 2, 3 & 4 to vacant spaces 7, 8, 9, 10, 11 & 12 so that the distance travelled is minimized, The matrix below shows the distance.

	1	2	3	4
7	4	7	3	7
8	8	2	5	5
9	4	9	6	9
10	7	5	4	8
11	6	3	5	4
12	6	8	7	3

18. The captain of a cricket team has to allot five middle batting positions to five batsmen. The average runs scored by each batman at these positions are given below:

		Batting positions				
		1	2	3	4	5
Batsman	P	40	40	35	25	50
	Q	42	30	16	25	27
	R	50	48	40	60	50
	S	20	19	20	18	25
	T	58	60	59	55	53

(Hint: Maximization type)

(i) Find the assignment of batsmen to positions, which would give the maximum number of runs.

(ii) If another batsman U with the following average runs in the batting positions are given below:

Batting positions	1	2	3	4	5
Average runs	45	62	38	50	49

is added to the team, should be included to play in the team? If so, who will be replaced by him?

19. A solicitors firm employs typists on hourly piece-rate basis for their daily work. There are five typists for service and their charges and speeds are different. According to an earlier understanding only one job is given to one typist and the typist is paid for full hours even if he works for a fraction of an hour. Find the least cost allocation for the following data:

Typists	Rate/hr (₹)	No. of Pages Type/hr
A	5	12
B	6	14
C	3	8
D	4	10
E	4	11

Job	No. of Pages
P	199
Q	175
R	145
S	198
T	178

(Hint: If A takes the job P, then the amount for A can be calculated as $(199/12) = 16.58 = 17$ hrs (i.e.) $17 \times 5 =$ ₹85)

20. ABC company has taken the third floor of a multistoried building for rent with a view to locate one of their zonal offices. There are five main rows in this floor to be assigned to five Managers. Each room has its own advantages and disadvantages. Some windows, some are closer to the washrooms or to the canteen or secretarial pool. The rooms are of all different sizes and shapes. Each of the five Managers was asked to rank their room preferences amongst the rooms 301, 302, 303, 304 and 305. Their preferences were recorded in a table as indicated below:

MANAGER				
M_1	M_2	M_3	M_4	M_5
302	302	303	302	301
303	304	301	305	302
304	305	304	304	304
	301	305	303	
		302		

Most of the Managers did not list all the five rooms since they were not satisfied with some of these rooms and they have left of these rooms from the list. Assuming that their preferences can be quantified by numbers, find out as to which Manager should be assigned to which rooms so that their total preferences ranking is a minimum.
[Hint: For each Manager rank it a 1, 2, 3, 4 & 5 and for a non-interested rooms put –.]

21. The Secretary of a school is taking bids on the City's four school bus routes. Four companies have made the bids as detailed in the following table:

	Bids (Amounts in ₹)			
	Route 1	Route 2	Route 3	Route 4
Company 1	4000	5000	–	–
Company 2	–	4000	–	4000
Company 3	3000	–	2000	–
Company 4	–	–	4000	5000

Suppose each bidder can be assigned only one route. Use the assignment model to minimize the school's cost of running the four bus routes.

ANSWERS TO THE EXERCISE PROBLEMS

1. Refer Section 5.2.

2.
Man	A	B	C	D	
Jobs	III	I	II	IV	Optimum time = 59 hrs

3. Refer Flowchart 5.1.

4.
Persons	A	B	C	D	E	
Jobs	5	1	4	3	2	Optimum value = 9

5.
Worker	1	2	3	
Job	A	B	C	Optimum cost = ₹11

6.
Machine	1	2	3	4	5	
Job	3	5	4	2	1	Optimum Return = ₹50

7.
Job	I	II	III	IV	
Men	1	3	2	4	Optimum value = 60

8. Travelling Salesman Model:
The sequence of processing is.

A→B→C→D→E→A

Minimum set-up cost = ₹15

9.
Jobs	A	B	C	D	E	
Subordinates	H_3	H_2	H_1	H_4	H_5	Optimum time = 45 hrs

10.
Machine	1	2	3	4	5	6	
Job	4	2	5	3	1	6	Optimum cost = ₹19

Alternate Solution

Machine	1	2	3	4	5	6
Job	4	1	5	3	2	6

Optimum cost = ₹19

11.

Representatives	A	B	C	D	E
Zones	V	II	I	III	V

Maximum Sales = ₹54,000

12.

Job	1	2	3	4	5
Machine	B	A	E	D	C

Maximum Profit = ₹191

13.

Salesman	A	B	C	D	E	F
Sales Region	6	1	4	3	2	5

Maximum earnings = 305

14.

Sales territory	A	B	C	D
Sales men	I	II	III	IV

Maximum Sales = 260

15.

Towns	A	B	C	D	E
Depots	e	c	b	a	d

Minimum distance travelled = 570 miles

16. The optimum route is

A – B – C – D – E – A

Optimum distance = 30

17.

Vacant space	7	8	9	10	11	12
Trucks	3	2	1	5	6	4

Optimum distance travelled = 12

18. (i)

Batsman	P	Q	R	S	T
Batting Positions	V	I	IV	III	II

Maximum number of runs = 232

(ii)

Batsman	P	Q	R	S	T	U
Batting Positions	V	I	IV	dummy	III	II

Maximum number of runs = 263

19.

Typist	A	B	C	D	E
Job	T	R	Q	P	S

Total Cost = ₹399

20.

Room Nos.	301	302	303	304	305
Manager	M_5	M_1	M_3	M_2	M_4

Minimum ranking is = 7

21.

Company	1	2	3	4
Route	1	2	3	4

Minimum cost = ₹15,000

REVIEW QUESTIONS

Short Questions

1. What is an assignment problem? Give an example. (Section 5.2)
2. Write down the structure of an assignment model. (Section 5.2)
3. Give any two areas of application of an assignment problem.
   i. Selection of operators for machine operation in production planning.
   ii. Selection of persons for sales area in marketing department.
4. Show that an assignment problem is a special case of transportation problem. (Section 5.3)
5. Define an unbalanced assignment model. (Section 5.9)
   Given matrix should be a rectangular one.
6. How can we convert the maximization objective into a minimization objective while applying the Hungarian method? (Refer Example 3)
7. Bring out the difference between a transportation model and an assignment model.

	Transportation Model	Assignment Model
i.	Minimization objective	Minimization objective
ii.	Must be balanced	Must be balanced
iii.	Not necessary	One to one relation
iv.	The solution indicates the number of units	The solution, indicates the one-to-one assignment to be transported.

8. What do you mean by restricted assignments?
   Consider an example,

$$\text{Job}\quad \begin{matrix} \\ A \\ B \\ C \\ \\ \end{matrix} \overset{\text{Person}}{\begin{bmatrix} 1 & 2 & 3 \\ 4 & - & 1 \\ 3 & 4 & 2 \\ 1 & 3 & 4 \end{bmatrix}}$$

   Here, the hyphen the restricted assignments That is, we can't allocate the Job-B to person-2. We should keep the hyphen as such and don't consider it for any computation,

9. Can we think of more than one optimal solution? How we can detect?

   Yes, we can think of more than one optimal solution for a given APP. While selecting the zeros for encircling, if any row/column is having more than one zero, it indicates that there exists an alternate optimum

10. Define the Travelling Salesman Problem. (Section 5.5)

11. State the difference between an assignment problem and a travelling salesman problem.
In the assignment problem after attaining the optimum solution, we stop as such. Whereas in the travelling salesman problem. We check the extra condition, whether the existing solution guarantees the visit to different places are once only. Also the starting and the returning points are one and the same.

State True or False

12. The APP is a special case of a TPP. (T)
13. The APP can be solved using TPP algorithm. (T)
14. An impossible assignment and the restricted assignments are one and the same. (T)
15. Assignment problem can be converted into an equivalent LPP. (T)
16. The given APP can be solved directly by the Hungarian Method. (F)
17. The given APP must be square. (T)
18. In the case of an unbalanced APP with maximization objective, first we have to convert into minimization model and introduce the number of dummy rows or columns required. (F)
19. Using the Hungarian Method we can solve only a minimization model. (T)

Fill in the Blanks

20. ______ method is used to solve APP. (Hungarian)
21. The algorithm used to solve APP is known as Hungarian method because ______.
(It was designed by the Hungarian Mathematicians)
22. The Hungarian method was designed by ______ and ______. (Khun, Flood)
23. Theorems related to the assignment model were solved by ______ and ______. (Konieg, Egervary)
24. The APP can be referred to by a Matrix form of order nxn known as ______. (Cost matrix)
25. In an APP, if the number of rows and columns are equal then the APP is said to be ______. (Balanced)

6

PERT - CPM

6.1 INTRODUCTION

A project comprises of interrelated activities, each contributing concurrently and or sequentially to the achievement of certain specific objectives. During the late 1950's, PERT (MALCOLM, ROSEBOOM, CLARK, and FRAZER 1959) and CPM (KELLY and WALKER 1959) were developed to provide a tool to enhance the effectiveness of the three phases planning, scheduling and controlling of a project management.

Basically, PERT and CPM represent one and the same technique. PERT assumes that for each activity the duration (activity time) as a random variable, whereas CPM uses known fixed activity timings.

The planning phase leads to the construction of a network that represents the activities of the project in the predetermined sequence and interdependencies in time for this end, the following concepts need to be introduced.

6.1.1 Activity

Each case of the project that has a definite beginning and end point in time is called an activity. Each activity on the project is represented by an arrow with its head indicating the direction in the project.

6.1.2 Activity Duration/Activity Time

The time required to perform an activity is called the activity duration time.

6.1.3 Event

The starting and end point of an activity in time is called event. An event has no time duration. A project can have one starting and one ending event. Normally, the events will be represented by small circles called nodes of a project network.

6.1.4 Network/Arrow Diagram of a Project

A graphical representation of the activities of a project in their proper sequence and interdependencies in time are called project network.

6.2 METHOD FOR CONSTRUCTION OF A NETWORK

***Example:* 1**

Construct a network diagram for the following situation.

A < D, E; B, D < F; C < G; B < H

Step 1: Tabulate the given relation as follows:

	Predecessor	A	B, D	C	B
Activities	RHS Successor	D, E	F	G	H

Since the activities, A, B and C are not there in row of successors, imply that they are independent and can be started simultaneously.

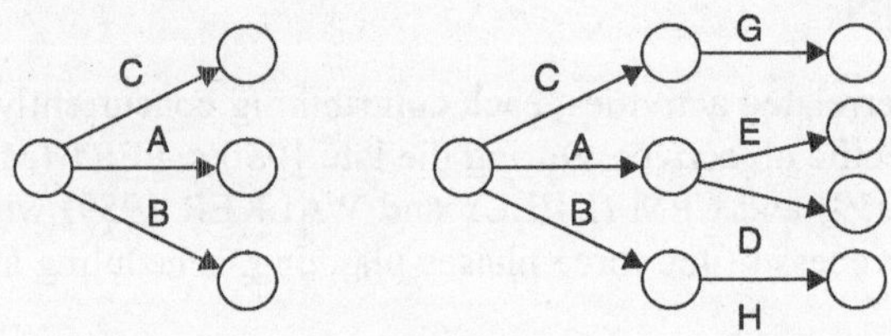

Step 2: Then start the successor activities which depend only on A, B, C as in Step 1.

Step 3: In Step 1 table remove the columns which contain the activities A, B and C. The resultant table is,

Activities in	LHS	B, D*
	RHS	F

As per the table, F succeeds B and D. * Whenever more than two activities are preceding one or more activities, we must seek the help of dummy activities.

Here, we have two options

Option 1: F can be linked after D, if so, link B and D using a dummy activity.
Option 2: F can be linked after B, if so, link D and B using a dummy activity.

Step 4: Using option 1, the network becomes,

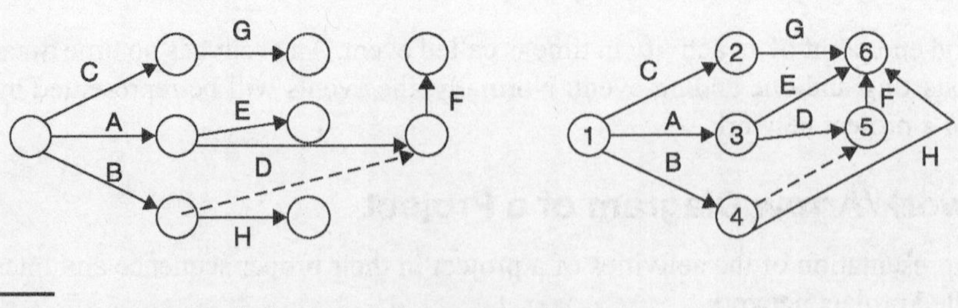

*These questions have appeared in Indian university examination papers.

Step 5: As per diagram in Step 4, indicates the activities E, F, G and H head nodes are the end nodes. Hence they can be combined as above. The above network is the required one.

6.3 NUMBERING THE NODES

***Example:* 2**

Consider the following network for numbering the nodes.

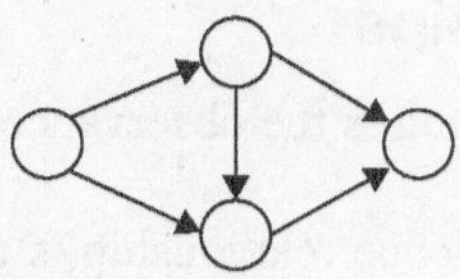

Step 1: Identify the node only with outgoing arrows.

Step 2: Number those nodes.

Step 3: Delete the outgoing arrows at the node already numbered.

Step 4: Go to Step 1 and repeat this process till all the nodes are numbered.

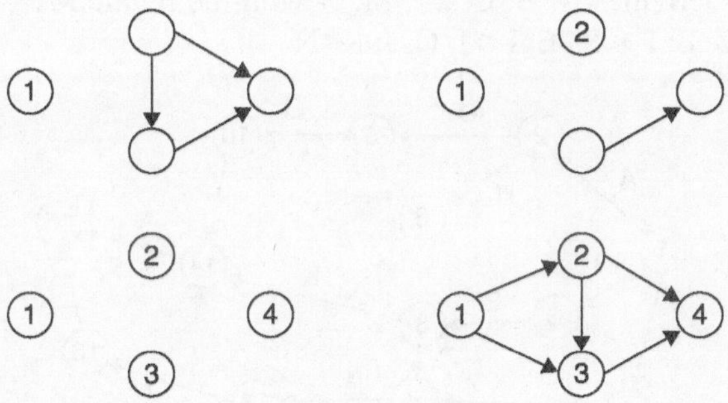

***Example:* 3**

A simple project consisting of five activities A, B, C, D and E. The restrictions are such that activity B cannot start until the completion of activity A. and activity E depends upon the completion of activity D, while activity C is independent of all other activities.

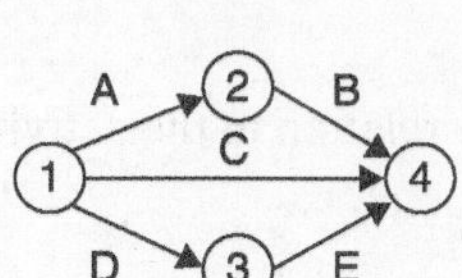

6.3.1 Dummy Activity

No two activities can have same starting and same ending points

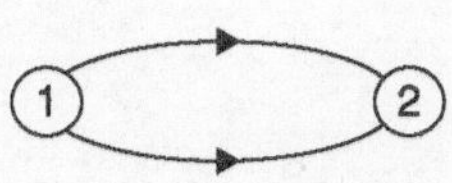

Not Allowed

The obvious deficiency of this diagram is that both A & B would be identified by 1-2. To overcome this, we introduce the dummy activity (represented in dotted line and the duration is "0")

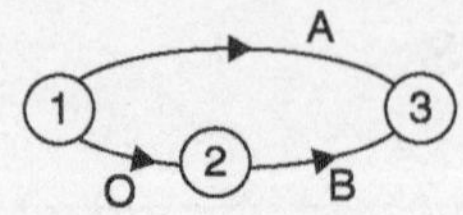

6.3.2 Precedence Relationships

To construct the network of a project requires the observance of some sequencing requirements of the activities.

Notation used to represent the precedence relationship is given by A < B ; where A and B are two activities. B can be started only after the completion of A.

Construction of a Network

***Example: 4**

A project consists of the activities A, B, C. ... , M, N with the relation A < D, H ; B < E ; C < I, F; D < G, H; L < M, E; I < L, E; F < K, E; I < J, G, J; K<N

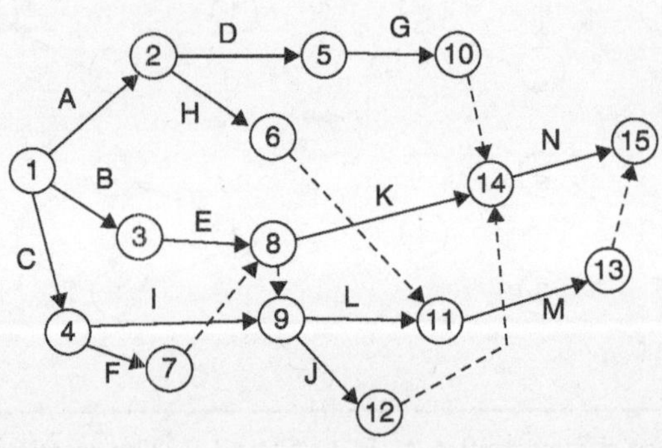

Remark 1: For example, E, I < L means that the activity L can be started after the completion of E and I. While representing it using network, we have to connect either E with I (or I with E) using dummy then L with I (or E with I).

Remark 2: Since, there is no precedence relation to the activities A, B and C, all three activities can be started simultaneously.

***Example: 5**

Draw a network for the following project:

(i) A is the start event and K is the end event.
(ii) J is the successor event to F.
(iii) C and D are the successor events to B.
(iv) D is the preceding event to G.

(v) E and F occur after event C.
(vi) E precedes F.
(vii) C restraints the occurrence of G and G which precede H.
(viii) H precedes J.
(ix) K succeeds event J.

Tabulate the given conditions:

Activity											
	Predecessor	A	B	C	D	E	F	G	H	J	K
	Successor	–	C, D	G, E, F	G	F	J	H	J	K	–

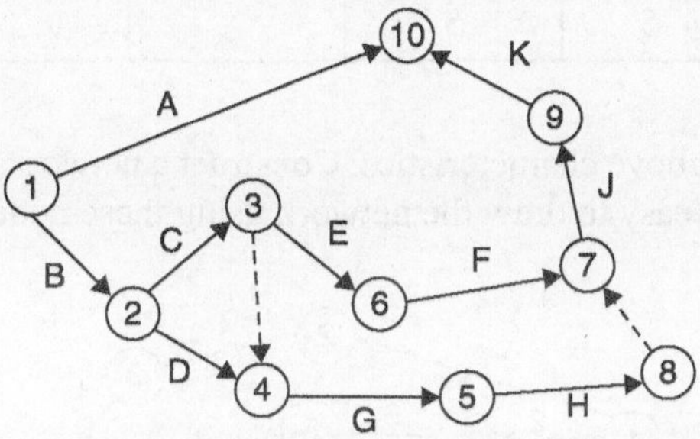

*Example: 6

Draw the network for the following:

(i) A, B and C are the first activities of the project, can start simultaneously.
(ii) A and B precedes D.
(iii) B precedes E, F and H.
(iv) F and C precedes G.
(v) E and H precedes I and J.
(vi) C, D, F and J precedes K.
(vii) K precedes L.
(viii) I, G. L are the terminal activities of the project.

Activity									
	Predecessor	A	B	A, B	F, C	E, H	C, D, F, J	K	I, G, L
	Successor	–	E, F, H	D	G	I, J	K	L	–

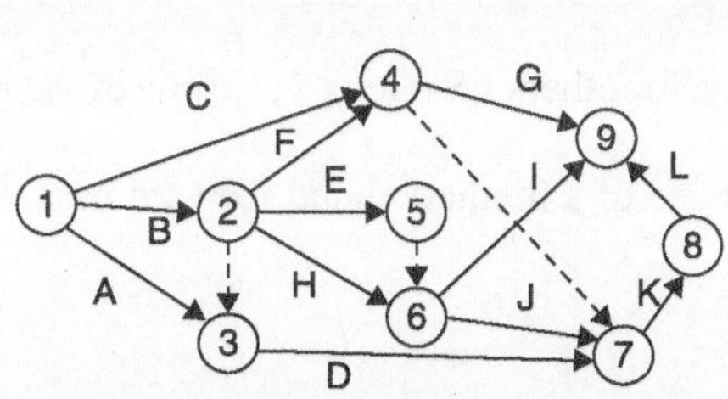

***Example:* 7**

Activity	Time	Activity	Time
1–2	2	4–8	8
1–4	2	5–6	4
1–7	1	6–9	3
2–3	4	7–8	3
3–6	1	8–9	5
4–5	5		

A project schedule has the above characteristics. Construct a network. Here, the activity contains the node numbers. Then it is very easy to draw the network using these nodes.

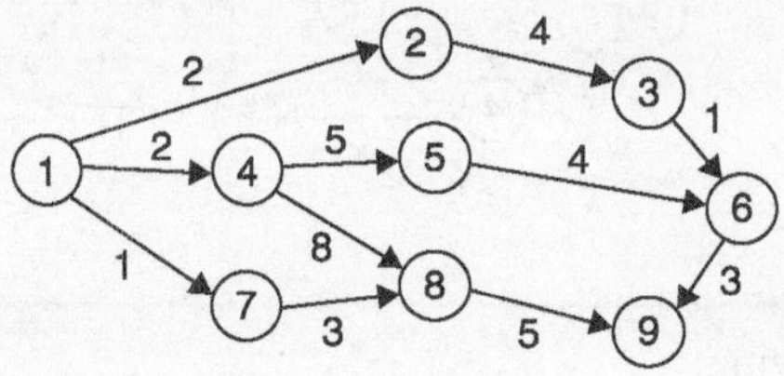

6.4 CRITICAL PATH METHOD (CPM)

CPM assumes constant time estimates. After the time estimates are produced for each activity the next step is to compute the (ES) Earliest Start and Earliest Completion (EC) time of each activity.

6.4.1 ES and EC Time of an Activity

ES time indicates the earliest possible time that a specific activity can begin, assuming that all of the preceding activities are also started at their ES timings. So, for any activity the EC is defined as EC = ES + activity time

- For the first activity, usually the earliest start time is "0"; then EC = 0 + activity time = activity time.
- Apart from the first activity, for others ES = max {EC time of the immediate predecessors (including the dummy)}
- To evaluate this we make use of a method called forward pass, i.e. move from the start node to the end node.

6.4.2 Latest Start (LS) and Latest Completion (LC) Time of an Activity

LS time for an activity is the time at which the activity should start in order not to delay the completion of the project beyond its LC time. So in general for any activity the LS is defined as LS = LC – Activity time.

- For the final activity, LC = EC
- For other activities,
  LC = Min{LS time of immediate successors (including the dummies)}
- To evaluate this we make use of a method called backward pass, i.e., move from the end node to the start node assuming the arrows are in the opposite direction.

6.4.3 Total Slack (TS)

We must be in a position to answer the question how long an activity can be delayed without disturbing the overall completion of the project The time that an activity can be delayed beyond its ES time without delaying the completion of the project is called TS (total slack or total float) time.

TS time for any activity can be evaluated using the relation, TS = LC – EC

6.4.4 Free Slack (FS)

This is the amount of time an activity can be delayed without disturbing the ES time of any other activity.

FS time for any activity can be evaluated using the relation,

$$\text{FS} = \begin{Bmatrix} \text{minimum ES time of all} \\ \text{its immediate} \\ \text{successors} \end{Bmatrix} - \begin{Bmatrix} \text{EC time of} \\ \text{the activity} \end{Bmatrix}$$

6.4.5 Independent Float (IF)

It refers the time by which the start of an activity can be delayed without affecting the earliest event time of the succeeding event. IF = FF – Tail event slack.

6.4.6 Critical Activity and Critical Path

Critical activity: If for an activity the total slack is zero, then it is called critical activity.

Critical Path: It is the longest time taking path in a network. Otherwise activities with zero total slack would constitute a path in a network called critical path. Normally in a network the critical path will be denoted by a thick line or double line. There is a chance for a given network to have more than one critical path with the same length.

The various steps to be followed to find the critical path for the given network are clearly depicted in the Flowchart 6.1.

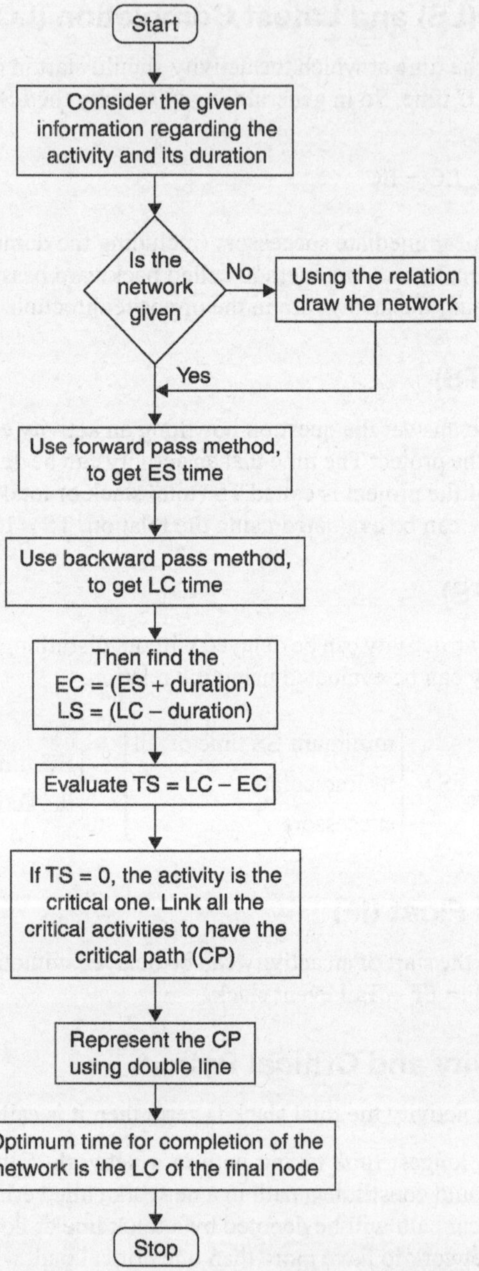

Flowchart 6.1 Critical path method

Example: 8

Find the critical path for the following scheduling problem.

Activity Code	Immediate Predecessor Activity	Duration
A	–	10
B	A	2
C	A	3
D	A	5
E	D	6
F	C	15
G	B	5
H	G	2
I	E, F, H	7
J	F, H	18

Step 1: Construct the network using the precedence conditions given

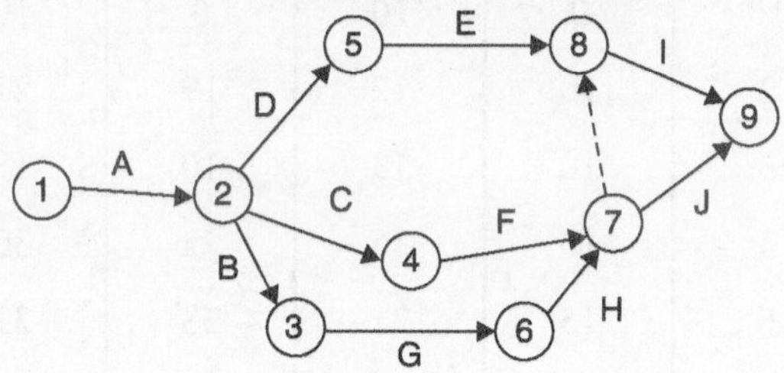

Step 2: Attach the duration of the activities and find the ES and LC timings using the forward and backward pass methods.

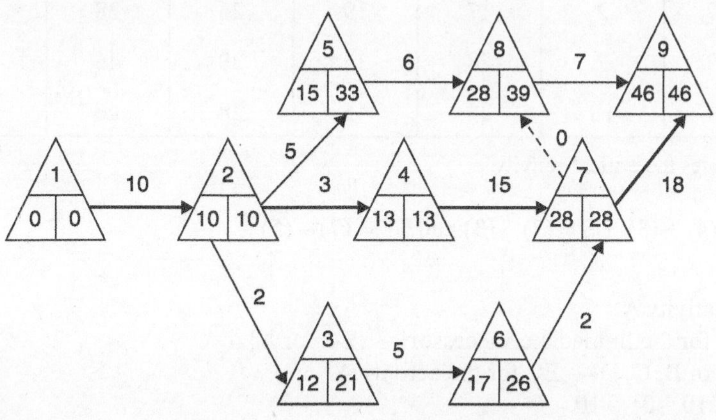

1. ES time for the initial activity is always "0".
2. Find all other ES timings using the formula and represent it in the network.

Remark 1: The ES for the (i, j)th activity is represented in the lower left half of the jth node.

Clearly the earliest completion time for the final activity (J) is 46 days. Thus, the project required 46 days to complete the project.

This implies that the latest completion time for the final activity (J) is 46 days. Also evaluate, the LC time for all activities, using the backward pass method.

Remark 2: The LC for the (i, j)th activity is represented in the lower right half of the jth node.

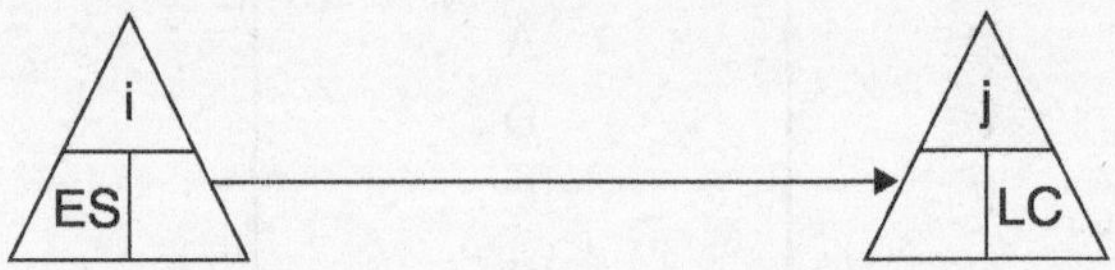

After evaluating the ES and LC timings, tabulate in the following table.

Activity	Node	Duration	Earliest		Latest		T F	F F
			S	C	S	C		
(1)	(2)	(3)	(4)	(5)	(6)	(7)	(8)	(9)
*A	1–2	10	0	10	0	10	0	0
B	2–3	2	10	12	19	21	9	0
*C	2–4	3	10	13	10	13	0	0
D	2–5	5	10	15	22	33	18	0
E	5–8	6	15	21	33	39	18	7
*F	4–7	15	13	28	13	28	0	0
G	3–6	5	12	17	21	26	9	0
H	6–7	2	17	19	26	28	7	9
I	8–9	7	28	35	39	46	11	0
*J	7–9	18	28	46	28	46	0	0

* In column-1 indicates the critical activity.

Remark 3: EC = (4) + (3); LS = (7) – (3) and TF = (7) – (5)

Remark 4:

(i) FF for the activity A
  = Min {ES for the immediate successor} – {EC for A}
  = Min {ES of B, C, D} – EC for the activity A
  = Min {10, 10, 10} – 10.
  = 10–10 = 0

(ii) FF for the activity H
= Min {J}–{EC of H}
= Min {28}–19 = 28–19 = 9

The optimal time required to complete the project is 46 days.
Then, the critical activities are (1–2), (2–4), (4–7) and (7–9). The critical path is 1–2–4–7–9
The thick line in the network indicates the critical path.

Example: 9

Activity	Time	Activity	Time
1–2	2	4–8	8
1–4	2	5–6	4
1–7	1	6–9	3
2–3	4	7–8	3
3–6	1	8–9	5
4–5	5	*	*

A project schedule has the above characteristics. Construct a PERT network. Find the critical path and duration of the project.

Step 1: Construct the network.

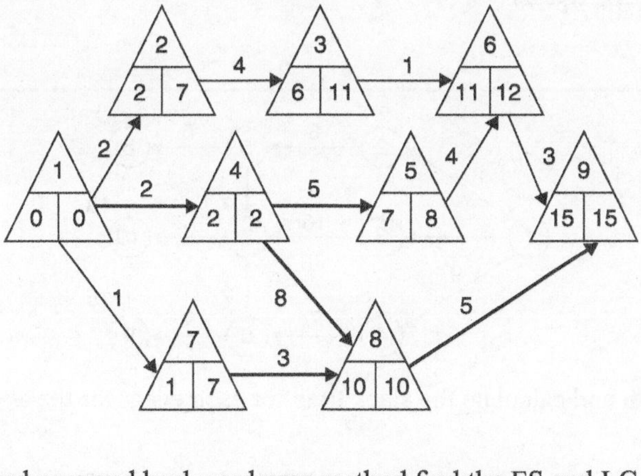

Step 2: Using the forward pass and backward pass method find the ES and LC.

Step 3: Tabulate the results and find the critical path using total float.

Node (1)	Duration (2)	Earliest S (3)	Earliest C (4)	Latest S (5)	Latest C (6)	Total Float (7)
1–2	2	0	2	5	7	5
*1–4	2	0	2	0	2	0
1–7	1	0	1	6	7	6
2–3	4	2	6	7	11	5
3–6	1	6	7	11	12	5
4–5	5	2	7	3	8	1
*4–8	8	2	10	2	10	0
5–6	4	7	11	8	12	1
7–8	3	1	4	7	10	6
*8–9	5	10	15	10	15	0

EC: (2) + (3)
LS: (6) – (2)
TF: (6) – (4)

Clearly the activity correspond to the "0" total float is called critical activity.
They are 1 – 4; 4 – 8; 8 – 9.
The optimal time required to complete the project is 15.
The critical path is 1→4→8→9

***Example:* 10**

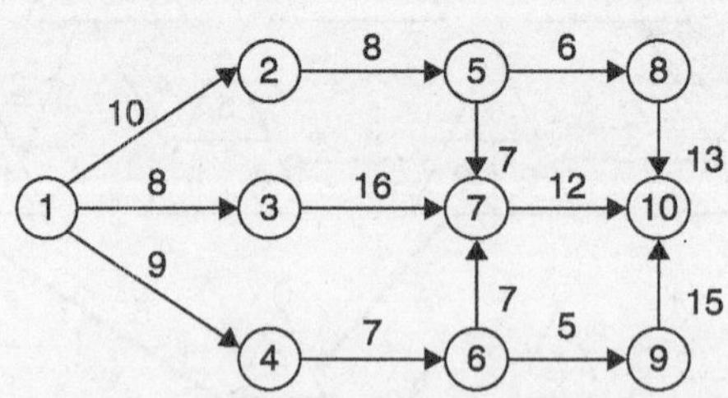

Find the critical path and calculate the slack time for each event for the above PERT diagram.

Step 1: Consider the network, and find the ES and LC time using the forward pass and backward pass methods.

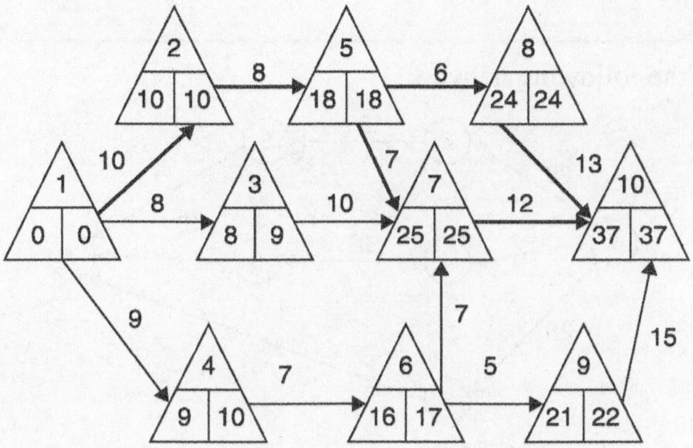

Step 2: Tabulate the results and find the critical path using total float.

Node (1)	Duration (2)	Earliest S (3)	Earliest C (4)	Latest S (5)	Latest C (6)	Total Float (7)
*1–2	10	0	10	0	10	0
1–3	8	0	8	1	9	1
1–4	9	0	9	2	10	1
*2–5	8	10	18	10	18	0
3–7	16	8	24	9	25	1
4–6	7	9	16	11	17	1
*5–7	7	18	25	18	25	0
*5–8	6	18	24	18	24	0
6–7	7	16	23	18	25	2
6–9	5	16	21	17	22	1
*7–10	12	25	37	25	37	0
*8–10	13	24	37	24	37	0
9–10	15	21	36	22	37	1

For the given problem we have two critical paths. First critical path is, 1→2→5→8→10, and the second critical path is 1→2→5→7→10. The total time required to complete the project is 37.

***Example:* 11**

Find the critical path of the following network

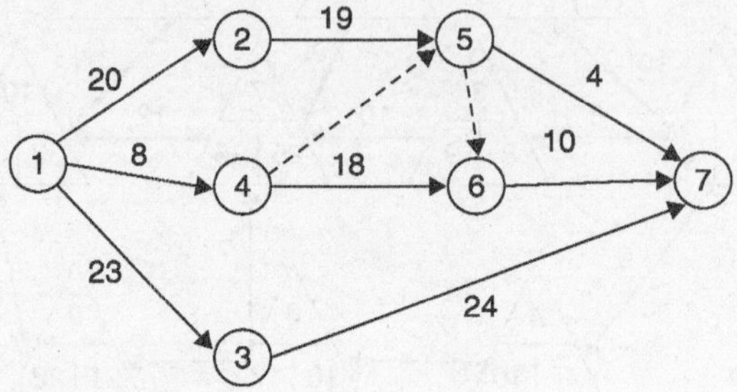

Step 1: Consider the network and find the ES and EC timings.

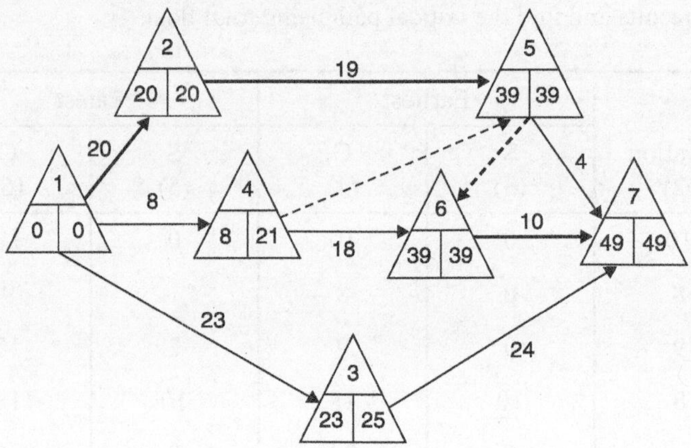

Step 2: Tabulate the results and find the critical path using total float.

Node (1)	Duration (2)	Earliest S (3)	Earliest C (4)	Latest S (5)	Latest C (6)	Total Float (7)
*1–2	20	0	20	0	20	0
1–3	23	0	23	2	25	2
1–4	8	0	8	13	21	13
*2–5	19	20	39	20	39	0
3–7	24	23	47	25	49	2
4–5	0	8	8	39	39	31
4–6	18	8	26	21	39	13
*5–6	0	39	39	39	39	0
5–7	4	39	43	45	49	6
*6–7	10	39	49	39	49	0

The critical path for the corresponding network is. 1 → 2 → 5→ 6 → 7.
The optimum time required to complete the Job is 49.

6.5 PROJECT EVALUATION REVIEW TECHNIQUE (PERT)

As stated earlier, this model assumes that the activity time is a random variable conforming to some probability distribution In accordance with each activity of a project having three different estimates:

t_o = Optimistic time, stands for the completion time for an activity, if every thing goes very well without any disturbance.

t_m = Most likely time, stands for the completion time for an activities, if we have some sort of disturbances.

t_p = Pessimistic time, stands for the completion time for an activity, if we have adverse situation leads to lot of delay.

These time estimates stated above are made on the basis of the level of available resources by experienced persons in the specific type of activities. Past records of similar situation can be very helpful for such estimations.

Once the estimates are known for an activity, they are converted into a single mean activity time, which is then used for calculating the critical path of the project.

$$\text{Clearly } t_o \leq t_m \leq t_p$$

The t_o, t_m and t_p are related using statistical concept to develop the expected time (t_e) for an activity.

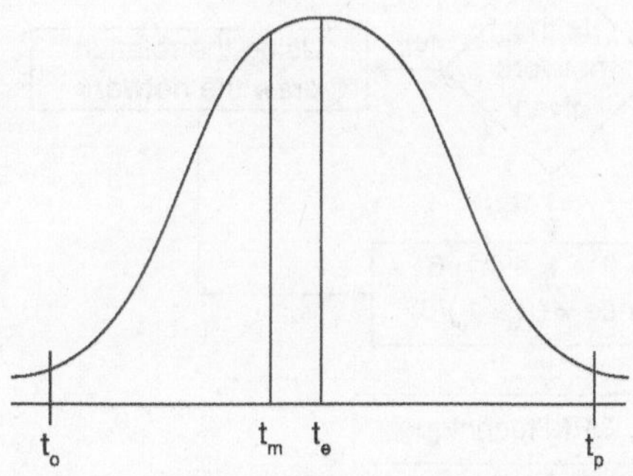

The basic assumption in the PERT model is that the three time estimates form the end points and the mode of β-distribution. The estimate of mean t_e of this distribution is derived by selecting the weighted arithmetic mean of mid point ($t_o + t_p$), that is ($t_o + t_p$)/2, and twice the modal value t_m, that is $2t_m$ such that

$$t_e = \frac{(t_o + t_p)/2 + 2t_m}{1+2} = \frac{t_o 4t_m + t_p}{6}$$

The estimate of the variance σ^2 of the distribution can be calculated by equating the range ($t_p - t_o$) to 6 standard deviations of the distribution This is justified by the fact that about 90% or more of any probability density function lies within 3 standard deviations of its mean. Hence,

$$t_p - t_o = 6\sigma$$

$$\sigma = \frac{t_p - t_o}{6}$$

$$\sigma^2 = ((t_p - t_o)/6^2)$$

These reduced estimates t_e and σ^2 approximately follow the normal distribution. Using the estimates t_e, we can find the expected time or mean completion time (μ) of the project and the critical path. Since the expected time t_e divides the area under the curve into two equal parts. This implies that probability of completion of the project is 0.5 (i.e., 50% chance for completion).

Also we can compute the probability of completion for a given time X. For which we make use of the concept of standard normal variate, $Z = (x - \mu)/\sigma$

Where $\sigma = \sqrt{\text{Sum of all variance of the critical activities}}$

Also, the project variance $= \sigma^2$

Since P (X) = P (Z), we can get the probability value using the standard normal table.

The PERT algorithm is clearly explained in the Flowchart 6.2.

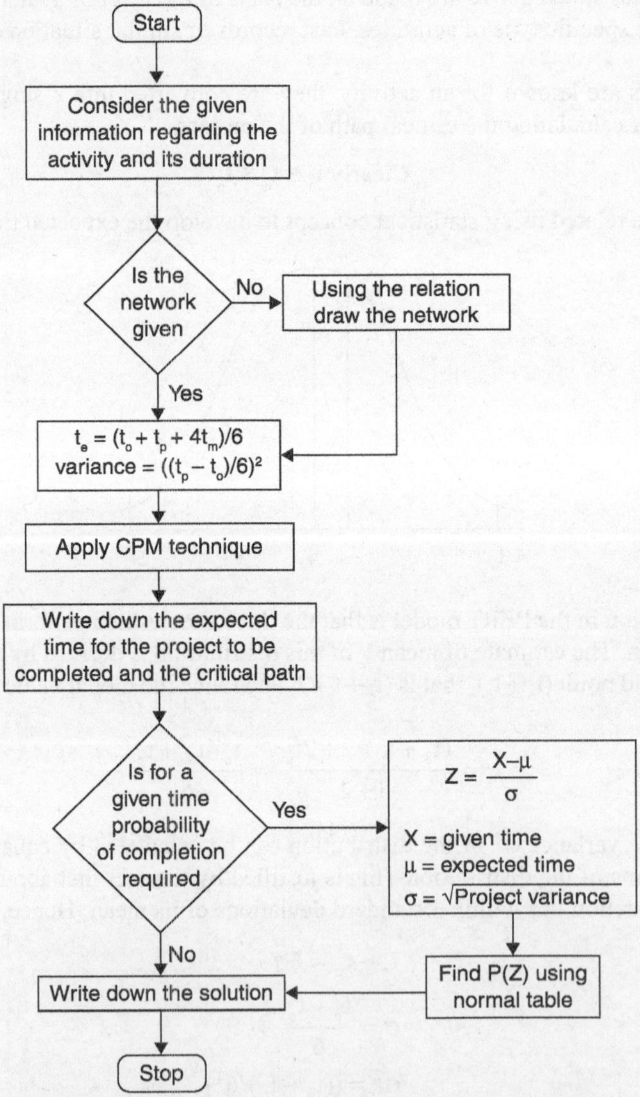

Flowchart 6.2 Project evaluation review technique

*Example: 12

For the following PERT network determine (i) Critical path (ii) probability of completing the project in 23 days.

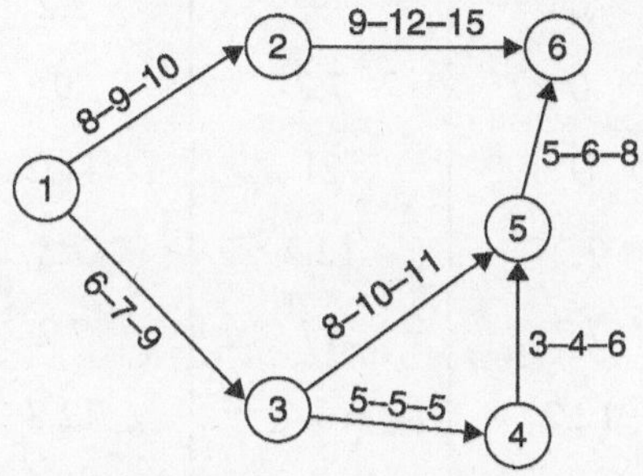

Step 1: Tabulate the activity timings and find the expected time and the activity variance.

Node	t_o	t_m	t_p	t_e	Variance
1–2	8	9	10	9	0.11
1–3	6	7	9	7.2	0 25
2–6	9	12	15	12	1
3–4	5	5	5	5	0
3–5	8	10	11	9.8	0.25
4–5	3	4	6	4.2	0.25
5–6	5	6	8	6.2	0.25

Step 2: Consider the Network and the t_e's to find the expected completion time using CPM.

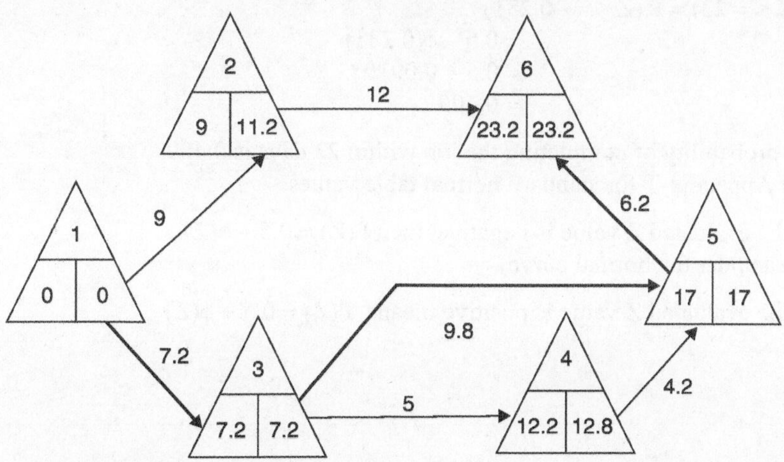

Node	t_e	Expected Early S	Expected Early C	Expected Latest S	Expected Latest C	TF
1–2	9	0	9	2.2	11.2	2.2
*1–3	7.2	0	7.2	0	7.2	0
2–6	12	9	21	11.2	23.2	2.2
3–4	5	7.2	12.2	7.8	12.8	0.6
*3–5	9.8	7.2	17	7.2	17	0
4–5	4.2	12.2	16.4	12.8	17	0.6
*5–6	6.2	17	23.2	17	23.2	0

The critical activities are 1 – 3, 3 – 5 and 5 – 6 and the critical path is 1 – 3 – 5 – 6. Expected time of completion = 23.2 days

Step 3: Project variance = σ^2 = sum of all the variances of critical activities.

$$\sigma^2 = 0.25 + 0.25 + 0.25 = 0.75$$
$$\sigma = \sqrt{0.75}$$
$$= 0.866$$

Step 4: To find the probability of completion of the jobs within 23 days (P(x <= 23)) can be evaluated using the relation

$$Z = (x - \mu)/\sigma$$
$$X = 23$$
$$\mu = 23.2$$
$$\sigma = 0.866$$
$$Z = (23 - 23.2)/0.866 = -0.231$$

i.e. $P(X <= 23) = P(Z <= -0.231)$

$$= 0.5 - \phi(0.231)$$
$$= 0.5 - 0.0910^*$$
$$= 0.409$$

i.e. the probability of completing the job within 23 days is 0.409.

* Refer Appendix-T for standard normal table values.

Remark 1: If the evaluated Z value is negative, then $P(Z) = 0.5 - \phi(Z)$ ($\phi(Z)$: is the area under the normal curve)

Remark 2: If the evaluated Z value is positive means. $P(Z) = 0.5 + \phi(Z)$

Example: 13

A project has the following characteristics:

Job i–j	Duration (days)		
	t_o	t_m	t_p
1–2	3	6	15
1–6	2	5	14
2–3	6	12	30
2–4	2	5	8
3–5	5	11	17
4–5	3	6	15
6–7	3	9	27
5–8	1	4	7
7–8	4	19	28

(i) Draw the network.
(ii) Calculate the length and variance.
(iii) Find the probability of completion within 42 days.

Step 1: Evaluate the expected time (t_e) and σ^2 for each activity.

Node	t_o	t_m	t_p	t_e	Variance
1–2	3	6	15	7	4
1–6	2	5	14	6	4
2–3	6	12	30	14	16
2–4	2	5	8	5	1
3–5	5	11	17	11	4
4–5	3	6	15	7	4
6–7	3	9	27	11	16
5–8	1	4	7	4	1
7–8	4	19	28	18	16

Step 2: Draw the network using the nodes relation with expected timings and find the critical path using CPM.

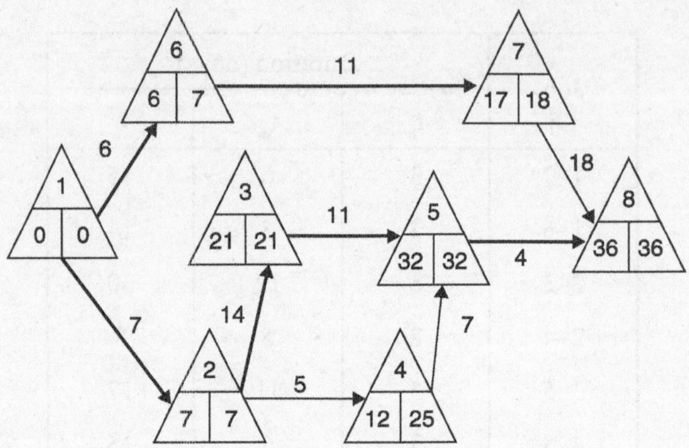

Step 3: Using the table as stated in example 1; we have the critical path as. 1→2→3→5→8. The expected completion time of the project is 36 days.

Step 4: The project variance = 4 + 16 + 4 + 1 = 25

$\sigma = \sqrt{25} = 5$

Given X = 42

$Z = (X - \mu)/\sigma = (42 - 36)/5$

$= 6/5 = 1.2$

i.e., $P(X \le 42) = P(Z \le 1.2)$

$= 0.5 + \phi(1, 2)$

$= 0.5 + 0\ 3849$

$= 0.8849$

The approximate probability of completing the job within 42 days is 0.8849.

*Example: 14

PERT calculations yield a project length T normally distributed with a mean of 60 days and SD of 4 days. Within how many days would you expect the project to be completed with probability 0.95? What is the probability of completing the project 4 days sooner than expected?

Given $\mu = 60$

$\sigma = 4$

To find the value of X

$P(X) = 0.95$

$= 0.5 + 0.45$

$= 0.5 + \phi(Z)$

$\Rightarrow \phi(Z) = 0.45$

From the normal table, Z = 1.65
We know that $Z = (X - \mu)/\sigma$

$$\Rightarrow \quad Z\sigma = X - \mu$$
$$X = \mu + Z\sigma$$
$$X = 60 + 4 \times 1.65$$
$$X = 66.6$$

i.e., The expected completion time of the project with the probability 0.95 is 66.6 days.

To find probability of completion if X = 66.6 – 4 = 62.6

$$Z = (62.6 - 60)/4$$
$$= 26/4$$
$$= 0.65$$
$$P(Z \le 0.65) = 0.5 + \phi(0.65)$$
$$= 0.5 + 0.2422 \quad \text{(using Normal table)}$$
$$= 0.7422$$

Hence, the probability of completing 4 days sooner than expected is 0.7422.

*Example: 15

A project has the following schedule. Draw the network to determine the critical path and the probability of completing the job in 30 days.

Activity	t_o	t_m	t_p
1–2	2	5	14
1–3	9	12	15
2–4	5	14	17
3–4	2	5	8
4–5	6	6	12
3–5	8	17	20

Step 1: Evaluate the expected time (t_e) and the variance.

Activity	t_o	t_m	t_p	t_e	σ^2
1–2	2	5	14	6	4
1–3	9	12	15	12	1
2–4	5	14	17	13	4
3–4	2	5	8	5	1
4–5	6	6	12	7	1
3–5	8	17	20	16	4

Step 2: Draw the network using the nodes relation with expected timings and find the critical path.

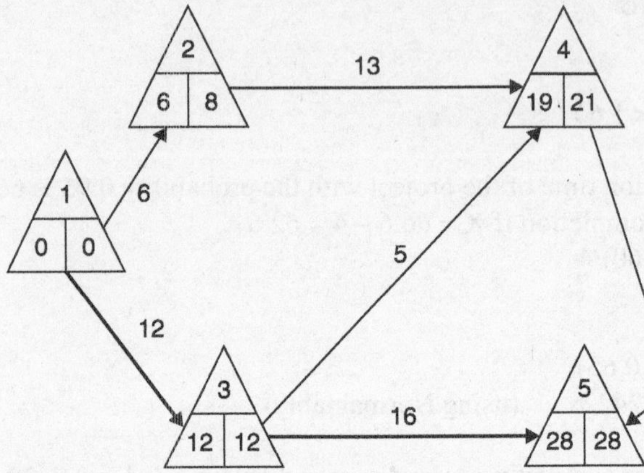

Step 3: Using the table as stated in example 1, we have the critical path as, 1→3→5

The expected completion time of the project is 28 days.

Step 4: The project variance = 1 + 4 = 5

$$\sigma = \sqrt{5} = 2.236$$

Given $X = 30$

$$Z = (X - \mu)/\sigma = (30 - 28)/2.236 = 0.895$$

$P(X \leq 30) = P(Z \leq 0.895) = 0.5 + \phi(0.895)$

$= 0.5 + 0.3159$ (Using Normal table)

$= 0.8159$

The probability of completing the job within 30 days is 0.8159.

6.6 PERT-COST

***Example:* 16**

A project comprises of the activities A, B, C, D, E, F, G and H. The cost-time data for these jobs are given below:

Job	Immediate Predecessor	Normal		Crash	
		Time (days)	Cost (₹)	Time (days)	Cost (₹)
A	–	3	140	2	210
B	–	6	215	5	275
C	–	2	160	1	240
D	A	4	130	3	180
E	A	2	170	1	250
F	A	7	165	4	285
G	B, D	4	210	3	290
H	C, E	3	110	2	160

(a) Draw the network
(b) If all the jobs are done in their normal timings, what will be the minimum cost schedule?
(c) If the project is to be completed in 8 days, determine the minimum cost.

6.6.1 Crashing

The reduction of time from its normal time for completing the specific activity is called Crashing (Compression). For example in this problem, the normal time for A is 3 days and the Crash time is 2, i.e. we want to reduce (Crash) one day (3 – 2 = 1).

Remark: Whenever we want to Crash, we have to pay extra amount.
Additional amount = (Number of days crashed) * (Cost slope)

6.6.2 Project Cost

It is the over all cost required to complete the project.
Project Cost = Sum of all activities of the normal cost

Step 1: Network for the given problem.

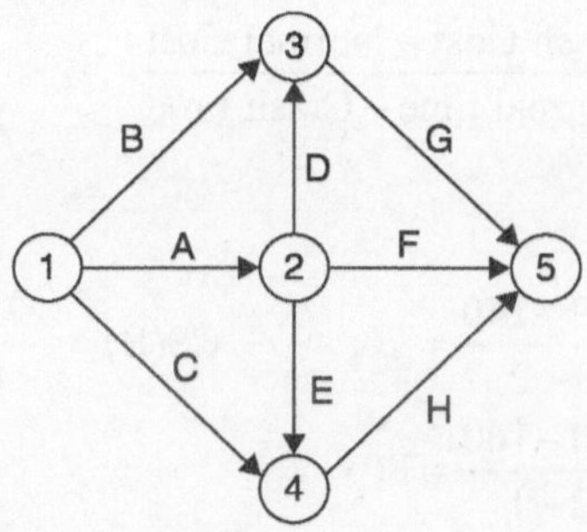

Step 2: Using the given normal duration in the network, find the normal optimum time.

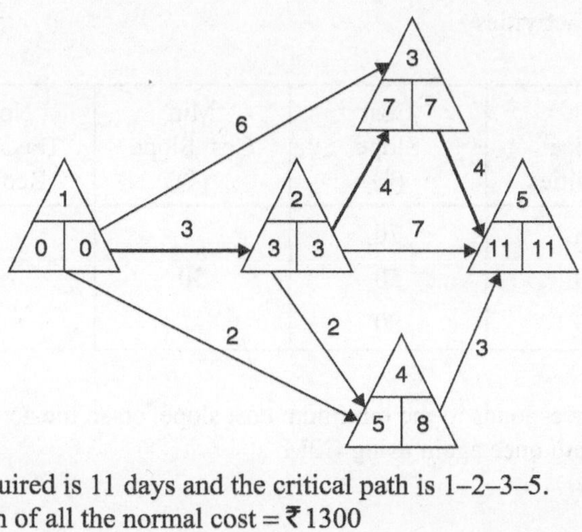

The optimum time required is 11 days and the critical path is 1–2–3–5.
The project cost = Sum of all the normal cost = ₹1300

Step 3: Before crashing the activities, we must find the cost slope for each activity.

Activity	Node	*Cost Slope (CS)
A	1–2	70
B	1–3	60
C	1–4	80
D	2–3	50
E	2–4	80
F	2–5	40
G	3–5	80
H	4–5	50

$$\text{*Cost Slope} = \frac{\text{Crash Cost} - \text{Normal Cost}}{\text{Normal time} - \text{Crash time}}$$

For example the cost slopes are

$$CS(A) = \frac{210-140}{3-2} = 70; \qquad CS(B) = \frac{275-215}{6-5} = 60$$

$$CS(C) = \frac{240-160}{2-1} = 80$$

Similarly find for all the activities.

Step 4: Start crashing the activities

Path	Critical Activities	Cost Slope (₹)	Min Cost Slope (₹)	No. of Days to be Reduced	Cost of Reduction (₹)
A-D-G	A	70			
	D	50	50	1	50
	G	80			

Since the activity D corresponds to the minimum cost slope, crash the activity D. After crashing the activity, find the critical path once again using CPM.

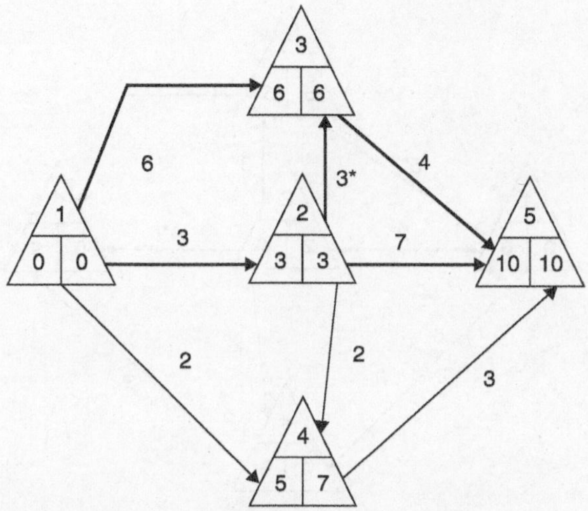

* The activity is crashed already, further crashing is not possible.
In this case we have three critical paths, they are

(i) 1–3–5 (ii) 1–2–5 (iii) 1–2–3–5

If we have more than one critical path, crash one activity in all the paths simultaneously.

Path	Critical Activities	Cost Slope (₹)	Min Cost Slope (₹)	No. of Days to be Reduced	Cost of Reduction (₹)
1–3–5	B	60	60	1	60
	G	80			
1–2–5	A	70			
	F	40	40	3	120
1–2–3–5	A	70	70	1	70
	D	–			
	G	80			

After crashing the activities B, F & A, find the critical path using CPM.

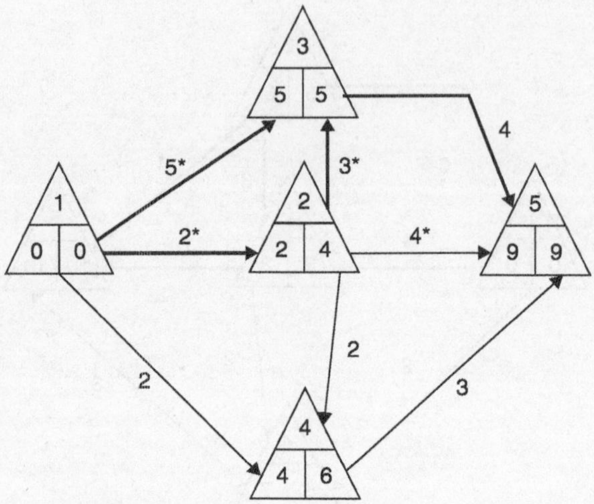

Path	Critical Activities	Cost Slope (₹)	Min Cost Slope (₹)	No. of Days to be Reduced	Cost of Reduction (₹)
1–2–3–5	A	–			
	D	–			
	G	80	80	1	80
1–3–5	B	–			
	G	80	–	–	–

After crashing the activity G, find the critical path using CPM.

Note: Since the activity G is taking place in both the paths, it can be reduced once only.

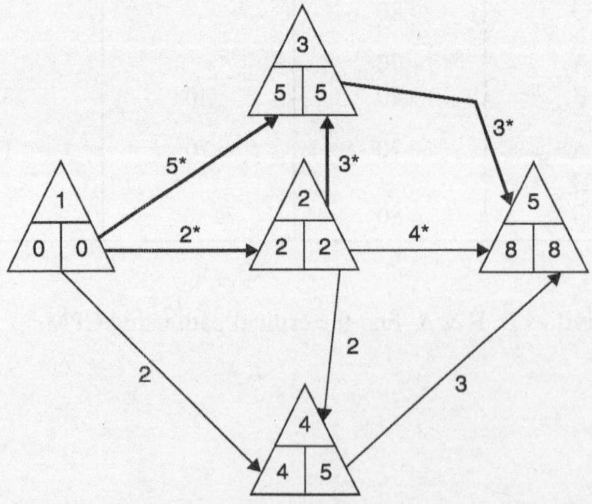

The latest critical path is 1–2–3–5 and 1–3–5. Since all the activities in both the paths are already crashed, we can not crash it further more. Hence, the optimum number of days required to complete the project is 8 days.

$$\begin{aligned}\text{Project cost} &= \text{Normal cost} + \text{Crash cost}\\ &= 1300 + 50 + 60 + 70 + 120 + 80\\ &= ₹1680\end{aligned}$$

Remark: When the independent cost is given, then

Total independent cost = Sum of all [(Length after crashing) * (Independent cost/day)]

Total normal cost = Sum of all normal cost

Total cost = (Total normal cost) + (Total crash cost) + (Total independent cost)

6.7 RESOURCE LEVELLING

These problems arise when there are sufficient resources to schedule all concurrent activities competing for the same resource types, but the pattern of resource utilization by the activities has undesirable features, such as frequent and significant changes in the amount of particular resource category. The objective of the leveling process is to smooth, as much as possible, the pattern of resource over time, without increasing the project duration beyond the critical path duration This is accomplished by careful rescheduling of activities within their avoidable slack time to give the most acceptable pattern.

***Example:* 17**

The following diagram gives the project's early start schedule The manpower requirement for each job is represented within brackets Use the resource leveling technique to reduce the peak resource requirement.

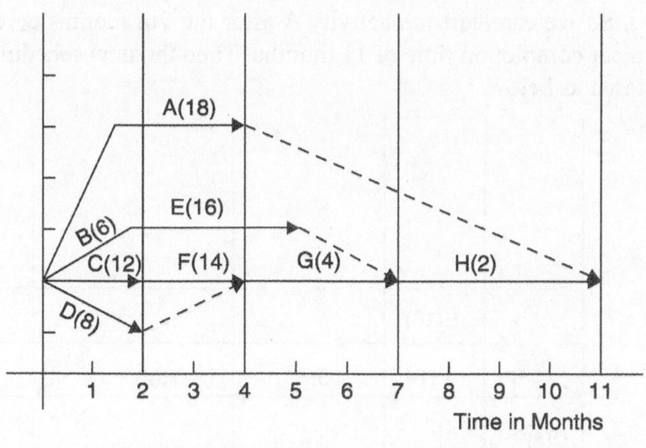

Draw the manpower loading chart for the given early start schedule of the projects.

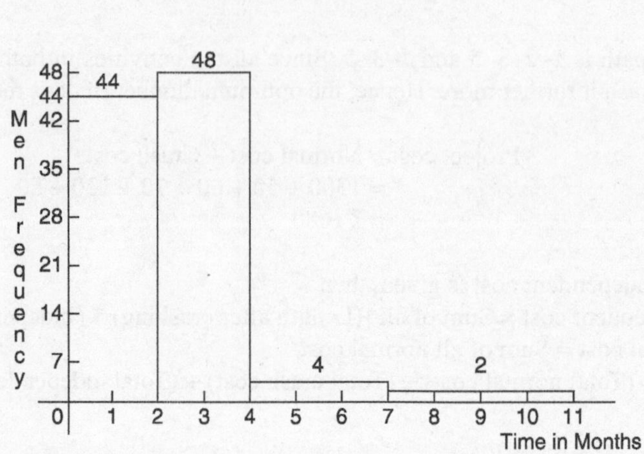

The man power loading chart shows that the peak manpower requirement is 48 men during the 3rd & 4th months when the activities A, E & F are active.

Using the given project early start schedule we can tabulate the activities along with their positive (>0) slack time.

Activity	Slack Time (months)
A	7
B	2
D	2
E	2

We can reschedule the activity A, B, D & E by considering its slack time without affecting the project completion time of 11 months in such a manner leveling the resources. We can not reschedule the activities C, F, G & H, because they are critical activities.

Since the activity A is having the highest slack, we can delay it. That is we can start the activity A after the peak period. So we can start the activity A after the 7th months certainly this delay is not going to disturb the project completion time of 11 months. Then the new schedule chart and manpower loading chart can be drawn as below.

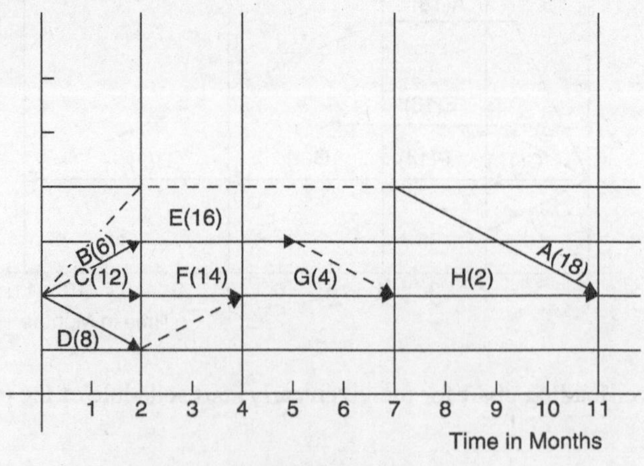

Schedule chart

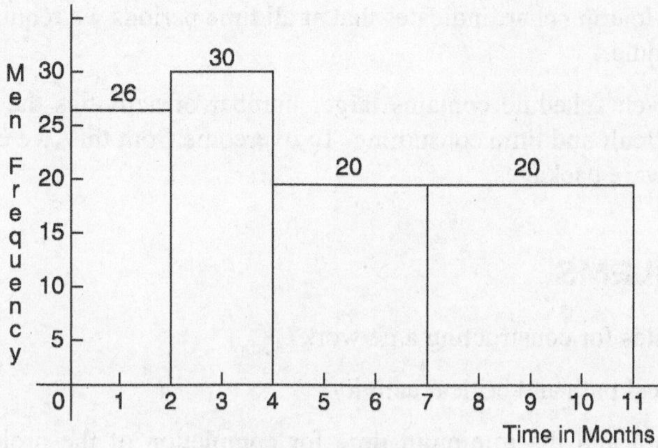

Manpower loading chart

For the next smoothening, we can delay the start of B by 2 months. Then the new schedule chart and manpower loading chart can be drawn as below.

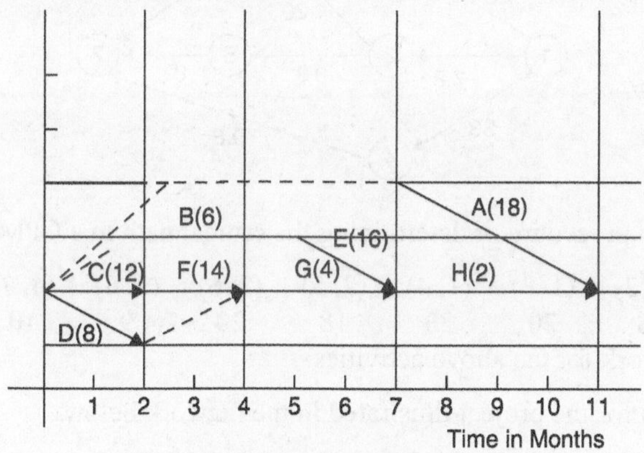

Schedule chart

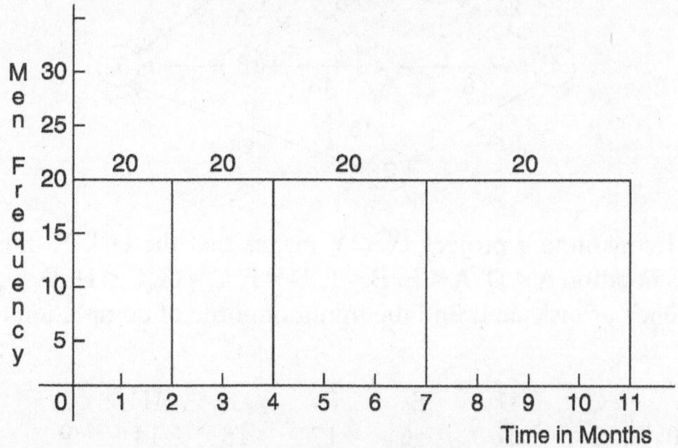

Manpower loading chart

The latest manpower loading chart indicates that at all time periods we require evenly 20 men. This is the best possible schedule.

Note: Whenever the given schedule contains larger number of activities the application of manual smoothening is very difficult and time consuming. To overcome from this, we can seek the help of the available computer software packages.

EXERCISE PROBLEMS

*1. (i) What are the rules for constructing a network?

(ii) Define the critical path and critical activity

*2. Find the critical path and the minimum time for completion of the project represented by the following network diagram.

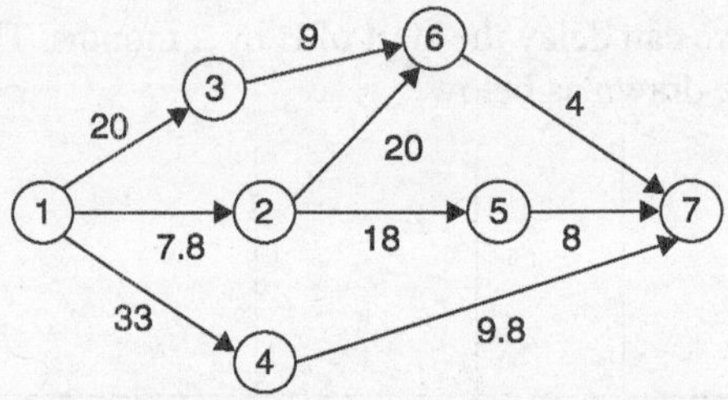

*3. Give the step by step procedure of determining the critical path in a CPM network.

*4.

Job (i, j)	(1, 2)	(1, 3)	(1, 4)	(2, 5)	(2, 6)	(3, 6)	(4, 7)	(5, 7)	(6, 7)
Duration Days	8	20	33	18	20	9	10	8	4

Draw the CPM network for the above activities.

*5. Find the critical path for the project illustrated in the network below.

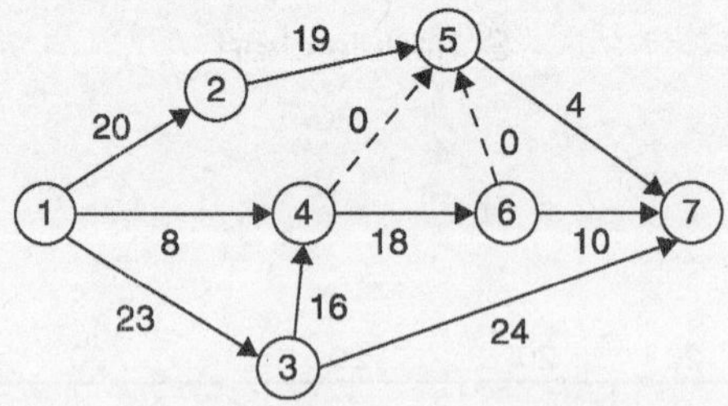

*6. Tasks A, B ... H, I constitute a project, X < Y means that the task X must be completed before Y begins. With this notation A < D, A < E, B < F, D < F, C < G, C < H, F < l, G < l. Draw a graph to represent the sequence of tasks and find the minimum time of completion of the project, when the time (in days).

Task	A	B	C	D	E	F	G	H	I
Time	8	10	8	10	16	17	18	14	9

7.

Activity	A	B	C	D	E	F	G	H	I	J
Immediate Predecessor	–	–	B	A, B	A, B	B	E, D, F	D, E	C	H, G

Draw the network for the above project, (use dummies wherever necessary).

*8.

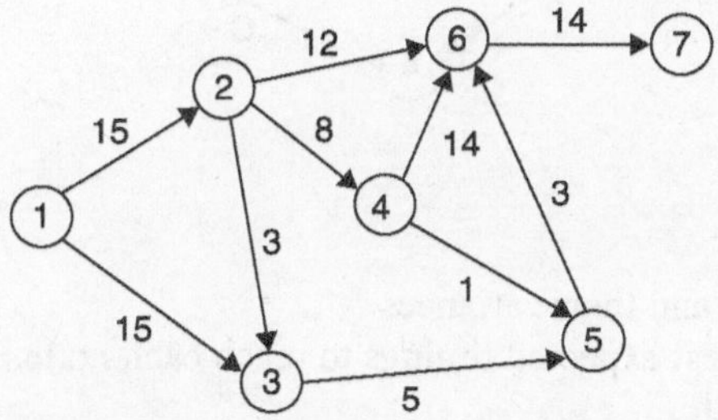

For the CPM network with the above activity, find the critical path and length of the critical path.

*9.

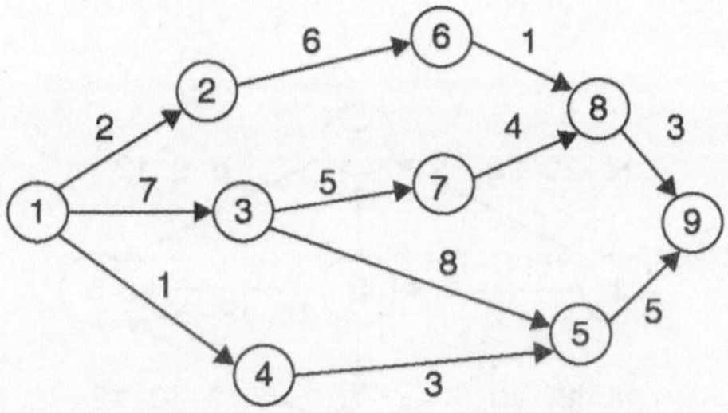

Find the critical path and calculate the slack time for each event of the above PERT diagram.

*10. The following table gives the activities of a construction project and duration.

Activity	1–2	1–3	2–3	2–4	3–4	4–5
Duration	20	25	10	12	5	10

(i) Draw the network for the project
(ii) Find the critical path

*11. A project is represented by the network given below and has the following data

Task	A	B	C	D	E	F	G	H	I
Latest time (weeks)	5	18	26	16	15	6	7	7	3
Greatest time (weeks)	10	22	40	20	25	12	12	9	5
Most likely time (weeks)	8	20	33	18	20	9	10	8	4

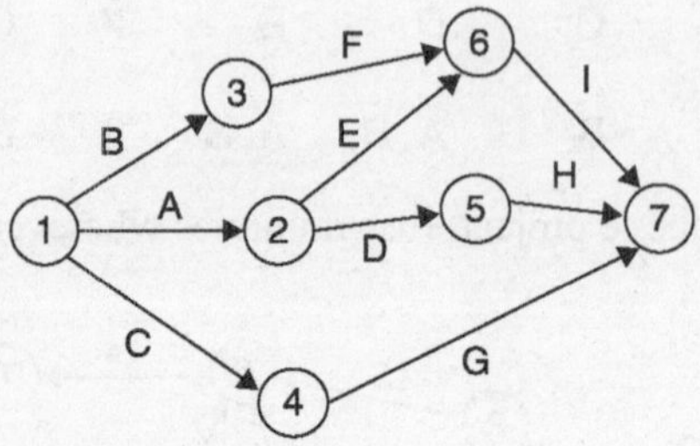

Determine the following:

(i) Expected task time and their variances
(ii) The earliest and latest expected timings to reach each node.
(iii) Find the critical path.
(iv) Find the probability of completing with in 41.5 weeks.

*12. The least, most likely and the greatest time of each task is shown in the following network diagram.

(i) Calculate the expected task times and their variances.
(ii) The critical path.

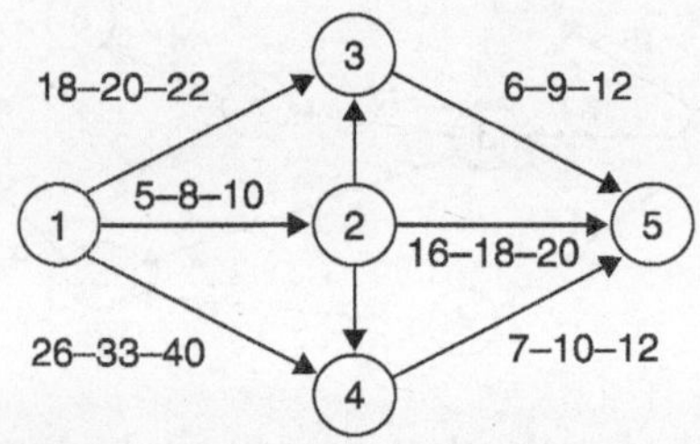

13.

Activity / Time (Weeks)	1–2	1–3	1–4	2–5	3–5	4–6	5–6
t_o	1	1	2	1	2	2	3
t_m	1	4	2	1	5	5	6
t_p	7	7	8	1	14	8	15

(i) Draw a network to the above data
(ii) Find also the expected duration, variance of each activities
(iii) Find the critical path.
(iv) Find the probability of completing the project 4 weeks earlier than the expected time.

*14.

Activity	(1, 2)	(1, 6)	(2, 3)	(2, 4)	(3, 5)	(4, 5)	(6, 7)	(5, 8)	(7, 8)
Optimistic Time	3	2	6	2	5	3	3	1	4
Most Likely Time	6	5	12	5	11	6	9	4	19
Pessimistic Time	15	14	30	8	17	15	27	7	28

Draw the network for the above activities whose three time estimates in days are given against them. Calculate the length and variance of the critical path.

*15.

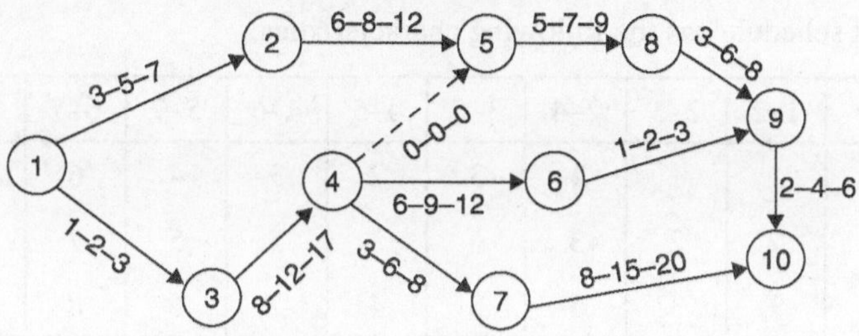

For the above PERT network, determine

(i) Critical path

(ii) Probability of finishing the project in 34.67 weeks.

*16. **M/s Real Value applications Ltd. provides you the following scenario with respect to the development of a consumer product during the Market Test Phase.**

Activity	Activity Code	Preceding Activities	Time Estimate (Weeks)
Design promotion campaign	A	–	3
Initial pricing analysis	B	–	1
Product design	C	–	5
Promotional cost analysis	D	A	1
Manufacture prototype models	E	C	6
Product cost analysis	F	E	1
Financial pricing analysis	G	B, D, F	2
Market test	H	G	8

Draw the network to this project and also calculate the slacks and interpret their meaning.

*17. Determine the earliest start time, latest completion time and critical path for the network given below:

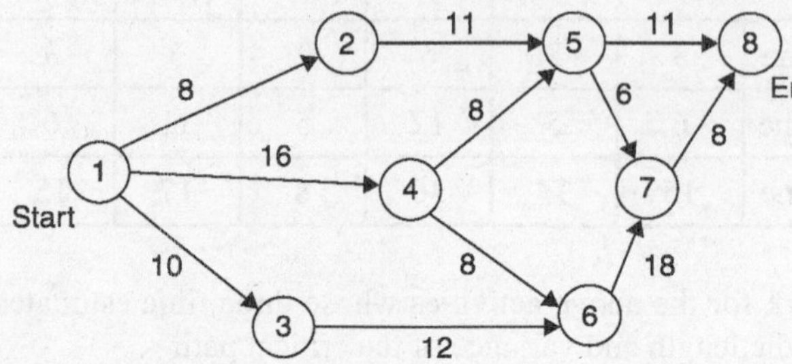

*18. A project schedule has the following characteristics:

Activity	1–2	2–3	2–4	3–5	4–5	4–6	5–7	6–7	7–8	7–9	8–10	9–10
t_o	1	1	1	3	2	3	4	6	2	4	1	3
t_m	2	2	3	4	3	5	5	7	4	6	2	5
t_p	3	3	5	5	4	7	6	8	6	8	3	7

Construct a PERT network and find the critical path. Also determine the probability that the project is completed in 30 weeks, given that the time in the above table is in number of weeks.

19. For the data given below:

Activity	Normal Duration (days)	Cost (₹)	Crash Duration (days)	Cost (₹)
1–2	5	170	4	240
1–3	9	310	6	550
2–3	6	80	4	200
2–4	10	130	8	230
3–4	6	110	4	290

Indirect cost for the project is ₹120 per day.

(i) Draw the network of the project.

(ii) Find the normal duration and cost of the project.

(iii) Find the cost of completing within 15 days. [Hint: Use the crashing idea]

*20.

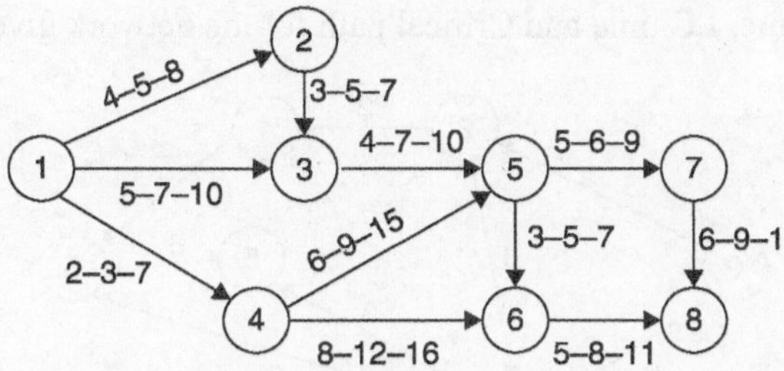

In the network given above, the three time estimates for the activities are indicated. Calculate the variance and the expected time for each activity.

21. Find the critical path for the following network.

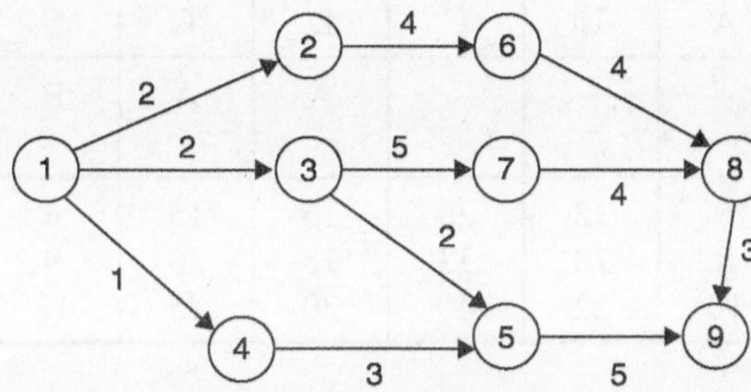

*22. A maintenance schedules has the following activities.

Activity	A	B	C	D	E	F	G	H	I	J	K	L
Predecessor Activity	–	–	A	B	B	C, D	E	E	G	H	I, J	F
Duration (Days)	5	4	3	5	4	6	3	4	6	5	8	6

K and L are the terminal activities.

(a) Construct the network.

(b) Find the critical path and the total duration for the completion of the project.

*23. Draw the network and find the following for the data given below:

(i) Critical path

(ii) Early start and late start for each activity.

Activity	A	B	C	D	E	F	G	H	I
Immediate Predecessors	–	–	–	A	B	C	D, E	B	H, F
Estimated Time (weeks)	3	5	4	2	3	7	8	7	9

*24. Determine the ES time, LC time and Critical path for the network given below:

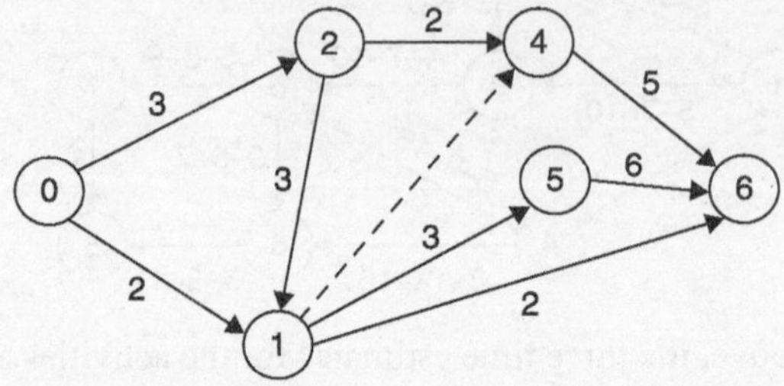

*25. A project is represented by the network shown below and has the following data.

Task	A	B	C	D	E	F	G	H	I
Immediately Proceeding Tasks	–	–	–	A	A	B	C	D	E, F
a (weeks)	5	18	26	16	15	6	7	7	3
m (weeks)	8	20	33	18	20	9	10	8	4
b (weeks)	10	22	40	20	25	12	12	9	5

(i) Draw the project network
(ii) Determine expected task time and their variance
(iii) Determine the critical path

26. The following table gives data on normal time and cost and crashed time and cost for a project

Activity	Time		Cost (₹)	
	Normal	Crash	Normal	Crash
1–2	9	4	1300	2400
1–3	15	13	1000	1380
2–3	7	4	7000	1540
2–4	7	3	1200	1920
2–5	12	6	1700	2240
3–6	12	11	600	700
4–5	6	2	1000	1600
5–6	9	6	900	1200

Find the optimum project time and corresponding minimum total project cost by crashing appropriate activities in proper order. The indirect cost per day is ₹400.

*27.

(i, j)	Normal Time	Crash Time	Normal Cost (₹1,000)	Crash Cost (₹1,000)
A(1, 2)	2	1	10	15
B(1, 3)	8	5	15	21
C(2, 4)	4	3	20	24
D(3, 4)	1	1	7	7
E(3, 5)	2	1	8	15
F(4, 6)	5	3	10	16
G(5, 6)	6	2	12	36

The above table provides cost and time estimates for seven activities of a project.

(i) Draw the project network corresponding to normal time.
(ii) Determine the critical path and the normal duration and cost of the project.
(iii) Crash the activities so that the project completion time reduces to 9 weeks.

ANSWERS TO THE EXERCISE PROBLEMS

1. Refer Sections 6.2; 6.4.6
2. Critical path: 1 – 4 – 7, Optimum time = 42.8
3. Refer Flowchart 6.1
4.

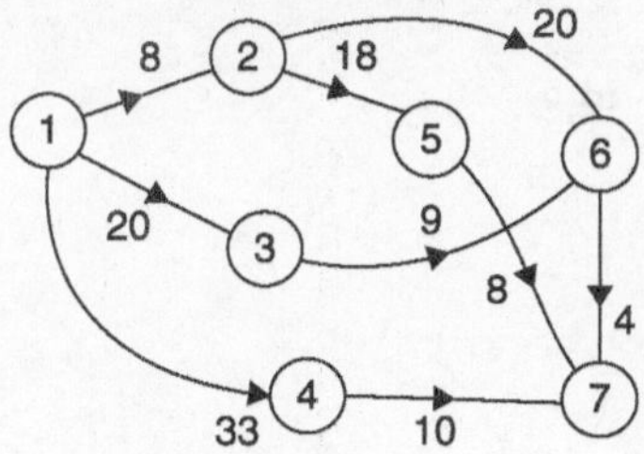

5. Critical path: 1 – 3 – 4 – 6 – 7, Optimum time = 67
6. Critical path: 1 – 2 – 3 – 7 – 8, Optimum time = 44 days
7.

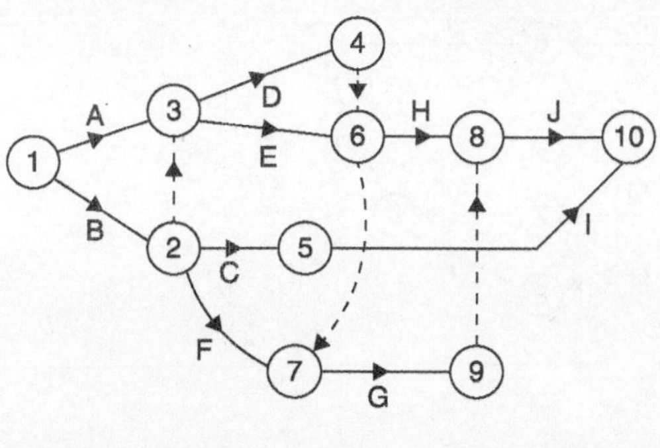

8. Critical path: 1 – 2 – 4 – 6 – 7, Optimum time = 51
9. Critical path: 1 – 3 – 5 – 9, Optimum time = 20
10. Critical path: 1 – 2 – 3 – 4 – 5, Optimum time = 45
11. Critical path: 1 – 4 – 7. Expected optimum time = 42.8 weeks,

    P (Completion in 41.5 weeks) = 0.3015
12. Critical path 1 – 4 – 5, Expected optimum time: 42.8
13. Critical path: 1 – 3 – 5 – 6, Expected optimum time = 17 weeks
14. Critical path: 1 – 2 – 3 – 5 – 8, Expected Optimum time = 36, Project variance = 21.5
15. Critical path: 1 – 3 – 4 – 7 – 9, P (completion in 34.67 weeks) = 0.5
16.

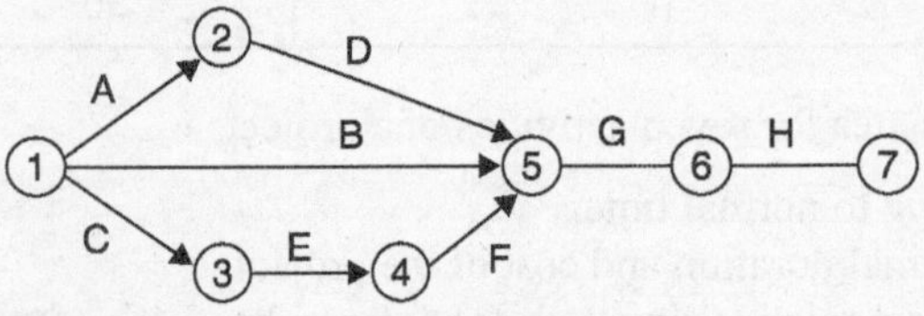

    Critical path = 1 – 3 – 4 – 5 – 6 – 7 Optimum time = 22
17. Critical path: 1 – 4 – 6 – 7 – 8 Optimum time = 50
18. Critical path = 1 – 2 – 4 – 6 – 7 – 9 – 10, Expected optimum time = 28 weeks,
    P(Completion in 30 weeks) = 0 9207
19. Normal duration = 17 days,
    Expected cost of completing the project within normal duration = 800 + 17* 120 = ₹2840
    Cost of completing within 15 days = ₹2720
20. Critical path = 1 – 2 – 3 – 5 – 7 – 8, Expected optimum time = 32.8
21. Critical path = 1 – 3 – 7 – 8 – 9, Optimum time = 14
22. Critical path:

    (i) 1 – 3 – 5 – 7 – 9 – 10
    (ii) 1 – 3 – 5 – 8 – 9 – 10

    Optimum time: 25

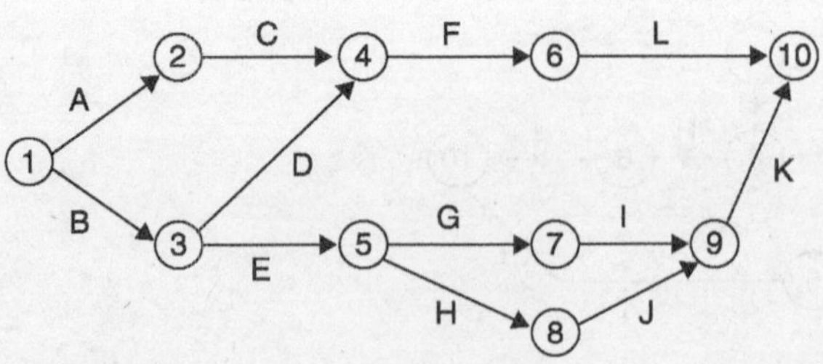

23. Critical path = 1 – 3 – 6 – 7, Optimum time = 21

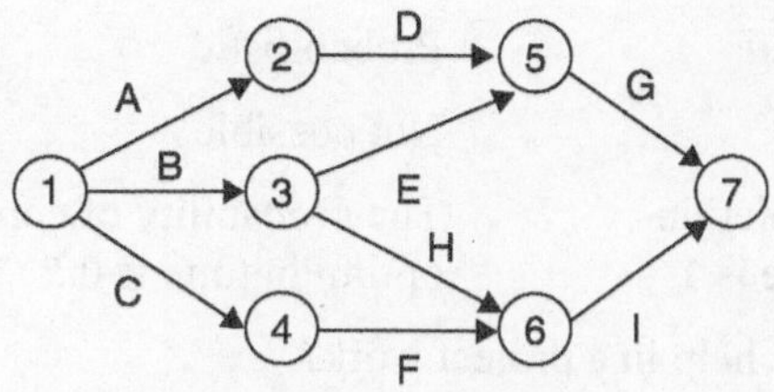

24. Critical path = 0 – 1 – 5 – 6, Optimum time = 11
25. Critical path = C – G, Expected optimum time = 42.8
26. Optimum project time = 27 days, Optimum cost ₹19750
27. Critical path = 1 – 3 – 5 – 6, Normal cost = 82*1000 = ₹82000

    Crash cost = ₹36000, Total cost = 82000 + 36000 = ₹118000

REVIEW QUESTIONS

Short Questions

1. Define the following terms:
   PERT, CPM, Activity, Activity duration. Node, Dummy activity and Critical path. (Section 6.1)
2. What do you mean by the precedence relation and how to express the idea that A precedes B.
   It refers to all the activities to be completed before starting the specific activity. We can represent the given case as: A < B.
3. State the rules and regulations to construct a given network. (Section 6.2)
4. How the nodes can be numbered? (Section 6.3)
5. Define t_0, ... t_p and t_m. (Section 6.5)
6. Why is CPM known as the deterministic type?
   All the activity durations are fixed values i.e., constant.
7. Define Crashing (Section 6.6.1)
8. Define Cost slope (Section 6.6.2)
9. What do you mean by resource levelling? (Section 6.7)
10. What is the difference between an activity and an event? (Sections 6.1.1 & 6.1.3)
11. Define independent float (Section 6.4.5)
12. Define Total float (Section 6.4.3)
13. Define Free float. (Section 6.4.4)

14. Differentiate between PERT and CPM

CPM	PERT
1. Activity timings are fixed	Probabilistic
2. Can be solved directly	Not possible
3. The probability of completion within the optimum time is 1.	The probability of completion within the expected optimum time is 0.5

15. How does network analysis help in a project work?

    (i) Optimum time for completing the project can be got.

    (ii) We can identify the critical activities.

16. Define an activity variance and project variance
    Activity variance refers to the individual activity defined as
    Variance = $((t_p - t_0/6)^2$; for each activity
    Project variance = Sum of all the variances of all critical activities

17. What is the role of a dummy activity in a network?
    Used to avoid starting and ending up of two/more activities from the same nodes.

State True or False

18. Two activities can start and end at the same nodes. (F)
19. The time estimates t_o, … t_m and t_n related to a PERT model approximately follows a Beta distribution. (T)
20. The converted, expected time estimates approximately follow normal distribution. (T)
21. For a given problem, the network is unique. (F)
22. The total float and the free float are one and the same (F)
23. A network can have more than one critical path (T)

Fill in the Blanks

24. PERT model was developed by ______, ______, ______, and ______. {6 1)
25. CPM model was developed by ______ and ______. (6.1)
26. CPM and PERT were developed to provide a tool to enhance the effectiveness of the three phases of ______, ______ and ______ of a project management. (6.1)
27. PERT assumes that for each activity the duration is a ______. (random variable)
28. TS refers to ______. (Total slack)
29. Total float refers to ______. (the maximum time when an activity can be delayed)
30. This formula used to find

    t_e = ____ $((t_o + 4t_m + t_p)/6)$
31. The compression of the activity timings in a PERT-COST model is called ______. (crashing)
32. The relationship between the activity in a network model is named as ______. (precedence relation)

7

Sequencing

7.1 INTRODUCTION

Sequencing can be viewed as a special kind of problem commonly encountered in production shops where various types of products are to be processed over various combinations of machines.

For example, let there are m jobs, each of which has to be processed one at a time, on each of n different machines. Also the time required for different machines and different jobs are given. The main objective is, to find the optimal sequence of jobs to be done using different machines in such way that the over all time required for completion of all jobs (Total Elapsed Time) must be minimum. Also the idle time (machine waiting time) of the machines must be minimized.

We can solve the given situation by using the method of enumeration. The difficulty with the enumeration method is that it requires large number of calculations. Thus for the above stated model of the jobs with n-machines, we have to compute $(m!)^n$ calculations, [for $m = 10$ and $n = 2$, the number of different sequences will be $(10!)^2$ (ie) 1316818944000]

As an alternative we have a very simple method devised by S.M. Johnson. Generally named as Johnson's rule.

7.1.1 Assumptions

1. A machine can process only one job at a time.
2. Each job, once started on a machine must be completed.
3. Only one machine is available in each type.
4. The time gap between the movement from one machine to another is negligible.
5. The jobs can be taken only in the specified order.

7.2 JOHNSON'S METHOD (RULE)

The various steps to be followed while using Johnson's rule are clearly depicted in the following Flowchart 7.1.

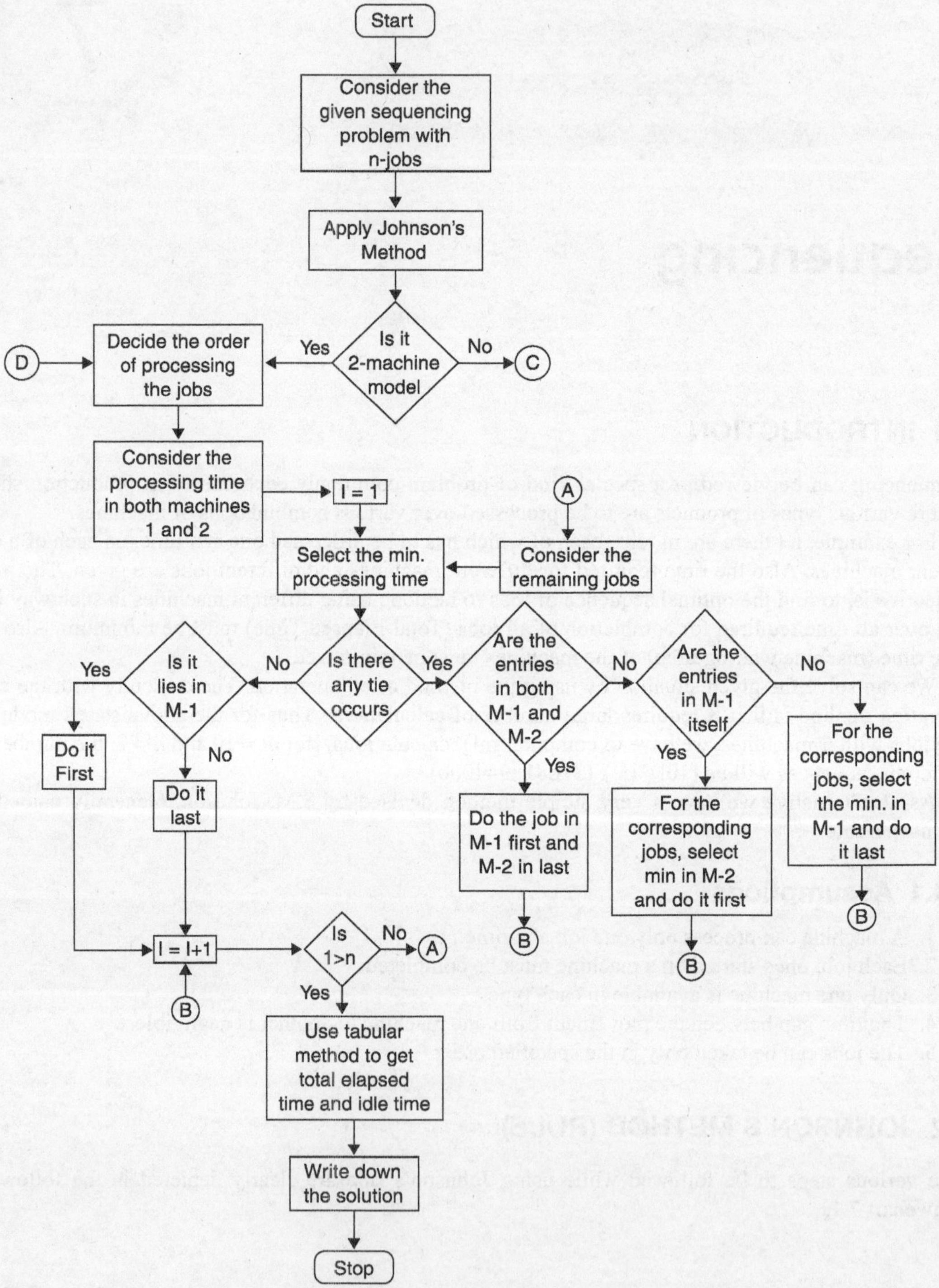

Flowchart 7.1 Johnson's method

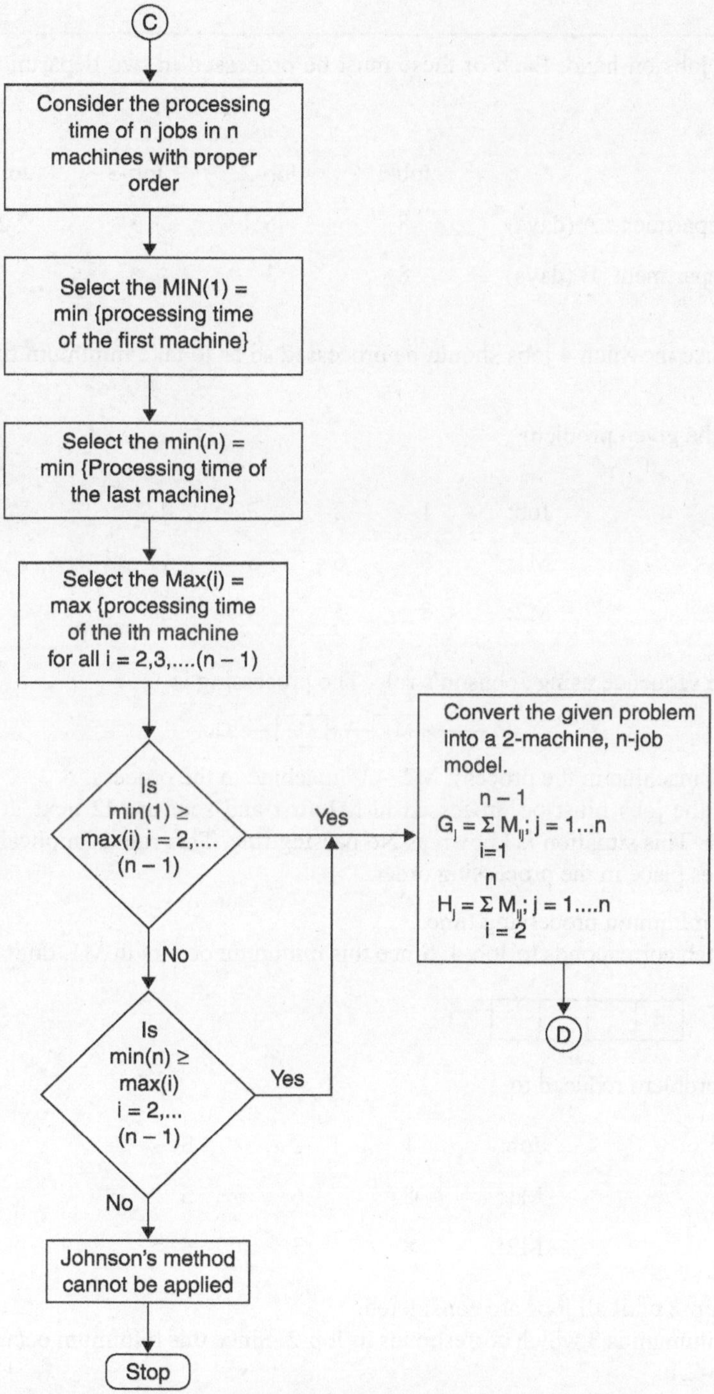

Flowchart 7.1 Continued

Example: 1

A company has 4 jobs on hand. Each of these must be processed in two departments, the sequential order is A→B.

	Job-1	Job-2	Job-3	Job-4
Department: A (days)	8	6	5	2
Department: B (days)	8	3	4	5

Find the sequence in which 4 jobs should be processed so as to take minimum time to finish all the 4 jobs.

Step 1: Consider the given problem

Job:	1	2	3	4
M1:	8	6	5	1
M2:	8	3	4	5

Find the sequence using Johnson's rule. The processing is

In → M1 → M2 → Out

M1 – 1st machine in the process; M2 – 2nd machine in the process.

Clearly the jobs must be processed in M1 first and sent to M2 next. It is not possible to bypass M1. This situation is known as No passing rule. This rule is applicable for any number of machines place in the processing order.

Step 2: Select the minimum processing time.

It is which corresponds to Job-4. Since this minimum occurs in M1, do it first. Delete Job-4.

Sequence:

4			

Then the problem reduced to

Job:	1	2	3
M1:	8	6	5
M2:	8	3	4

Step 3: Repeat Step-2 until all jobs are considered.

Next minimum is 3 which corresponds to Job-2. Since this minimum occurs in M2, do it last. Delete job-2.

Sequence:

4			2

Then the problem reduced to

Job:	1	3
M1:	8	5
M2:	8	4

Next minimum is 4 which corresponds to job-3. Since this minimum occurs in M2, do it last. Delete job-3.

4		3	2

The the problem reduced to

JOB	1
M1	8
M2	8

Since only one job is left out, we can fill it in the gap available in the sequence.

Hence the optimal sequence is

4	1	3	2

Step 4: Find the optimal elapsed time and the idle time for the machines, based on the sequence got in step 3.

Job Sequence	Processing Time		M1		M2		Idle Time in M2
	M1	M2	In Time	Out Time	In Time	Out Time	
4	1	5	0*	0 + 1 = 1	1	1 + 5 = 6	1
1	8	8	1	1 + 8 = 9	9	9 + 8 = 17	3
3	5	4	9	9 + 5 = 14	17	17 + 4 = 21	–
2	6	3	14	14 + 6 = 20	21	21 + 3 = 24	–

The optimal elapsed time is 24 hrs. [The time needed to complete all jobs]

Idle time for M1 = 1 + 3 = 4 hrs

Idle time for M2 = 24 – 20 = 4 hrs

* Start time must be '0'.

***Example: 2**

A ready-made garments manufacturer has to process 7 items through two stages of production, viz., cutting and sewing. The time taken for each of these items at the different stages is given below in appropriate units.

Item		1	2	3	4	5	6	7
Processing Time	Cutting	5	7	3	4	6	7	12
	Sewing	2	6	7	5	9	5	8

*These questions have appeared in Indian university examination papers.

Find an order in which these items are to be processed through these stages so as to minimize the total processing time (i.e make span or time for completion).

Step 1: Cutting-M1 – 1st machine in the sequence
Sewing-M2 – 2nd machine in the sequence.

Consider the given problem,

	Item	1	2	3	4	5	6	7
M1	Cutting	5	7	3	4	6	7	12
M2	Sewing	2	6	7	5	9	5	8

Find the sequence using Johnson's rule and represent it in the box.

Step 2: Select the minimum processing time. It is 2. Which corresponds to job-1. Since this minimum occurs in M2, do it Last. Delete job-1.

Step 3: Repeating the same process. Minimum from the rest is 3. Which corresponds to job-3. Since this minimum occurs in M1, do it first. Delete job-3. Minimum from the rest is 4 which corresponds to job-4. Since this minimum occurs in M1, do it first. Delete job-4. Minimum from the rest is 5 which corresponds to job-6. Since this minimum occurs in M2, do it last. Delete job-6. Minimum from the rest is 6 which corresponds to job-5 and job-2. Since this minimum occurs in M1 of job-5 and M2 of job-2, do the job first and last respectively.

Finally we are left with job-7. Hence the required optimal sequence is

Sequence:

3	4	5	7	2	6	1

Step 4: Find the optimal elapsed time and the idle time for machines, based on the sequence got in step-3.

Job Sequence	Processing Time		M1		M2		Idle time in M2
	M1	M2	In Time	Out Time	In Time	Out Time	
3	3	7	0	3	3	10	3
4	4	5	3	7	10	15	–
5	6	9	7	13	15	24	–
7	12	8	13	25	25	33	1
2	7	6	25	32	33	39	–
6	7	5	32	39	39	44	–
1	5	2	39	44	44	46	–

The optimal elapsed time is 46 hours.
Idle time for cutting = 46 – 44 = 2 hrs
Idle time for sewing = 3 + 1 = 4 hrs

*Example: 3

A company has six jobs which go through three machines, X, Y and Z in the order XYZ. The processing time in minutes for each job on each machine is as follows.

Job	Machine		
	X	Y	Z
1	18	7	19
2	12	12	12
3	29	11	23
4	36	2	47
5	43	6	28
6	37	12	36

What should be the sequence of the jobs such that the total make span is minimized?

Step 1: Since, the number of machines used in this model is more than two, it is not possible for us to apply Johnson's method directly. That is, the conversion is required.

Step 2: Select minimum in machine-1 (X), machine-3 (Z) and maximum in machine-2 (Y).

Min (X) = 12 Min (Z) = 12 Max (Y) = 12

Since, Min (x) ≥ Max (Y) for corresponding entries, we can convert the given three machine problem into two machine problem.

$$\text{Define}\quad \begin{array}{ll}\text{Machine-G}: & X+Y \\ \text{Machine-H}: & Y+Z\end{array}\Bigg\}\text{add the corresponding entries}$$

		1	2	3	4	5	6
Machine	G	25	24	40	38	49	49
	H	26	24	44	49	34	48

Step 3: The optimum sequence is,

2	1	4	3	6	5

Note: Since the minimum 24 occurs in both G & H corresponding to job-2 it can be taken either first or last. Here it is taken as the first.

Step 4: Find the optimal elapsed time and idle time for the machines, based on the optimum sequence got in step-3.

Job Sequence	Processing Time			X		Y		Z		Idle	
	X	Y	Z	In	Out	In	Out	In	Out	Y	Z
2	12	12	12	0	12	12	24	24	36	12	24
1	18	7	19	12	30	30	37	37	56	6	1
4	36	2	47	30	66	66	68	68	115	29	12
3	29	11	23	66	95	95	106	115	138	27	–
6	37	12	36	95	132	132	144	144	180	26	6
5	43	6	28	132	175	175	181	181	209	31	1

The optimal elapsed time is 209 hrs
Idle time for X = 34 hrs
Idle time for Y = 131 + (209 – 181) = 159 hrs
Idle time for Z = 44 hrs

Example: 4

Solve the following sequencing problem, giving an optimal solution when passing is not allowed:

Machine \ Job	A	B	C	D	E
M1	10	12	8	15	16
M2	3	2	4	1	5
M3	5	6	4	7	3
M4	14	7	12	8	10

The processing method is

In → M1 → M2 → M3 → M4 → Out

Step 1: Before applying Johnson's method, we have to verify the condition.

Min (1) ≥ Max(j); j = 2, 3
(or)
Min (4) ≥ Max(j); j = 2, 3

Here, Min (1) is 8; Max (2) = 5
Min (4) is 7; Max (3) = 7

Clearly min (1) ≥ max (2) and
min (1) ≥ max (3)
implies that, Johnson's method can be applied.

Step 2: Reduce the four machine problem into the two machine problem, using the evaluation.

G: M1 + M2 + M3 } add the corresponding entries.
H: M2 + M3 + M4 }

Machine \ Job	A	B	C	D	E
G	18	20	16	23	24
H	22	15	20	16	18

Step 3: The optimal sequence is

C	A	E	D	B

Step 4: Using the table, evaluate the total elapsed time and the idle time for different machines.

Job Sequence	Processing Time				M1		M2		M3		M4		Idle Time		
	M1	M2	M3	M4	In	Out	In	Out	In	Out	In	Out	M2	M3	M4
C	8	4	4	12	0	8	8	12	12	16	16	28	8	12	16
A	10	3	5	14	8	18	18	21	21	26	28	42	6	5	–
E	16	5	3	10	18	34	34	39	39	42	42	52	13	13	–
D	15	1	7	8	34	49	49	50	50	57	57	65	10	8	5
B	12	2	6	7	49	61	61	63	63	69	69	76	11	6	4

The optimal elapsed time is 76 hrs.
Idle time for M1 = 76 – 61 = 15 hrs
Idle time for M2 = 11 + 10 + 13 + 6 + 8 + (76–63) = 61 hrs
Idle time for M3 = 12 + 5 + 13 + 8 + 6 + (76–69) = 51 hrs
Idle time for M4 = 16 + 5 + 4 = 25 hrs

*Example: 5

Determine an optimum sequence to process the various types of fan blades each day from the following information so as to minimize the total elapsed time:

Types of Fan Blades	Number to be Processed/Day	Processing Time On	
		Machine-A (Minutes)	Machine-B (Minutes)
1	4	4	8
2	6	12	6
3	5	14	16
4	2	20	22
5	4	8	10
6	3	18	2

Also work out the total elapsed time for an optimum sequence. What is the total idle time on Machine-A and on Machine-B?

Step 1: In this case we are going to process more than one job of the same type. First consider the processing time and get the optimal sequence.

	1	2	3	4	5	6
M-A	4	12	14	20	8	18
M-B	8	6	16	22	10	2

1	5	3	4	2	6

Sequence	No. of Jobs	Job	Machine-A In	Machine-A Out	Machine-B In	Machine-B Out	Idle Time B
1	4	1	0	4	4	12	4
		1	4	8	12	20	–
		1	8	12	20	28	–
		1	12	16	28	36	–
5	4	5	16	24	36	46	–
		5	24	32	46	56	–
		5	32	40	56	66	–
		5	40	48	66	76	–
3	5	3	48	62	76	92	–
		3	62	76	92	108	–
		3	76	90	108	124	–
		3	90	104	124	140	–
		3	104	118	140	156	–
4	2	4	118	138	156	178	–
		4	138	158	178	200	–
2	6	2	158	170	200	206	–
		2	170	182	206	212	–

Sequence	No. of Jobs	Job	Machine-A		Machine-B		Idle Time B
			In	Out	In	Out	
		2	182	194	212	218	–
		2	194	206	218	224	–
		2	206	218	224	230	–
		2	218	230	230	236	–
6	3	6	230	248	248	250	12
		6	248	266	266	268	16
		6	266	284	284	286	16

The optimal idle time = 286 minutes
Idle time on A = 286 – 284 = 2 minutes
Idle time on B = 4 + 12 + 16 + 16 = 48 minutes

7.3 GRAPHICAL METHOD

If we want to process two jobs using m machines we can make use of the graphical method using two-dimension. Normally the X-axis refers to job-1 processing time and the Y-axis refers to the processing time of job-2. Erect a square or rectangle using the corresponding machine processing time on two different jobs. Draw a bisector line, based on that and decide the optimal sequence.

Example: 6

Use graphical method to determine the minimum time needed to process two jobs on four machines A,B,C and D. The technological order for these jobs on machines is as follows:

Job 1:	A	B	C	D
Job 2:	D	B	A	C

Processing time (in hours) are given as follows:

	A	B	C	D
Job 1:	4	6	7	3
Job 2:	4	7	5	8

Step 1: Represent the given information in a two-dimensional graph. The blocks represent the clashes/ overlaps which are to be avoided. To the possible, extent, move diagonally, which represents work on both the jobs simultaneously.

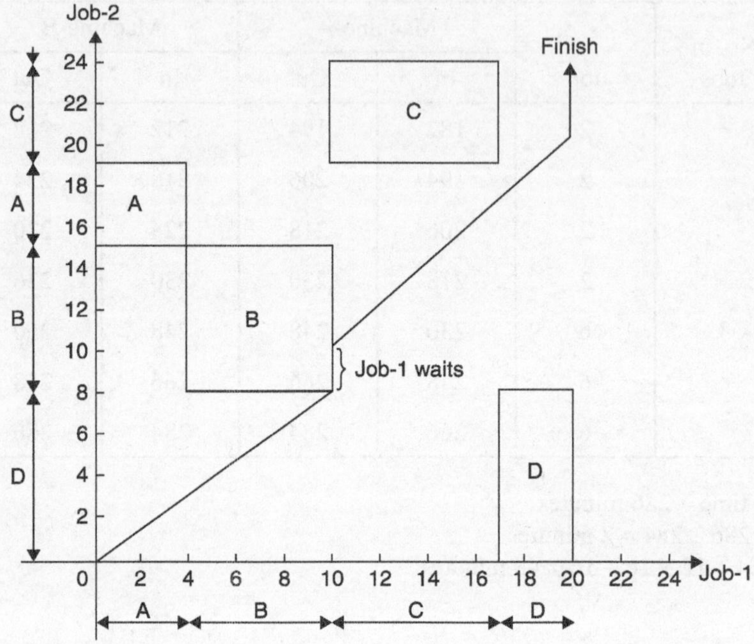

The above figure clearly indicates that the jobs can go together upto the first 10 hrs, after that the job-1 should wait for 2 hrs. This is because the machine needed to process job-1 is not available. After this break, the work can go together for the next 10 hrs. At the end of this, job-1 is completed and job-2 is to be continued for 2 more hours. Till then, job-1 waits. On the whole job-1 waits for 4 hrs. The optimal total elapsed time for completion of both jobs is 24 hrs.

Processing time of job-1 + idle time for job-1 = 20 + 4 = 24 hrs or
Processing time of job-2 + idle time for job-2 = 24 + 0 = 24 hrs

*Example: 7

Use the graphical method to minimize the time needed to process the following on the machines shown, i.e., for each machine find the job which should be done first. Also calculate the total time elapsed to complete both the jobs.

Job 1	Sequence	A	B	C	D	E
	Time	3	4	2	6	2
Job 2	Sequence	B	C	A	D	E
	Time	5	4	3	2	6

Represent the given information in a two-dimensional graph. The blocks represent the clashes/over laps which are to be avoided. To the possible extent, move diagonally, which represents work on both the jobs simultaneously.

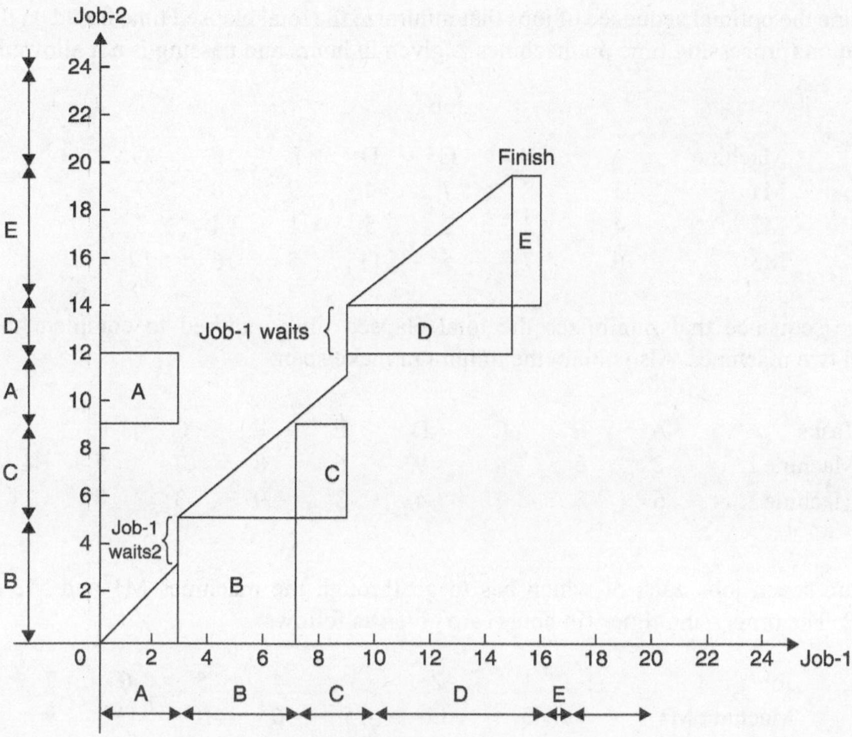

The figure clearly indicates that the jobs can go together upto the first 3 hrs, after that the job-1 should wait for 2 hrs. After this break the jobs can go together upto 13 hrs, then the job-1 should wait for 3 hrs. After the second break the work can go together till the completion.

Processing time of job-1 + idle time of job-1 = 17 + 2 + 3 = 22 hrs
Processing time of job-2 + idle time of job-2 = 20 + 2 = 22 hrs
Hence the optimal elapsed time for the completion of both the jobs is 22 hrs.

EXERCISE PROBLEMS

*1. Six jobs go first over machine I and then over machine II. Order of the completion of jobs has no significance. The following table gives the machine times in hours for six jobs and two machines:

Job No.	1	2	3	4	5	6
Time on Machine I	5	9	4	7	8	6
Time of Machine II	7	4	8	3	9	5

Find the sequence of jobs that minimizes the total elapsed time to complete the jobs.

*2. Determine the optimal sequence of jobs that minimize the total elapsed time based on the following information (processing time on machines is given in hours and passing is not allowed).

Job

Machine	A	B	C	D	E	F	G
M1	3	8	7	4	9	8	7
M2	4	3	2	5	1	4	3
M3	6	7	5	11	5	6	12

*3. Find the sequence that minimizes the total elapsed time required to complete the following tasks on two machines. Also obtain the minimum make span.

Tasks :	A	B	C	D	E	F	G	H	I
Machine I :	2	5	4	9	6	8	7	5	4
Machine II :	6	8	7	4	3	9	3	8	11

*4. There are seven jobs each of which has to go through the machines M1 and M2 in the order M1, M2. The processing times (in hours) are given as follows:

Job :	1	2	3	4	5	6	7
Machine M1 :	3	12	15	6	10	11	9
Machine M2 :	8	10	10	6	12	1	3

Determine the sequence of these jobs that will minimize the total elapsed time T.

*5. A book binder has one printing press, one binding machine and the manuscripts of a number of different books. The time required to perform the printing and binding operations for each book are known. Determine the order in which the books should be processed in order to minimize the total time required to process all the books. Find also the total time required.

Processing Time in Minutes

Book	1	2	3	4	5
Printing Time	40	90	80	60	50
Binding Time	50	60	20	30	40

6. Alongwith problem-5, suppose that an additional operation is added to the process described as in problem-5, namely finishing. The time required for operations is given below.

Finishing Time in Minutes

Book	1	2	3	4	5
Finishing time	80	100	60	70	110

What is the order in which the books should be processed? Find also the minimal total elapsed time. (Hint: Include this finishing operation as a third process along with first operation printing and second operation Binding. Then do this problem as a three machine process problem)

*7. Suppose a third stage of production is added along with the processes stated in Example-2, namely pressing & packing, with the processing time for these items as follows:

Item	1	2	3	4	5	6	7
Time required for pressing & packing	10	12	11	13	12	10	11

Find an order in which these seven items are to be processed so as to minimize the time taken to process all the item through all the stages.
[Hint: Very similar to problem-6]

8. Find an optimal sequence for the following sequencing problem of four jobs and five machines when passing is not allowed having the processing time (in hours) as given below:

Machine\Job	1	2	3	4
M1	6	5	4	7
M2	4	5	3	2
M3	1	3	4	2
M4	2	4	5	1
M5	8	9	7	5

Find the optimum total elapsed time.
[Hint: Verify the condition; min (1) ≥ max (j); j = 2, 3, 4 (1)

(or)

min (5) ≥ max (j); j = 2, 3, 4 (2)

Here the 2nd condition is satisfied. Then convert the given problem in to a two machine problem and hence solve it.

G: M1 + M2 + M3 + M4} for the corresponding timings
H: M2 + M3 + M4 + M5}

9. We have 4 jobs each of which has to go through the machines. M1, M2, M6 in the same order.

Job \ Machine	Processing Time (in hours)					
	M1	M2	M3	M4	M5	M6
A	18	8	7	2	10	29
B	17	6	9	6	8	19
C	11	5	8	5	7	15
D	20	4	3	4	8	12

Determine a sequence of these four jobs that minimizes the total elapsed time.

ANSWERS TO THE EXERCISE PROBLEMS

1. 3 – 1 – 5 – 6 – 2 – 4; Elapsed time: 42 hrs
2. A – D – G – B – F – C – E; Elapsed time: 59 hrs
3. A – I – C – B – H – F – D – E – G (or) A – I – C – H – B – F – D – E – G; Elapsed time = 61 hrs
4. 1 – 4 – 5 – 3 – 2 – 7 – 6; Elapsed time = 67 hrs
5. 1 – 2 – 5 – 4 – 3; Elapsed time = 340 hrs
6. 4 – 1 – 5 – 2 – 3; Elapsed time = 510 hrs
7. 1 – 4 – 3 – 6 – 2 – 5 – 7; Elapsed time = 86 hrs
8. 1 – 3 – 2 – 4; Elapsed time = 43 hrs
9. C – A – B – D; Elapsed time = 112 hrs

REVIEW QUESTIONS

Short Questions

1 Define Sequencing (Section 7.1)

2. Define the following:

   (i) Idle time (ii) Elapsed time (Section 7.1)

3. State the assumptions related to a sequencing problem (Section 7.1.1)
4. Explain how an machine (n > 2) can be converted into a two machine problem (Section 7.7)
5. Highlight the term 'no passing rule' (Section 7.4)
6. Discuss Johnson's algorithm (Section 7.2)
7. How we can make use of the Graphical method to solve a sequencing problem (Section 7.12)

State True or False

8. Johnson's Method can be applied for all cases (F)
9. While applying Johnson's Method, if there is any tie occurs at the time of selecting a minimum entry, we have to make use of a special tie breaking methodology. (T)

Fill in the Blanks

10. Sequencing can be viewed as a special kind of problem commonly encountered in the ______. (production shop)
11. If we can solve a sequencing model using the method of enumeration, we have to compute ______ number of calculations. ($(m!)^n$)
12. Sequencing Method was designed by ______. (S.M. Johnson)
13. Apart from Johnson's Method Sequencing problem can also be solved by using the ______ (Graphical Method)
14. The optimum time required to complete all jobs using sequencing is known as ______. (Elapsed time)

8

Queuing Theory

8.1 INTRODUCTION

The term 'Queue' refers to a waiting line. Almost every individual spent some part of his life waiting in the line in order to get some kind of service. Even this statement holds good for every object also.

For example,

(i) People wait at Railway Counters
(ii) People wait at Ration shops
(iii) People wait at Banks
(iv) People wait at Petrol Pumps
(v) Trucks wait at the Weigh-Bridge
(vi) Broken machine waits for repair
(vii) Officers of an organization may wait for a car. All these cause queuing situations.

Queuing theory was originated from the work of A.K. Erlang, an engineer in Copenhagen Telephone Exchange who studied the reason for the delay of telephone operators and published his findings as a paper in 1909, under the title "The Theory of probabilities and Telephone Conversation".

8.2 SOME QUEUING TERMINOLOGIES

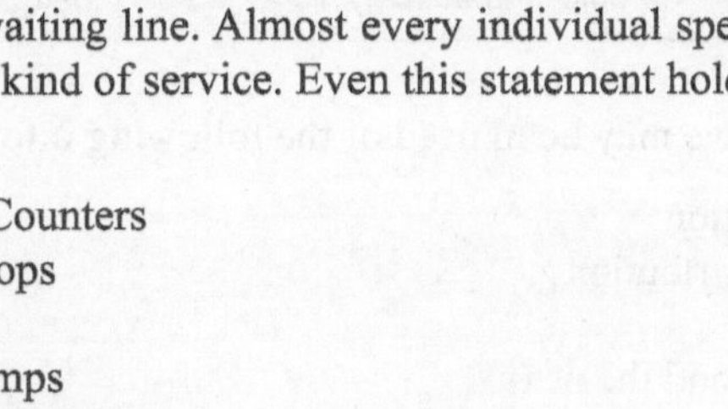

8.2.1 The Input/Arrival Process

The input process is usually called the arrival process. Arrivals are called customers. Customers arrive at a service station for service. They do not come at regular intervals but arrivals into the system occur according to some chance of mechanism. Often arrival occurs at random and is independent of what has previously occurred.

Customers may arrive for service individually or in groups. Customers visiting a bank can be taken as an example of a single arrival case. On the other hand, families visiting a restaurant, is an example of bulk or group arrival. The nature of arrivals may be constant or may follow probability distributions like poisson distribution, etc. The population of the customers requiring service may be finite or infinite.

The nature of the customers who enters into the Queue can be viewed as follows:

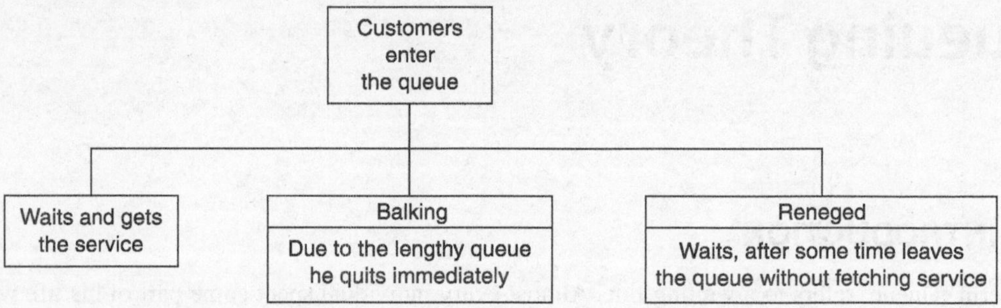

Based on the input process, we may be in need of the following information:

1. Pattern of arrival distribution
2. Pattern of inter-arrival distribution
3. Mean arrival rate and
4. Mean arrival time of one and the next.

8.2.2 Queue Discipline

It describes the method used to determine the order in which the customers are served. The most common discipline is the FCFS discipline (First Come First Served), in which customers are served in the order of their arrival.

Under the LCFS discipline (Last Come First Served), the most recent arrivals are the first to enter service. Sometimes the order in which the customers arrive has no effect on the order in which they are served. This would be the case if the next customer to enter service is chosen at random from those customers waiting for service. Such a situation is referred to as SIRO discipline (Service In Random Order).

Lastly, we consider Priority Queuing disciplines. A priority discipline classifies each arrival into one of several categories. Each category is then given a priority level, and within each priority level, customers enter service on an FCFS pattern. Priority disciplines are often used in emergency rooms to determine the order in which customers receive treatment.

8.2.3 Service Mechanism

This is nothing but a description of the service operation. The time that elapse while one unit is being served is called its service time. The service time of successive customers is taken to be independent random variables and their probability distribution is known as the service time distribution. The number of persons who can be served per unit of time is called service rate.

8.2.4 Service Channel

The service mechanism may be of single or multi channels. In a multichannel service mechanism, the channels may be arranged either in parallel or in series or as a combination of both (called Tandem).

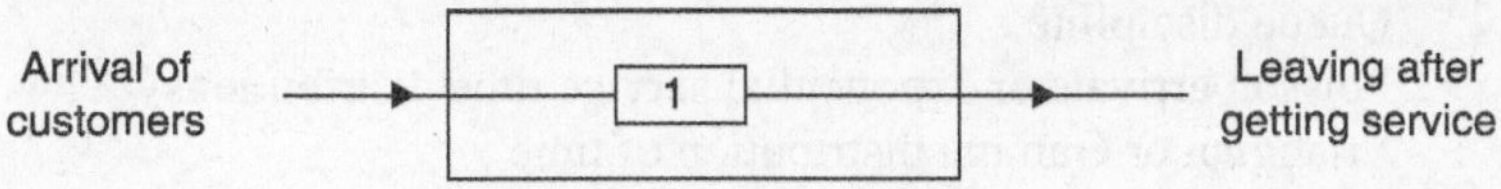

Single channel system

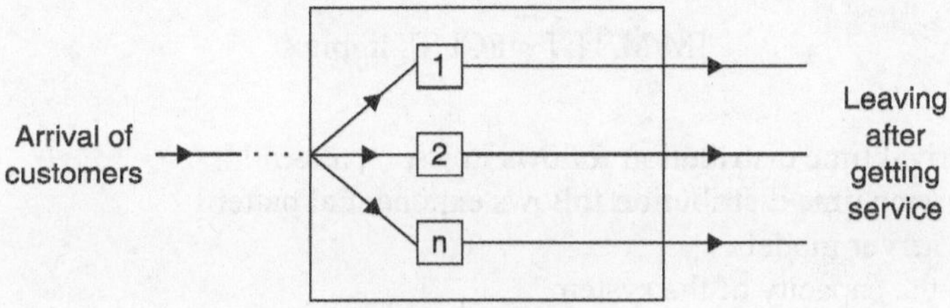

Multi channel system

8.2.5 Maximum Capacity of the Queue

The maximum queue length is normally infinite. But occasionally we have queues with finite number of customers. Finite Queue : Capacity of a Bus

Notations Used

n	:	Number of customers in the system, both waiting and in service.
λ	:	Average number of customers arriving per unit of time.
C	:	Average number of customers being served per unit of time.
λ/μ	:	Traffic Intensity (ρ)
C	:	Number of parallel service channels (Servers)
E(S)	:	Average number of customers in the system both waiting and in service.
E(Q)	:	Average number of customers in the Queue only.
W(S)	:	Average waiting time of a customer in the system both waiting and in service
W(Q)	:	Average waiting time of a customer in the Queue
P_0	:	Probability of no customers in the system
P_n	:	Probability of n customers in the system
$P_s(> n)$	:	Probability of more than n customers in the system
$P_Q(> n)$	:	Probability of more than n customers in the Queue
$P_s(w \geq t)$	:	Probability of the waiting time of a customer in the system greater than or equal to t units of time.
$P_Q(w \geq t)$	:	Probability of the waiting time of n customer in the queue greater than or equal to t units of time.

8.2.6 Classification of Queues

The characteristics of a queuing situation can be summarized using the following format:

$$[a/b/c] : [d/e]$$

Where a : The mode of arrival pattern
b : The mode of service pattern
c : Total number of parallel servers (C = 1, 2...∞)
d : Capacity of the system
e : Queue discipline
M : Poisson arrivals or exponential service time distributions
E_K : Erlangian or Gamma distribution of time
G_1 : General Input distribution
G : General service time distribution.

For example:

[M/M/1]: [∞/FCFS], implies

that

1. M : The arrival time distribution follows Poisson pattern
2. M : The service time distribution follows exponential pattern
3. 1 : Single server model
4. ∞ : Refers the capacity of the system
5. FCFS : The Queue discipline is based on First Come First Served basis

8.2.7 Methods Used to Solve a Queuing Situation

The various conditions to be verified before selecting a model are clearly explained in the Flowchart 8.1.

8.3 MODEL : 1 SINGLE SERVER MODEL WITH INFINITE QUEUE (M/M/1): (∞/FCFS)

This model deals with a queuing system having single service channel, input pattern based on Poisson, service pattern based on exponential, infinite capacity and the discipline is First Come First Served basis. We consider the situation when there is no customer (unit) in the system initially. That is, n = 0 initially. Then moving from this state n = 0 to n = 1; at a rate of λ from 0 to 1 and at a rate μ from state 1 to 0.

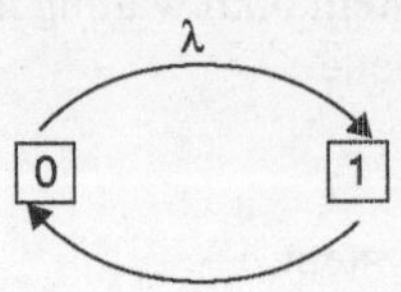

$$\lambda P_0 = \mu P_1 \qquad \text{(I)}$$

(I) is called a balanced equation because Rate in = Rate out. Extending to (n + 1) states, we have

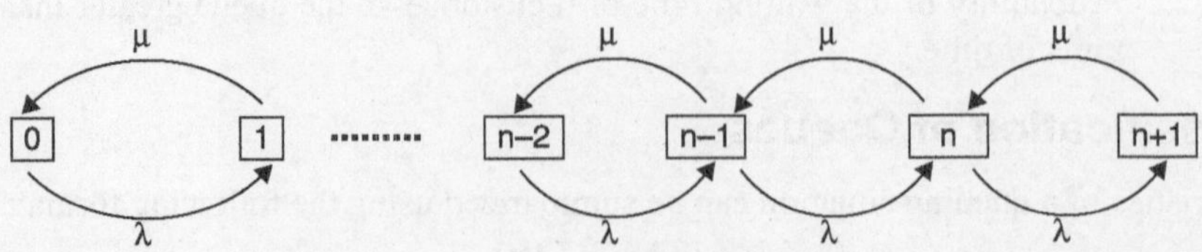

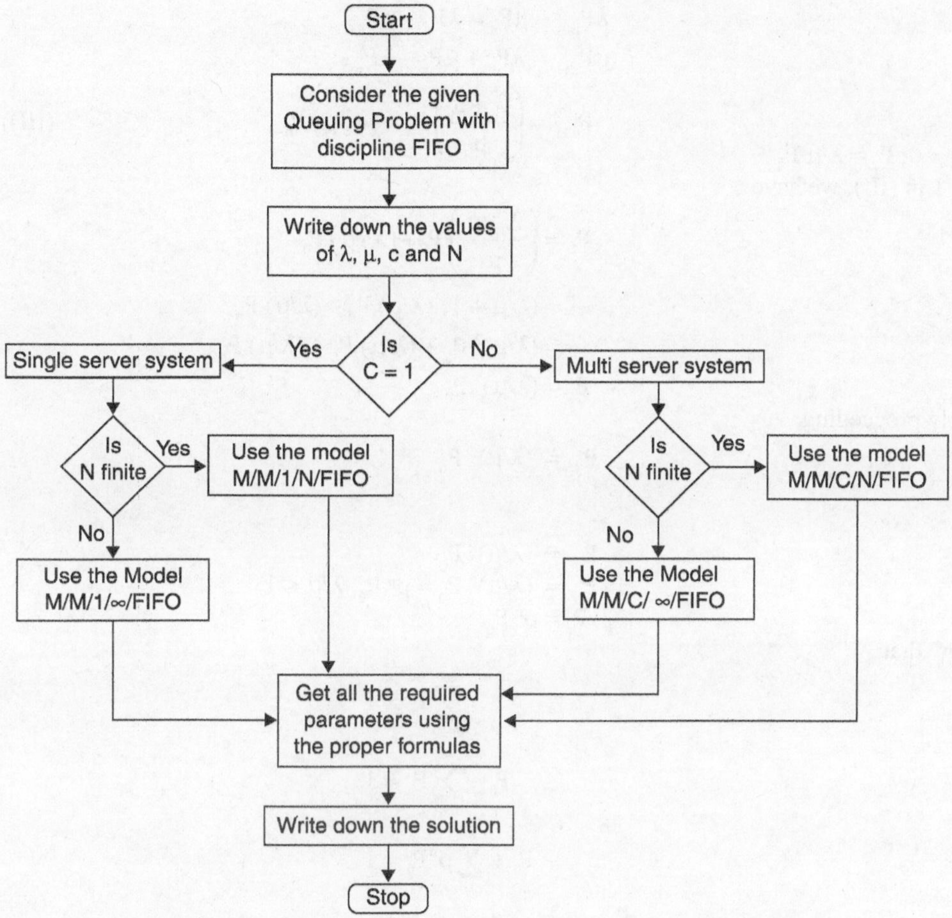

Flowchart 8.1 Methods used to solve a queuing situation

implies that

$$
\begin{aligned}
\lambda P_1 &= \mu P_2 \\
\lambda P_2 &= \mu P_3 \\
&\vdots \\
&\vdots \\
\lambda P_{n-1} &= \mu P_n \\
\lambda P_n &= \mu P_{n+1} \qquad \text{(II)}
\end{aligned}
$$

Consider the last two equations of system II,

$$
\begin{aligned}
\lambda P_{n-1} - \mu P_n = 0 \\
\lambda P_n - \mu P_{n+1} = 0
\end{aligned}
$$

Since RHS of both equations are zero, equating we have,

$$\lambda P_{n-1} - \mu P_n = \lambda P_n - \mu P_{n+1}$$

$$\mu P_{n+1} = \lambda P_n + \mu P_n - \lambda P_{n-1}$$

$$P_{n+1} = \left(\frac{\lambda+\mu}{\mu}\right) P_n - (\lambda/\mu) P_{n-1} \qquad \text{(III)}; \quad n \geq 1$$

when $n = 0$; $P_1 = \lambda/\mu\, P_0$
put $n = 1$ in (III), we have

$$P_2 = \left(\frac{\lambda+\mu}{\mu}\right) P_1 - (\lambda/\mu) P_0$$

$$= (\lambda/\mu + 1)\,(\lambda/\mu)\, P_0 - (\lambda/\mu)\, P_0$$

$$= (\lambda/\mu)^2 P_0 + (\lambda/\mu) P_0 - (\lambda/\mu) P_0$$

$$P_2 = (\lambda/\mu)^2 P_0$$

Similarly proceeding,

$$P_3 = (\lambda/\mu)^3 P_0$$

$$\vdots$$

$$P_n = (\lambda/\mu)^n P_0$$

$$\Rightarrow P_n = (\lambda/\mu)^n P_0 = \rho^n P_0; \ \lambda/\mu < 1$$

Hence,

$$P_n = \rho^n P_0$$

we know that;

$$\sum_{n=0}^{\infty} P_n = 1$$

$$P_0 + \sum_{n=1}^{\infty} P_n = 1$$

$$P_0 + \sum_{n=1}^{\infty} \rho^n P_0 = 1$$

$$P_0\, [1 + \rho + \rho^2 + \ldots + \rho^n + \ldots] = 1$$

Since

$$(1 - \rho)^{-1} = [1 + \rho + \rho^2 + \ldots \rho^n + \ldots] \qquad \text{(a)}$$

$$P_0\, [1 - \rho]^{-1} = 1$$

We have,

$$P_0\left(\frac{1}{1-\rho}\right) = 1$$

$$P_0 = (1 - \rho)$$

Using P_0 value in P_n,

$$P_n = (1 - \rho) \cdot \rho^n$$

$$P_s(\geq n) = \sum_{k=n}^{\infty} P_k = \sum_{k=n}^{\infty} \rho^k \cdot P_0$$

$$= P_0 \sum_{k=n}^{\infty} \rho^k = (1-\rho) \sum_{k=n}^{\infty} \rho^k$$

$$= (1-\rho) \sum_{k=n}^{\infty} \rho^n \cdot \rho^{(k-n)}$$

$$= (1-\rho)\rho^n \sum_{k=n}^{\infty} \rho^{(k-n)}$$

$$= (1-\rho)\,\rho^n[1+\rho+\rho^2+....] \qquad \text{[using (a)]}$$

$$= (1-\rho)\,\rho^n \cdot \left(\frac{1}{1-\rho}\right)$$

$$= P_s\,(\geq n) = \rho^n$$

$$P_s(>n) = \sum_{k=n+1}^{\infty} P_k$$

$$= \sum_{k=n+1}^{\infty} \rho^k\, P_0$$

$$= (1-\rho)\rho^{n+1} \cdot \sum_{k=n+1}^{\infty} \rho^{k-(n+1)}$$

$$P_S(>n) = \rho^{n+1}$$

$$P_Q(>n) = P_S\,(>n+1)$$

$$= \rho^{(n+1)+1} = \rho^{n+2}$$

$$P_Q(>n) = \rho^{n+2}$$

Average number of customers in the system (E(S))

$$E(S) = \sum_{n=0}^{\infty} n\, P_n$$

$$= \sum_{n=0}^{\infty} n \cdot (1-\rho)\rho^n$$

$$= (1-\rho)\sum_{n=0}^{\infty} n \cdot \rho^{n-1} \cdot \rho$$

$$= (1-\rho)\rho \sum_{n=0}^{\infty} \frac{d}{d\rho}(\rho^n)$$

$$= (1-\rho)\rho \frac{d}{d\rho}\left(\sum_{n=0}^{\infty} \rho^n\right)$$

$$= \rho(1-\rho)\,\frac{d}{d\rho}\,((1-\rho)^{-1}) \qquad \text{[using (a)]}$$

$$= \rho\,(1-\rho)\,(-1) \cdot (1-\rho)^{-2}\,(-1)$$

$$= \frac{\rho\,(1-\rho)}{(1-\rho)^2} = \frac{\rho}{1-\rho}$$

$$E(S) = \frac{\rho}{1-\rho}$$

Average number of persons in the Queue: E(Q)

$$
\begin{aligned}
E(Q) &= \sum_{n=1}^{\infty} (n-1)P_n \\
&= \sum_{n=1}^{\infty} nP_n - \sum_{n=1}^{\infty} P_n \\
&= \left(\frac{\rho}{1-\rho}\right) - \left[\sum_{n=0}^{\infty} P_n - P_0\right] \\
&= \left(\frac{\rho}{1-\rho}\right) - [1-P_0] = \left(\frac{\rho}{1-\rho}\right) - [1-(1-\rho)] \\
&= \frac{\rho}{1-\rho} - \rho = \frac{\rho - \rho\,(1-\rho)}{1-\rho} \\
&= \frac{\rho - \rho + \rho^2}{1-\rho}
\end{aligned}
$$

$$E(Q) = \frac{\rho^2}{1-\rho} = \rho \cdot \frac{\rho}{1-\rho}$$

$$E(Q) = \rho \times E(S)$$

Average waiting time in the system W(S)

$$
\begin{aligned}
W(S) &= \frac{1}{\lambda} \cdot E(S) \\
&= \frac{1}{\lambda} \cdot \frac{\rho}{1-\rho} \\
&= \frac{1}{\lambda} \cdot \frac{(\lambda/\mu)}{(1-(\lambda/\mu))} = \frac{1}{\lambda} \cdot \frac{\lambda}{\mu} \frac{1}{((\mu-\lambda)/\mu)} \\
&= \frac{1}{\lambda} \cdot \frac{\lambda}{\mu} \cdot \frac{\mu}{\mu-\lambda} = \frac{1}{\mu-\lambda}
\end{aligned}
$$

$$W(S) = \frac{1}{\lambda} E(S) = \frac{1}{\mu-\lambda}$$

Average waiting time in a Queue W(Q)

$$
\begin{aligned}
W(Q) &= \frac{1}{\lambda} \cdot E(Q) = \frac{1}{\lambda} \cdot \frac{\rho^2}{(1-\rho)} \\
&= \rho \cdot \left(\lambda \cdot \frac{\rho}{(1-\rho)}\right) = \rho \cdot \frac{1}{(\mu-\lambda)}
\end{aligned}
$$

$$W(Q) = \frac{1}{\lambda} \cdot E(Q) = \rho * W(S) = \frac{\rho}{(\mu-\lambda)} = \frac{\lambda}{\mu\,(\mu-\lambda)}$$

Probability of waiting time is more than or equal to the units of time in a given system: $P_s(w \geq t)$
We know that,

$$P(t) = \begin{cases} 1-\rho;\ t = 0 \\ \rho(\mu-\lambda)\, e^{-(\mu-\lambda)}t;\ t > 0 \end{cases}$$

Then
$$P(t/>0) = \frac{P(t)}{t(t>0)}$$

$$= \frac{\rho(\mu-\lambda)\, e^{-(\mu-\lambda)t}}{\rho}$$

$$P(t/t>0) = (\mu-\lambda)\, e^{-(\mu-\lambda)t}$$

$$P(w \geq t) = \int_t^{\infty} P(t/t>0)\, dt$$

$$= \int_t^{\infty} (\mu-\lambda)e^{-(\mu-\lambda)t}dt = (\mu-\lambda)\int_t^{\infty} e^{-(\mu-\lambda)t}\, dt$$

$$= (\mu-\lambda)\left(\frac{e^{-(\mu-\lambda)t}}{-(\mu-\lambda)}\right)_t^{\infty} = \frac{-(\mu-\lambda)}{(\mu-\lambda)}.(e^{-(\mu-\lambda)t})_t^{\infty}$$

$$= -[0 - e^{-(\mu-\lambda)t}] = e^{-(\mu-\lambda)t}$$

$$P_S(w \geq t) = e^{-(\mu-\lambda)t}$$

Similarly proceeding, we have

$$P_Q(W \geq t) = \rho\, P_S(W \geq t) = \rho\, e^{-(\mu-\lambda)t}$$

Formulas

1. $\rho \quad = \lambda/\mu\ (0 < \rho < 1)$
2. $P_0 \quad = 1 - \rho$
3. $P_s(n) \quad = (1-\rho)\,\rho^n$
4. $P_s(\geq n) \quad = \rho^n$
5. $P_Q(\geq n) \quad = \rho^{n+1}$
6. $E(S) \quad = \rho/(1-\rho)$
7. $E(Q) \quad = \rho \times E(S) = \rho^2/(1-\rho)$
8. $W(S) \quad = 1/(\mu-\lambda)$
9. $W(Q) \quad = \rho \times W(S) = \rho/(\mu-\lambda) = 1/(u(\lambda-\mu))$
10. $P_s(w \geq t) = e^{-(\lambda-\mu)t}$
11. $P_Q(w \geq t) = \rho \times P\,(w \geq t)$

*Example: 1

1. A T.V. repairman finds that the time spent on his jobs has an exponential distribution with mean 30 minutes. If he repairs sets in the order in which they came in and if the arrival of sets is approximately Poisson with an average rate of 10 per 8 hour-day, what is repairman's expected idle time each day? How many jobs are ahead of the average set just brought in?

Given: One day = 8 hrs
Arrival rate = 10 per day
Mean service time/Job = 30 mts.
$\lambda = 10/8$ per hr. [hr:hour]
$\mu = 1/30$ per mt. [mt:minute]

Since, λ is in hours and μ is in minutes. Convert μ in hours.
$\mu = 1/30 \times 60$ per hr
$\mu = 2$ per hr

We know that (wkt), $\rho = \lambda/\mu = (10/8)/(2) = 10/8 \times 2 = 5/8$
$\rho = 5/8$

To find,

(i) Idle time each day $[P_0]$
(ii) Average number of sets in the system (E(S))

(i) $P_0 = 1 - \rho = 1 - 5/8 = 3/8$
The idle time per day = 3/8*8 = 3 hrs

(ii) $\text{wkt}: E(S) = \dfrac{\rho}{1-\rho}$; $E(S) = \dfrac{(5/8)}{(1-5/8)} = \dfrac{(5/8)}{(3/8)} = 5/3 = 1.67$

The average number of sets in the system is 2 sets approximately.

*Example: 2

Arrivals at a telephone booth are considered to be Poisson, with an average time of 10 minutes between one arrival and the next. The length of phone call is assumed to be distributed exponentially with a mean of 3 minutes.

(i) What is the probability that a person arriving at a booth will have to wait?
(ii) Find the average number of person in the system.
(iii) What is the mean queue size?

Given:
Time between the arrivals = 10 mts.
Service time per customer = 3 mts.

⇒ Arrival rate = 1/10 per mt.
Service rate = 1/3 per mt.
$\rho = \lambda/\mu = (1/10)/(1/3) = 3/10 = 0.3$

*These questions have appeared in Indian university examination papers.

To find:

(i) Probability that a person arriving at a booth will have to wait.
That is, a person who arrives at the booth will have to wait, if one or more persons are in the system.
ie) $P_s(\geq 1)$

(ii) Average number of persons in the system, ie) E(S)

(iii) The mean queue size ie) E(Q)

(i) wkt: $P_s(\geq n) = \rho^n$

ie) $P_s(\geq 1) = \rho^1 = \rho$ (here, n =1)

$= 0.3$

$P_s(\geq 1) = 0.3$

(ii) wkt : $E(S) = \dfrac{\rho}{1-\rho} = \dfrac{0.3}{1-0.3}$

$= 0.3/0.7 = 3/7$

$E(S) = 3/7$

(iii) wkt: $E(Q) = \rho \cdot E(S)$

$= 0.3 * 3/7 = 0.9/7$

$= 9/70$

Hence the solution is,

(i) Probability of an arrival will have to wait is 0.3
(ii) Average number of persons in the system is (3/7)
(iii) Average number of persons in the Queue is (9/70)

Example: 3

An airline is planning to open a ticket desk at a new shopping complex staffed by one ticketing agent. Requests for tickets and information takes about 15 per hour on Poisson distribution. At exponential rate service time is assumed to be 3 minutes per request. Find the

(i) System usage.
(ii) % of time agent is idle.
(iii) Number of customers expected to be waiting to be served.
(iv) Average time customers will spend in the system.

Given:
Arrival rate = λ = 15 per hr
Service time = 3 mts per customer
Implies that,
Service rate = μ = 1/3 per mt.
Since, λ is in hour, μ can be converted into hour.
That is, $\mu = 1/3 * 60 = 20$/hr

To find:

(i) System usage (ρ)
(ii) % of time agent is idle (P_0)
(iii) Number of customers expected to be waiting to be served E(Q)
(iv) Average time customers will spend in the system. W(S)

(i) wkt: $\rho = \lambda/\mu = 15/20 = 3/4$
(ii) wkt: $P_0 = 1 - \rho = 1 - 3/4 = 1/4 = 0.25$
ie) 0.25 * 100; 25% of the time the agent is idle.

(iii) wkt: $E(Q) = \dfrac{\rho^2}{1-\rho} = \dfrac{(3/4)^2}{(1-(3/4))}$

$= \dfrac{(9/16)}{(1/4)} = \dfrac{9}{16} \times 4 = 9/4$ persons

(iv) wkt: $W(S) = \dfrac{1}{\mu - \lambda} = \dfrac{1}{20-15} = 1/5$ hr

Hence, the solution is,

(i) System usage is 0.75
(ii) Percentage of idle time is 25%
(iii) Number of customers expected to be waiting to be served is 2.25 persons
ie) Approximately 3 persons
(iv) Average time customers will spend in the system is (1/5) hr.

*Example: 4

In a railway marshalling yard, goods trains arrive at a rate of 30 trains per day. Assuming that the inter-arrival time follows an exponential distribution and service time distribution is also exponential with an average of 36 minutes, calculate the following:

(i) Average number of trains in the yard
(ii) The probability that the queue size exceeds 10.

Given:

Arrival rate = λ = 30 trains per day
Service time per train = 36 mts
ie) service rate = μ = 1/36 trains per mt.
Since λ is in days; convert μ in terms of day
That is, $\mu = 1/36 \times 24 \times 60 = 40$ trains
$\rho = \lambda/\mu = 30/40 = 3/4$

To find:

(i) Average number of trains in the yard. E(S)
(ii) The probability that the queue size exceeds 10. $P_Q(>10)$

(i) wkt: $E(S) = \dfrac{\rho}{1-\rho}$

$= \dfrac{3/4}{(1-3/4)} = \dfrac{3}{4} \times \dfrac{4}{1} = 3$ trains

(ii) wkt: $P_Q(>n) = \rho^{n+2}$

ie) $P_Q(>10) = \rho^{10+2} = \rho^{12} = (3/4)^{12}$

$= 0.0317$

Hence the solution is,

(i) Average number of trains in the yard is 3.
(ii) The probability that the Queue size exceeds 10 is 0.0317.

Example: 5

A person repairing radios finds that the time spent on the radio sets has an exponential distribution with mean 20 minutes. If the radios are repaired in the order in which they come in and their arrival is approximated Poisson with an average rate of 15 for an 8-hour day, what is the repairman's expected idle time each day? How many jobs are ahead of the average set just brought in?

Given:
Service time per radio = 20 mts
Service rate = $\mu = 1/20$ per mt
1 day = 8 hrs
Arrival rate = $\lambda = 15$ per day
$\mu = 1/20 * 60 = 3$ per hr
$\lambda = 15/8$ per hr
$\rho = \lambda/\mu = (15/8)/(3) = 15/3 \times 8 = 5/8$

To find:

(i) Repairman's expected idle time per day (P_0)
(ii) Number of jobs ahead of the average set just brought in ($E(S)$)

(i) wkt: $P_0 = 1 - \rho = 1 - 5/8 = 3/8$
expected idle time per day =3/8 * 8 = 3 hrs

(ii) wkt: $E(S) = \dfrac{\rho}{1-\rho} = \dfrac{(5/8)}{(1-5/8)}$

$= \dfrac{(5/8)}{(3/8)} = 5/3 = 1.67$ sets

Hence the solution is,

(i) Repairman's expected idle time per day is 3 hrs.
(ii) Approximately two jobs are ahead of the average set just brought in.

Example: 6

A university MBA DLP programme has a student project adviser. The adviser takes on an average 30 minutes per student. The service time per student is approximately exponential. Students arrival can be approximated to Poisson distribution. The average number of arrivals per day is 10 (Ten). Adviser works for 7 hrs and 30 minutes per day. Find the average utilization of the adviser. Under steady state conditions how many students are available in the system? What is the average waiting time per students?

Given:
Service time per student = 30 mts.
Service rate = μ = 1/30 per mt
Arrival rate = λ = 10 per day
1 day = 7.30 hrs = 450 mts
λ = 10/450 per mt

$$\rho = \frac{\lambda}{\mu} = \frac{(10/450)}{(1/30)} = \frac{300}{450} = \frac{2}{3}$$

To find:

(i) The average utilization of the adviser (ρ)
(ii) Number of students available in the system (E(S))
(iii) Average waiting time per student (W(S))

(i) The average utilization of the adviser is 2/3*450 = 300 mts.
(ii) wkt: E(S) = $\rho/1-\rho$

$$= \frac{(2/3)}{(1-2/3)} = 2 \text{ students}$$

(iii) wkt: W(S) = $1/\lambda$.E(S)

$$= \frac{1}{(10/450)} * 2 = \frac{2 \times 450}{10} = 90 \text{ mts}$$

Hence the solution is,

(i) The average utilization of the adviser is 300 mts.
(ii) Number of students in the system is 2.
(iii) Average waiting time per student is 90 mts.

***Example: 7**

At a public telephone booth in a post office, arrivals are considered to be Poisson with an average inter-arrival time of 12 minutes. The length of a phone call may be assumed to be distributed exponentially with an average of 4 minutes. Calculate the following:

(a) What is the probability that a fresh arrival will not have to wait for the phone?
(b) What is the probability that an arrival will have to wait for more than 10 minutes before the phone is free?
(c) What is the average length of queues that form from time to time?

Given:
Arrival rate = λ = 1/12 per mt
Service rate = μ = 1/4 per mt

$$\rho = \frac{\lambda}{\mu} = \frac{(1/12)}{(1/4)} = \frac{4}{12} = \frac{1}{3}$$

To find:

(a) The probability that a fresh arrival will not have to wait for the phone. [P_s ($\geq$ 1)]
(b) The probability that an arrival will have to wait for more than 10 mts [P_Q (w >10 mts)]
(c) The average length of Queue E(Q)

(a) wkt: $P_s(\geq n) = \rho^n$
That is, $P_s(\geq 1) = \rho^1 = \rho$ (Here n =1)
$= 1/3$

(b) wkt: $P_Q(w \geq t) = \rho * e^{-(\mu-\lambda)t}$
$= (1/3) * e^{-[1/4-1/12]10}$
$= (1/3) * e^{-[1.6667]}$
$= e^{-[1.6667]}/3$
$= 0.063$

(c) wkt: $E(Q) = \dfrac{\rho^2}{1-\rho}$
$= \dfrac{(1/3)^2}{(1-1/3)} = \dfrac{1}{9} * \dfrac{3}{2}$
$= 1/6$

Hence, the solution is

(a) The probability that an arrival will not have to wait for phone is 1/3.
(b) The probability that an arrival will have to wait for more than 10 mts before the phone is free is 0.063.
(c) The average number of persons in the Queue is (1/6)

***Example: 8**

A repairman is to be hired to repair machines that break down, following a Poisson process, with an average rate of four per hour. The cost of a nonproductive machine is ₹9 per hour. The company has the option of choosing either a fast or a slow repairman. The fast repairman charges ₹6 per hour and will repair machines at an average rate of 7 per hour. The slow repairman charges ₹3 per hour and will repair machines at an average rate of 5 per hour. Which repairman should be hired?

Given,
The arrival rate = λ = 4 per hr
Cost of nonproductive machine = ₹9 per hr
Fast repairman's service rate = 7 per hr
Fast repairman's wage rate = ₹6 per hr
Slow repairman's service rate = 5 per hr
Slow repairman's usage rate = ₹3 per hr
(TC) Total cost/hr = (Number of machines in system per hr) * 9 + wage cost per hr

Note that, the TC must be evaluated based on the fast and slow repairman respectively and select the best one.

wkt: $\rho = \lambda/\mu$ and $E(S) = \rho/1 - \rho$

Repairman	Arrival Rate (λ)	Service Rate (μ)	ρ	No. of Machine/hr in the System	Cost of Non-Productive Machine/hr (₹)	Wage Cost/hr (₹)	TC/hr (₹)
Fast	4	7	4/7	$\frac{(4/7)}{(1-4/7)}=\frac{4}{7}\times\frac{7}{3}=4/3$	4/3*9 = 12	6	18
Slow	4	5	4/5	$\frac{(4/5)}{(1-4/5)}=4$	4*9 = 36	3	39

Since the total cost for the fast repairman is less than the slow repairman, the fast repairman should be hired.

*Example: 9

The mean rate of arrival of planes at an airport during the peak period is 20 per hour but the actual number of arrivals in any hour follows a Poisson distribution. The airport can land 60 planes per hour on an average in good weather or 30 per hour in bad weather, but the actual number landed in any hour follows a Poisson distribution with the respective averages. When there is congestion, the planes are forced to fly over the field in the stack awaiting the landing of other planes that arrived earlier.

(i) How many planes would be flying over the field in the stack on an average in good weather and in bad weather.
(ii) How long a plane would be in the stack and in the process of landing in good and bad weather.

Given,
Arrival rate = λ = 20 per hr
Service rate = μ = 60 per hr; in good weather
30 per hr; in bad weather

$$\rho = \begin{cases} 20/60 = 1/3; \text{ in good weather} \\ 20/30 = 2/3; \text{ in bad weather} \end{cases}$$

To find the

(i) Number of planes that would be flying over the field in the stack
   (a) in good weather [E(Q)]
   (b) in bad weather [E(Q)]
(ii) Waiting time of a plane in the stack.
   (a) in good weather
   (b) in bad weather

(i) (a) $E(Q)=\frac{\rho^2}{1-\rho}=\frac{(1/3)^2}{(1-(1/3))}=\frac{1}{9}\cdot\frac{3}{2}=\frac{1}{6}$ planes in good weather

(b) $E(Q)=\frac{\rho^2}{1-\rho}=\frac{(2/3)^2}{(1-(2/3))}=\frac{4}{9}\cdot 3=\frac{4}{3}$ planes in bad weather

(ii) The distribution of stack of landing time together is given by t

$$\int_0^t (\mu - \lambda)\, e^{-(\mu-\lambda)t} dt = 0.95$$

That is,

$$(\mu - \lambda)\int_0^t e^{-(\mu-\lambda)t} dt = 0.95$$

$$(\mu - \lambda)\left[\frac{e^{-(\mu-\lambda)t}}{-(\mu-\lambda)}\right]_0^t = 0.95$$

$$-[e^{-(\mu-\lambda)t} - 1] = 0.95$$

$$1 - e^{-(\mu-\lambda)t} = 0.95$$

$$e^{-(\mu-\lambda)t} = 0.05$$

$$\log_e e^{-(\mu-\lambda)t} = \log_e (0.05)$$

$$-t(\mu - \lambda) = \log_e (0.05) = -2.9957$$

$$t = \frac{2.9957}{(\mu - \lambda)}$$

(a) in good weather

$$t = \frac{2.9957}{60-20} = \frac{2.9957}{40} = 0.0749 \text{ hr}$$

$$t = 0.0749 \text{ hr (or) } t = 4.5 \text{ mts}$$

(b) in bad weather

$$t = \frac{2.9957}{30-20} = 0.29957 \text{ hr}$$

$$t = 0.29957 \text{ hr or } t = 17.97 \text{ mts}$$

Hence the solution is,

(i) The number of planes flying over the field in the stack is
(1/6) in good weather and
(4/3) in bad weather

(ii) The waiting time of a plane in the stock and in the process of landing is
(4.5 mts) in good weather
and (17.97 mts) in bad weather

***Example: 10**

Let an average of 96 patients per 24 hour a day require the service of an emergency clinic. Also on an average, a patient requires 10 minutes of active attention. Assume that the facility can handle only one emergency at a time. Suppose that it costs the clinic ₹100 per patient treated to stay an average servicing time of 10 minutes and that each minute of decrease in this average time would cost ₹10 per patient treated. How much money would have to be budgeted by the clinic to decrease the average size of the Queue from 4/3 patient (one and third patients) to 1/2 a patient.

Given:
1 day = 24 hrs
Arrival rate = λ = 96 per day
Service time = 10 mts per patient
Service rate = μ = 1/10 per mt
Convert λ into mts.

$$\lambda = \frac{96}{24} \text{ per hr}$$

$$= \frac{96}{24 \times 60} \text{ per mt}$$

$$= 1/15 \text{ per mt}$$

wkt: $$\rho = \lambda/\mu = \frac{(1/15)}{(1/10)} = \frac{10}{15} = \frac{2}{3}$$

Cost to the patient at present = ₹100
Reduction cost/mt = ₹10
The aim of the Hospital is to reduce the service time as for as possible to reduce E(Q) = 1/2 by fixing the arrival rate as it is. That is the present E(Q) = 4/3, this must be reduced to E(Q) = 1/2.

wkt: $$E(Q) = \frac{\rho^2}{1-\rho}$$

$$\text{ie)} \quad \frac{\rho^2}{1-\rho} = \frac{1}{2}$$

$$2\rho^2 = 1 - \rho$$
$$2\rho^2 + \rho - 1 = 0$$
$$2\rho^2 + 2\rho - \rho - 1 = 0$$
$$2\rho(\rho + 1) - (\rho + 1) = 0$$
$$(2\rho - 1)(\rho + 1) = 0$$
$$2\rho - 1 = 0 \quad \text{or} \quad \rho + 1 = 0$$

Case (i): $2\rho - 1 = 0$
$2\rho = 1$
$\rho = 1/2$

Case (ii): $\rho + 1 = 0$
$\rho = -1$
Since ρ is negative, discard it.

Hence,

$$\rho = 1/2$$
$$\lambda/\mu = 1/2$$
$$\mu = 2\lambda$$
$$\mu = 2\,[1/15] = 2/15/\text{mt}$$

The new service rate is 2/15/mt
That is the,
New servicing time = 15/2 mts

Actual service time	= 10 mts
New service time	= 7.5 mts
Reduction in service time	= 2.5 mts

Budget:

Actual cost	= ₹100.00
Cost of reduction in service time (2.5 × 10)	= ₹ 25.00
Total cost	= ₹125.00

Hence the hospital should budget ₹125 per patient.

*Example: 11

Arrivals at a Telephone booth are considered to be Poisson, with an average time of 10 minutes between one arrival and the next. The duration of the phone call is assumed to be distributed exponentially, with a mean of 3 minutes.

(a) What is the probability that a person arriving at the booth will have to wait?
(b) The telephone department will install a second booth when convinced that an arrival would except waiting for at least for 3 minutes for phone. By how much should the flow of arrivals increase in order to justify a second booth?
(c) Find the average number of units in the system.
(d) Estimate the fraction of a day that the phone will be in use.
(e) What is the probability that it will take him more than 10 minutes altogether to wait for phone and complete his call?

Given:
Arrival rate = $\lambda = 1/10$ per mt
Service rate = $\mu = 1/3$ per mt

$$\rho = \lambda/\mu = \frac{(1/10)}{(1/3)} = 0.3$$

To find:

(a) The probability that a person arriving at the booth will have to wait. $P_s (\geq 1)$
(b) If W(Q) = 3, find the new arrival rate λ.
(c) The average number of units in the system (E(S)).
(d) The fraction of the day that the phone will be in use (ρ).
(e) The probability that it will take him more than 10 minutes altogether to wait for phone and complete his calls. [P (w ≥ 10)]

(a) wkt: $P_s (\geq 1) = \rho^1 = (0.3)^1 = 0.3$

(b) wkt: $W(Q) = \dfrac{\rho}{\mu - \lambda} = 3$, find new λ. μ is fixed.

$$= \frac{\rho}{\mu(1-\rho)} = 3$$

$$\rho = 3\,\mu(1-\rho) = 3.1/3(1-\rho)$$
$$\rho = 1 - \rho$$
$$2\rho = 1$$

$$\rho = 1/2$$
$$\lambda/\mu = 1/2$$
$$\lambda = 1/2\mu = 1/6$$
$$\lambda = 0.17 \text{ mt}$$

The arrival rate must increase to 0.17 per mt.

(c) wkt: $E(S) = \dfrac{\rho}{1-\rho} = \dfrac{0.3}{(1-0.3)} = 3/7$

$= 0.43$ customers

(d) wkt: $\rho = \lambda/\mu = 0.3$

(e) wkt: $P(w \geq t) = e^{-(\mu-\lambda)t}$

$P(w \geq 10) = e^{-(1/3-1/10)\,10}$

$= e^{-[7/30]*10}$

$= e^{-(7/3)}$

$= 0.09697$

$= 0.10$

Hence the solution is,

(a) The probability that a person arriving at the booth will have to wait is 0.3
(b) If the arrival rate increases to 0.17 per minute then it needs a second booth.
(c) The average number of units in the system is 0.43.
(d) The fraction of the day that the phone will be in use is 0.3.
(e) The probability that it will take him more than 10 minutes altogether to wait for phone and complete his calls is 0.10.

***Example:* 12**

A typist at an office receives on an average 22 letters per day for typing. The typist works 8 hours a day and it takes on an average 20 minutes to type a letter. The company has determined that the cost of a letter waiting to be mailed (opportunity) is 80 paise per hour and the equipment operating cost plus the salary of the typist will be ₹40 per day.

(i) What is the typists' utilization rate?
(ii) What is the average number of letters waiting to be typed?
(iii) What is the average waiting time needed to have a letter typed?
(iv) What is the total daily cost of waiting letters to be mailed?

Given:
Arrival rate of letters = $\lambda = 22$ per day
Time required to type a letter = 20 mts.
1 day = 8 hrs.
Opportunity cost to mail a letter per hr = ₹0.80

$$\mu = 1/20 \text{ /mt} = 1/20 \times 60 \times 8 = 24 \text{ per day}$$
$$\lambda = 22 \text{ per day}$$
$$\rho = \lambda/\mu = 22/24 = 11/12$$

To find:

(a) The typists utilization rate (ρ).
(b) The average number of letters waiting to be typed [E(Q)].
(c) The average waiting time needed to have a letter typed [W(Q)].
(d) The total cost of waiting letters to be mailed.

(a) $\rho = 11/12$

(b) $E(Q) = \frac{\rho^2}{1-\rho} = \frac{(11/12)^2}{(1-11/12)} = 10.08$ letters per day

(c) $W(Q) = \frac{\rho}{\mu-\lambda} = \frac{(11/12)}{24-22} = 11/24$

$= 0.458$ day $= 3.7$ hrs

(d) Opportunity cost per day to mail a letter = 0.8 × 8 = ₹6.4
Number of letters to be mailed E(S) = 11 per day
Total daily cost of waiting letters to be mailed = (Total opportunity cost/day) + wage/day
= 11 × 6.4 + 40 = ₹110.40

Hence the solution is,

(a) The typists' utilization rate is 0.92
(b) The average number of letters waiting per day is 10.08.
(c) The average waiting time for a letter in the queue is 3.7 hrs.
(d) The total cost of the waiting letters to be mailed is ₹110.40.

*Example: 13

In the Central Railway station 15 computerised reservation counters are available. A customer can book his/her ticket in any train on any day in any one of these computerised reservation counters. The average time spent per customer by each clerk is 5 minutes. Average arrivals per hour during the three types of activity periods have been calculated and customers have been surveyed to determine how long they are willing to wait during each type of period.

Type of Period	Arrivals Per Hour	Customer Acceptable Waiting Time
Peak	110	15 mts
Normal	60	10 mts
Low	30	5 mts

Making suitable assumptions on this queuing process, how many counters should be kept open during each type of period.

Let X be the number of single server system that can be thought of

$$\text{wkt: } W(Q) = \rho/\mu-\lambda$$

For further analysis refer table given in next page.

***Example: 14**

A warehouse has only one loading dock manned by a three person crew. Trucks arrive at the loading dock at an average rate of 4 trucks per hour and the arrival rate is Poisson distributed. The loading of a truck takes 10 minutes on an average and can be assumed to be exponentially distributed. The operating cost of a truck is ₹20 per hour and the members of the loading crew are paid @ ₹6 each per hour. Would you advise the truck owner to add another crew of three persons?

Given:
Arrival rate = λ = 4 per hr
Service time/truck = 10 mts.
Crew size = n
Service rate (μ) = 1/10 × 60 = 6 per hr
Operating cost (OC) = ₹20 per hr
Wage of a crew/person (WC) = ₹6 per hr

To find:

(a) Expected waiting time of a truck in the system (W(S))
(b) (i) Cost of waiting time = W(S) * λ * (OC)
(ii) Cost of crew = n * WC

Type of Period	Arrival Rate λ/hr	Services Rate μ/hr	$W(Q) = \rho/(\mu - \lambda)$	Given W(Q) in hrs	Evaluation of X	Number of Single Server Counters Required
1. Peak	110/X	12	$\frac{(110/12X)}{12-(110/X)}$	0.25	$\frac{(110/12X)}{12-(110/X)} = 0.25$ => X = 12.22 = 13	13
2. Normal	60/X	12	$\frac{(60/12X)}{(12-60/X)}$	0.17	$\frac{(60/12X)}{(12-60/X)} = 0.17$ => X = 7.45 = 8	8
3. Low	30/X	12	$\frac{(30/12X)}{(12-30/X)}$	0.081	$\frac{(30/12X)}{(12-30/X)} = 0.08$ => X = 5.10 = 6	6

Hence,

(i) During peak hours 13 single server counters must be kept open.
(ii) During normal hours 8 single server counters must be kept open.
(iii) During low hours 6 single server counters must be kept open.
(iv) Total cost = Cost waiting time + Cost of crew.

Case (i): When n = 3
wkt: W (S) = $1/\mu - \lambda = 1/6 - 4 = 0.5$ hr
Total cost = 0.5*4*20 + 3*6 = ₹ 58 per hr

Case (ii): When n = 6; $\lambda = 4$ per hr
and $\mu = 2*6 = 12$ per hr
W(S) = $1/\mu - \lambda = 1/12 - 4 = 1/8$ hr
Total cost = 1/8*4*20 + 6*6 = ₹ 46 per hr
Since the total coast is minimum of ₹ 46, when the crew size is 6.

Conclusion: The truck owner can add three more persons.

8.4 MODEL : 2 SINGLE SERVER MODEL WITH FINITE QUEUE (M/M/1): (N/FCFS)

This model differs from the Model: 1 in the sense that Maximum number of customers in that system is limited to N.

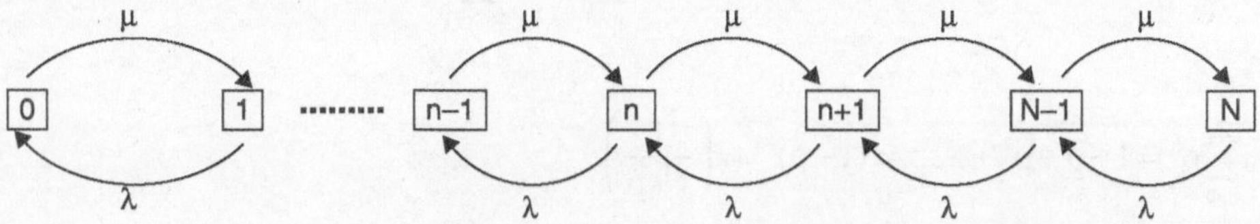

we have,

$$\lambda P_0 = \mu P_1 \tag{1}$$

$$(\lambda + \mu)\, P_n = \lambda P_{n-1} + \mu P_{n+1};\ 1 \le n \le N - 1 \tag{2}$$

and

$$\lambda P_{N-1} = \mu P_N \tag{3}$$

(1) implies that

$$P_1 = (\lambda/\mu)\, P_0$$

Put n = 1 in (2),

$$(\lambda + \mu)\, P_1 = \lambda P_0 + \mu P_2$$

$$\mu P_2 = (\lambda + \mu)\, P_1 - \lambda P_0$$

$$= (\lambda + \mu)\, (\lambda/\mu P_0) - \lambda P_0$$

$$= (((\lambda + \mu)/\mu) - 1)\, \lambda P_0$$

$$= \frac{\lambda + \mu - \mu}{\mu}.\ \lambda P_0$$

$$= (\lambda/\mu) \cdot \lambda P_0$$

$$\mu P_2 = \frac{\lambda^2}{\mu} P_0$$

$$P_2 = (\lambda/\mu)^2\, P_0$$

Proceeding similarly, we can have

$$P_n = (\lambda/\mu)^n\, P_0; \qquad n \le N - 1$$

Also,

$$P_n = (\lambda/\mu)^n\, P_0; \qquad n \le N \text{ holds.}$$

To find P_0:

wkt:
$$\sum_{n=0}^{N} P_n = 1$$

Case (i): If $\rho \neq 1$

$$\sum_{n=0}^{N} \rho^n P_0 = 1 \tag{4}$$

$$P_0 \sum_{n=0}^{N} \rho^n = 1$$

$$P_0 = \frac{1}{\sum_{n=0}^{N} \rho^n}$$

$$= \frac{1}{\sum_{n=0}^{\infty} \rho^n - \sum_{n=N+1}^{\infty} \rho^n} \tag{5}$$

Consider,

(i) $\sum_{n=0}^{\infty} \rho^n = 1 + \rho + \rho^2 + ... = (1-\rho)^{-1} = \left(\frac{1}{1-\rho}\right)$

(ii) $\sum_{n=N+1}^{\infty} \rho^n = \sum_{n-(N+1)=0}^{\infty} (\rho^{N+1}) \cdot \rho^{n-(N-1)}$

$$= \rho^{N+1} \sum_{n-(N+1)=0}^{\infty} \rho^{n-(N+1)}$$

$$= \rho^{N+1} \cdot (1-\rho)^{-1} = \frac{\rho^{N+1}}{1-\rho}$$

$$\sum_{n=0}^{\infty} \rho^n - \sum_{n=N+1}^{\infty} \rho^n = \frac{1}{1-\rho} - \frac{\rho^{N+1}}{1-\rho}$$

$$= \frac{1-\rho^{N+1}}{1-\rho}$$

Using in (5),

$$P_0 = \frac{1}{\left(\frac{1-\rho^{N+1}}{1-\rho}\right)} = \frac{1-\rho}{1-\rho^{N+1}}$$

Case (ii): If $\rho = 1$; (4) reduced to

$$\sum_{n=0}^{\infty} P_0 = 1 \tag{6}$$

$$P_0 + P_0 + \ldots\ldots (N+1) \text{ time} = 1$$

$$(N+1)\,P_0 = 1$$

$$P_0 = \frac{1}{N+1};\ \text{if } \rho = 1$$

$$P_0 \begin{bmatrix} \dfrac{1-\rho}{1-\rho^{N+1}}; & \text{if } \rho \neq 1 \\ \dfrac{1}{N+1}; & \text{if } \rho = 1 \end{bmatrix}$$

To find E(S):

Case (i): $\rho \neq 1$

$$E(S) = \sum_{n=0}^{N} n\rho_n = \sum_{n=0}^{N} n \cdot \rho^n\, P_0$$

$$= P_0 \sum_{n=0}^{N} n\rho^n$$

$$= \rho\, P_0 \sum_{n=0}^{N} n\rho^{n-1}$$

$$= \rho\, P_0 \sum_{n=0}^{N} \frac{d}{d\rho}(\rho^n)$$

$$= \rho\, P_0 \frac{d}{d\rho}\left(\sum_{n=0}^{N} \rho^n\right)$$

$$= \rho\, P_0 \frac{d}{d\rho}\left[\sum_{n=0}^{\infty} \rho^n - \sum_{n=N+1}^{\infty} \rho^n\right]$$

$$= \rho\, P_0 \frac{d}{d\rho}\left[\frac{1}{1-\rho} - \frac{\rho^{N+1}}{1-\rho}\right]$$

$$= \rho\, P_0 \frac{d}{d\rho}\left[\frac{1-\rho^{N+1}}{1-\rho}\right]$$

$$= \rho \frac{1-\rho}{1-\rho^{N+1}} \cdot \left(\frac{(1-\rho)\,[-(N+1)\cdot\rho^N] - (1-\rho^{N+1})\,(-1)}{(1-\rho)^2}\right)$$

$$= \frac{\rho}{(1-\rho^{N+1})\,(1-\rho)}[(N+1)\,\rho^{N+1} - (N+1)\rho^N + 1 - \rho^{N+1}]$$

$$E(S) = \frac{\rho\,[1-(N+1)\,\rho^N + N\rho^{N+1}]}{(1-\rho)\,(1-\rho^{N+1})}$$

Case (ii): $\rho = 1$

$$E(S) = \sum_{n=0}^{N} nP_0 = P_0 \sum_{n=1}^{N}$$

$$= P_0 \cdot \frac{N(N+1)}{2}$$

$$= \frac{1}{N+1} \cdot \frac{N(N+1)}{2} = N/2$$

$$E(S) = N/2$$

To find E(Q):

Case (i): $\rho \neq 1$

$$E(Q) = \sum_{n=1}^{N} (n-1) P_n$$

$$= \sum_{n=1}^{N} nP_n - \sum_{n=1}^{N} P_n$$

$$= \sum_{n=0}^{N} nP_n - \left[\sum_{n=0}^{N} (P_n) - P_0\right]$$

$$= E(S) - [1 - P_0]$$

$$= E(S) - \left[1 - \frac{1-\rho}{1-\rho^{N+1}}\right]$$

$$= E(S) - \left[\frac{1-\rho^{N+1} - 1 + \rho}{1-\rho^{N+1}}\right]$$

$$= E(S) - \left[\frac{\rho\left(1-\rho^{N}\right)}{1-\rho^{N+1}}\right]$$

Case (ii): $\rho = 1$

$$E(Q) = E(S) - (1 - P_0)$$

$$= N/2 - \left[1 - \frac{1}{N+1}\right]$$

$$= \frac{N}{2} - \left[\frac{N+1-1}{N+1}\right]$$

$$= \frac{N}{2} - \frac{N}{N+1}$$

$$= N/2 * \left[1 - \frac{2}{N+1}\right] = \frac{N}{2}\left[\frac{N+1-2}{N+1}\right]$$

$$= \frac{N(N-1)}{2(N+1)}$$

Formulas:

1. $\rho = \lambda/\mu$
2. $P_0 = \begin{cases} \dfrac{1-\rho}{1-\rho^{N+1}}; & \rho \neq 1 \\ \dfrac{1}{N+1}; & \rho = 1 \end{cases}$

3. $P_n = \begin{cases} \rho^n P_0; & \rho \neq 1 \\ P_0; & \rho = 1 \end{cases}$

4. $E(S) = \dfrac{\rho\,[1-(N+1)\rho^N + N\,.\,\rho^{N+1}]}{(1-\rho)\,(1-\rho^{N+1})}$

5. $E(Q) = \begin{cases} \dfrac{\rho^2[1-N\rho^{N-1}+(N-1)\,\rho^N]}{(1-\rho)\,(1-\rho^{N+1})} \\ \text{(or)} \\ E(S)-\rho\left[\dfrac{(1-\rho^N)}{(1-\rho^{N+1})}\right] \end{cases}$

6. $W(S) = E(S)/\lambda'$
7. $W(Q) = E(Q)/\lambda'$

where $\lambda' = \lambda\,(1 - P_N)$ is the mean rate of customers entering the system.

Example: 15

If for a period of 2 hours in the day (8 to 10 am) trains arrive at the yard every 20 minutes but the service time continues to remain 36 minutes, then calculate for this period.

(i) The probability that the yard is empty.
(ii) Average number of trains in the system; on the assumption that the line capacity of the yard is limited to 4 trains only.

Given:

$$N = 4$$
$$\lambda = 1/20 \text{ per mt.}$$
$$\mu = 1/36 \text{ per mt.}$$
$$\rho = \lambda/\mu = (1/20)/(1/36) = 36/20 = 9/5$$
$$\rho = 9/5\ (\neq 1)$$

To Find:

(i) The probability that the yard is empty (P_0)
(ii) Average number of trains in the system (E(S))

(i) wkt: $P_0 = \dfrac{1-\rho}{1-\rho^{N+1}}$ (if $\rho \neq 1$)

$$= \frac{1-(9/5)}{1-(9/5)^5} = \frac{-0.8}{-17.89568} = 0.0447$$

(ii) wkt: $E(S) = \dfrac{\rho\,[1-(N+1)\rho^N + N\,.\,\rho^{N+1}]}{(1-\rho)\,(1-\rho^{N+1})}$

$$= \frac{(9/5)\,[1-(4+1)\,(9/5)^4 + 4(9/5)^{4+1}]}{(1-9/5)\,(1-(9/5)^{4+1})}$$

$$= \frac{43.370496}{14.316544}$$

$$= 3.0294 \text{ trains.}$$

Hence, the solution is,

(i) The probability that the yard being empty is 0.0447
(ii) Expected number of trains in the system is 3.0294

*Example: 16

Assume that the goods trains are coming to the yard at the rate of 30 trains per day and suppose that the inter-arrival time follows an exponential distribution. The service time for each train is assumed to be exponential with an average of 36 minutes. If the yard can admit 9 trains at a time (there being 10 lines one of which is reserved for shunting purposes), calculate the probability that the yard is empty and find the average queue length.

Given:

$N = 9$
Arrival rate $= \lambda = 30$ trains/day
Service time/train = 36 mts.
That is service rate $= \mu = 1/36$ per mt.
$\mu = 1/36 \times 60 \times 24 = 40$ trains/day
$\rho = \lambda/\mu = 30/40 = 3/4$
To find (i) P_0 (ii) $E(Q)$

(i) Since $\rho \neq 1$;

$$P_0 = \frac{1-\rho}{1-\rho^{N+1}} = \frac{(1-(3/4))}{(1-(3/4)^{10})} = \frac{0.25}{0.9437} = 0.265$$

(ii) wkt: $E(Q) = \dfrac{\rho^2[1 - N\rho^{N-1} + (N-1)\,\rho^N]}{[1-\rho][1-\rho^{N+1}]}$

$$= \frac{(3/4)^2\,[1-9(3/4)^8 + 8*(3/4)^9]}{[1-3/4][1-(3/4)^{10}]}$$

$$= \frac{0.393598}{0.23592} = 1.668$$

Hence the solution is,

(i) The probability of the yard being empty is 0.265
(ii) The average number of trains in the queue is 1.668.

*Example: 17

Customers arrive at a one-window drive-in bank according to Poisson distribution with mean 10 per hour. Service time per customer is exponential with mean 5 minutes. The car space in front of the window, including that for the service can accommodate a maximum of 3 cars.

(a) What is the probability that an arriving customer can drive directly to the space in front of the window?
(b) Number of cars waiting in the queue.

Given:
$N = 3$
Arrival rate $= \lambda = 10/hr$
Service time per car = 5 mts
Service rate $(\mu) = 1/5$ per mt
$\mu = 1/5 \times 60 = 12/hr$
$\rho = \lambda/\mu = 10/12 = 5/6\ (\rho \neq 1)$

To find:

(a) The probability that an arriving customer can drive directly to the space in front of the window $[P_0 + P_1 + P_2]$
(b) Number of cars waiting in the Queue E(Q)

(a) $$P_0 + P_1 + P_2 = P_0 + \rho P_0 + \rho^2 P_0$$
$$= (1 + \rho + \rho^2) P_0$$
$$= (1+\rho+\rho^2)\frac{(1-\rho)}{1-\rho^{(3+1)}}$$
$$= \left(1+\frac{5}{6}+\left(\frac{5}{6}\right)^2\right)\frac{(1-(5/6))}{[1-(5/6)^4]}$$
$$= \frac{0.42130}{0.51775} = 0.81371$$

(b) $$E(Q) = \frac{\rho^2[1-N\rho^{N-1}+(N-1)\rho^N]}{[1-\rho][1-\rho^{N+1}]}$$
$$= \frac{(5/6)^2[1-(5/6)^2 * 3 + 2*(5/6)^3]}{[1-(5/6)][1-(5/6)^4]}$$
$$= \frac{0.05144}{0.08629}$$
$$= 0.5961$$

Hence the solution is,

(a) The probability that an arriving customer can drive directly to the space in front of the window is 0.81371
(b) Number of cars waiting in the Queue is 0.5961.

8.5 MODEL : 3 MULTI-SERVER MODEL WITH INFINITE QUEUE (M/M/C): (∞/FCFS)

This model is the generalization of the previous models. That is, here instead of one server, we are going to consider more than one server ($C \geq 2$). The arrival rate and the servicing rate are one and the same for all the servers.

$$\lambda_n = \lambda \text{ for all n}$$

$$\mu_n = \begin{cases} n\mu & 1 \le n < c \\ c\mu & n \ge c \end{cases}$$

$$\rho = \lambda/\mu$$

Formulas

1. $P_0 = \left[\sum_{n=0}^{c-1} (\rho^n/n!) + (1/C!)\,\rho^c \cdot \frac{1}{(1-\rho/c)}\right]^{-1}; \rho/c < 1$

2. $P_n = \begin{cases} \frac{\rho^n P_0}{n!} & 1 \le n < C \\ \frac{1}{C^{(n-c)}.C!}\rho^n P_0; & n \ge C \end{cases}$

3. Probability that an arrival has to wait is given by,

$$P(n \ge c) = \frac{\rho^c \cdot P_0}{C!(1-\rho/C)}$$

4. Probability that an arrival enters the service without waiting is

$$= 1 - P\,(n \ge c)$$

$$= 1 - \frac{\rho^c \cdot P_0}{C!(1-\rho/c)}$$

5. Average Queue length $E(Q) = \frac{\rho^{c+1} \cdot P_0}{(C-1)!(C-\rho)^2}$
6. Average number of persons in the system $E(S) = \rho + E(Q)$
7. Average waiting time of an arrival in the Queue $W(Q) = E(Q)/\lambda$
8. Average waiting time of an arrival in the system $W(S) = W(Q) + 1/\mu$

***Example: 18**

A super market has two girls ringing up sales at the counters. If the service time for each customer is exponential with mean 4 minutes and if people arrive in a Poisson fashion at the rate of 10 per hour.

(i) What is the probability of having to wait for service?
(ii) What is the expected percentage of idle time of each girl?
(iii) If a customer has to wait what is the expected length of his waiting time.

Given:

$C = 2$
Arrival rate = 10 per hour
Service time = 4 mts/customer
That is, service rate = μ = 1/4 per mt.
$\mu = 1/4 * 60 = 15$ per hour
$\rho = \lambda/\mu = 10/15 = 2/3$
$C = 2$

To find:

(a) The probability of having to wait for service [P(n ≥ C); C = 2]
(b) Expected percentage of idle time for each girl,
(c) Expected length of customers waiting time.

$$P_0 = \left[\sum_{n=0}^{1}(\rho^n/n!) + \frac{1}{2!}\rho^2\frac{1}{(1-\rho/2)}\right]^{-1}$$

$$= \left[1+\rho+\frac{\rho^2}{2}\frac{1}{(1-\rho/2)}\right]^{-1}$$

$$= \left[1+\frac{2}{3}+\frac{(2/3)^2}{2}\frac{1}{(1-1/3)}\right]^{-1}$$

$$= \left[1+\frac{2}{3}+\frac{2}{9}\cdot\frac{3}{2}\right]^{-1}$$

$$= \left[1+\frac{2}{3}+\frac{1}{3}\right]^{-1}$$

$$= [2]^{-1}$$

$$P_0 = 1/2$$

(a) $$P(n \geq 2) = \frac{\rho^2}{2!(1-\rho/2)}.P_0$$

$$= \frac{(2/3)^2}{2.(1-(1/3))}.\frac{1}{2}$$

$$= \frac{4}{2\times 9}\times\frac{3}{2}\times\frac{1}{2}$$

$P(n \geq 2) = 1/6$

(b) Expected idle time for girls $= 2.P_0 + 1.P_1$

$$= 2.\frac{1}{2}+\frac{1}{3} \qquad [P_1 - \rho P_0 = 1/3]$$

$$= 1 + (1/3)$$

$$= 4/3$$

Expected idle time for each girl = (4/3)/2 = 2/3

[Note: Both counters may be free or one counter may be free]

(c) $$E(Q) = \frac{\rho^{c+1}\cdot P_0}{(C-1)!\,(C-\rho)^2}$$

$$= \frac{(2/3)^3\cdot(1/2)}{(2-2/3))^2}$$

$$= \frac{(2/3)^3\cdot(1/2)}{(4/3)^2}$$

$$= \frac{(2/3)^3 \cdot (1/2)}{4 \cdot (2/3)^2}$$

$$= \frac{(2/3) \cdot (1/2)}{4} = 1/12$$

Average waiting time of customers = W(Q) = E(Q)/λ

$$= \frac{(1/12)}{10} = \frac{1}{120} \text{ hr} = 1/2 \text{ mt}$$

Hence the solution is,

(a) The probability of having to wait for service is 1/6.
(b) Expected percentage of idle time for each girl is 0.67.
(c) Expected length of customers waiting time in the Queue is (1/2) minute.

***Example: 19**

If the mean arrival rate is 24 per hour, consider from the customer's point of view, the time spent in the system, whether 3 channels in parallel with mean service rate of 10 per hour is better or worse than a single channel with mean service rate of 30 per hour.

Given:
N = 3
Arrival rate = λ = 24 per hour
Service rate = μ = 10 per hour C = 3

$\lambda/\mu = 24/10 = 2.4$

To find W(S) when C = 3 and when C = 1. Compare it.

(i) wkt: W(S) = E(Q) + 1/μ; if C = 3

$$W(Q) = \frac{E(Q)}{\lambda} = \frac{\rho^{c+1} P_0}{\lambda (C-1)! (C-\rho)^2}$$

$\rho = 2.4$
$C = 3$

$$P_0 = \left[\sum_{n=0}^{2} (\rho^n / n!) + \frac{1}{3!}\rho^3 \frac{1}{(1-\rho/3)} \right]^{-1}$$

$$= \left[1 + \rho + \frac{\rho^2}{2} + \frac{1}{6}\rho^3 \frac{3}{(3-\rho)} \right]^{-1}$$

$$= \left[1 + 2.4 + \frac{(2.4)^2}{2} + \frac{1}{6}(2.4)^3 \frac{3}{(3-2.4)} \right]^{-1}$$

$$= (17.56)^{-1} = 0.057$$

$$E(Q) = \frac{(2.4)^4 P_0}{(20)(2!)(3-2.4)^2} = 2.627$$

$$W(S) = \frac{2.627}{24} + \frac{1}{10} = 0.2095 \text{ hr} = 12.6 \text{ mts}$$

(ii) wkt: $W(S) = \frac{1}{\mu - \lambda}$; if C = 1

$$= \frac{1}{30-24} = 0.17 \text{ hr} = 10 \text{ mts}$$

Since the average time customer spends in a multichannel (c = 3) is more than a single channel, a single channel with mean service rate of 30 per hour is better.

8.6 MODEL : 4 MULTI-SERVER MODEL WITH FINITE QUEUE (M/M/C): (N/FCFS)

This model is very similar to the model-3 except that the size of the system is limited to N, where N > C.

Formulas

1. $\rho = \lambda/\mu$
2. $P_0 \left[\sum_{n=0}^{C-1} (\rho^n/n!) + \sum_{n=C}^{N} \frac{\rho^n}{C!C^{n-c}} \right]^{-1}$
3. $P_n = \begin{cases} (\rho^n/n!)P_0 & 0 \le n \le C \\ \frac{\rho^n P_0}{C!C^{n-c}} & C \le n \le N \end{cases}$
4. $E(Q) = \frac{\rho^c \, t \, P_0}{C!(1-t)^2} \quad [1 - t^{N-C+1} - [1-t](N-C+1)t^{N-C}] \qquad \text{where } t = \rho/c$
5. $E(S) = E(Q) + C - P_0 \sum_{n=0}^{C-1} \frac{C-n}{n!} \rho^n$
6. $W(S) = E(S)/\lambda'$
7. $W(Q) = E(Q)/\lambda'$ where $\lambda' = \lambda (1 - P_N)$

***Example: 20**

A barber shop has two barbers and three chairs for waiting customers. Assume that customers arrive in a Poisson fashion at a rate of 5 per hour and each barber services customers according to an exponential distribution with mean of 15 minutes. Further, if a customer arrives and there are no empty chairs in the shops he will leave. What is the probability that the shop is empty? What is the expected number of customers in the shop.

Given:

N = 3
C = 2
$\lambda = 5$/hr
$\mu = 4$/hr ie) [1/15 × 60 = 4]
$\rho = \lambda/\mu = 5/4 = 1.25$ ($\rho > 1$)

To find:

(a) P_0
(b) E(S)

$$t = \rho/c = 5/8$$
$$1 - t = 3/8;\ c = 2$$
$$N - C + 1 = 2;\ N - C = 1;\ P_0 = 0.2841;\ \rho = 5/4$$

$$E(Q) = \frac{\rho^c t P_0}{C!(1-t)^2} \quad [1 - t^{N-C+1} - (1-t)(N-C+1)t^{N-C}]$$

$$= \frac{(5/4)^2 (5/8)(0.2841)}{2!(3/8)^2} [1 - (5/8)^2 - (3/8) * 2 * (5/8)]$$

$$= 0.9865 * 0.1406$$
$$= 0.1387$$

$$E(S) = E(Q) + C - P_0 \sum_{n=0}^{C-1} \frac{C-n}{n!} \rho^n$$

$$= 0.1387 + 2 - 0.2841\ [2 + (5/4)]$$
$$= 1.225$$

Hence the solution is,

(a) The probability that the shop is empty 0.2841.
(b) The expected number of customers in the shop is 1.225 persons.

EXERCISE PROBLEMS

*1. A drive in a bank window has a mean service time of 2 minutes, while the customers arrive at a rate of 20 per hour. Assuming that these represent rates with a Poisson distribution, find

(i) what percentage of time will the teller be idle?
(ii) after driving in, how long will it take the average customer to wait in line and be served?
(iii) what fraction of customers will have to wait in line?

*2. An automatic car wash with a single service boy takes 5 minutes per car. On a typical Saturday, cars arrive an average of 8 per hour at Poisson distribution. Find the

(i) Average number of cars in the line.
(ii) Average time cars spend in line and service.

*3. In a bank every 15 minutes one customer arrives for cashing the cheque. The staff in the only payment counter takes 10 minutes for serving a customer on an average. State the suitable assumptions and find:

(i) The average queue length.
(ii) The increase in the arrival rate in order to justify a second counter (when the waiting time of a customer is at least 15 minutes the Management will increase on more counter)

4. In a super market the average arrival rate of customers is 5 every 30 minutes. The average time it takes to list and calculate the customer's purchases at the cash desk is 4.5 minutes, and this time is exponentially distributed.
   (i) How long will the customer expect to wait for service at the cash desk?
   (ii) What is the chance that the queue length will exceed 5?
   (iii) What is the probability that the Cashier is working.

   [Hint: $\lambda = 5/30$/mt $\mu = 1/4.5$/mt (a) W(Q) (b) $P_Q (> 5)$ (c) ρ]

*5. At what average rate must a clerk at a super market work in order to ensure a probability of 0.9 that the customer will net longer than 12 minutes? It is assumed that there is only one counter, at which customers arrive in a Poisson fashion at an average rate of 15 per hour. The length of service by the clerk has an exponential distribution.
   [Hint: $\lambda = 1/4$/mt; $P (\geq 12) = 0.1$ => service rate is 2.48 mts]

*6. At a one-man barber shop, customers arrive according to Poisson distribution with a mean arrival rate of 5 per hour and his hair cutting time was exponentially distributed with an average hair cut taking 10 minutes. It is assumed that because of his excellent reputation, customers were always willing to wait. Calculate the following
   (i) Average number of customers in the shop and the average number of customers waiting for a hair cut. [E(S) & E(Q)]
   (ii) The percent of time an arrival can walk right in without having to wait. (P_0)
   (iii) The percentage of customers who have to wait prior to getting into the barber's chair. ($P_s \geq 1$)

*7. In a railway marshalling yard, goods trains arrive at a rate of 30 trains per day. Assuming that the inter arrival time follows an exponential distribution and the service time distribution is also exponential with an average 36 minutes. Calculate the following:
   (i) The mean queue length
   (ii) The probability that the Queue size exceeds 10. If the input of trains increases to an average 33 per day, what will be the change in (i) & (ii).
   [Hint: Case (i): Take $\lambda = 30$/day & $\mu = (1/36) \times 24 \times 60 = 40$/day find (i) & (ii)
   Case (ii): Take $\lambda = 33$/day & $\mu = 40$/day find (i) & (ii)]

*8. Customers arrive at a sales counter manned by a single person according to a Poisson process with a mean of 20 per hour. The time required to serve a customer has an exponential distribution with a mean of 100 seconds. Find the average waiting time of a customer. (Hint: $\lambda = 20$/hr $\mu = (1/100) \times 60 \times 60 = 36$/hr find W(S) & W(Q))

9. A bank plans to open a single server service drive-in banking facility at a particular centre. It is estimated that 28 customers will arrive each hour on an average. If on an average, it requires 2 minutes to process a customer's transaction, determine.
   (i) The proportion of time when the system will be idle.
   (ii) On an average how long the customer will have to wait before reaching the service.
   (iii) The length of the driveway required to accommodate all the arrivals, on the average, if 20 feet of drive is required for each car that is waiting for service.

   [Hint: i) P_0 ii) W(Q) iii) E(Q) & Total driveway required = E(Q)*13 feet (given for one car the driveway required is 13 feet]

*10. Past records indicate that of the five machines that a factory owns, breakdown occur at random and the average time between the breakdown is 2 days. Assuming that the repairing capacity of the workman is one machine a day and the repairing is distributed exponentially, determine the following:

(i) the probability that the service facility will be idle
(ii) the probability of various number of machines (0 through 5) to be, and being repaired.
(iii) the expected length of the queue.

*11. Patients arrive at a clinic according to Poisson at a rate of 30 patients per hour. The waiting room does not accommodate more than 14 patients. Examination time per patient is exponential with mean rate of 20 per hour.

(a) Find the effective arrival rate at the clinic.
(b) What is the probability that an arriving patient will not have to wait?
(c) What is the expected waiting time until a patient is discharged from the clinic?

[Hint: (a) $\lambda' = \lambda [1 - P_N]$ (b) P_0 (c) W (S)]

*12. Consider a single server queuing system with a Poisson input, exponential service times. Suppose the mean arrival rate is 3 calling units per hour the expected service time is 0.25 hours and the maximum permissible calling unit in the system is two. Calculate the expected number in the system.

*13. A barbershop has space to accommodate only 10 customers. He can serve only one person at a time. If a customer comes to his shop and finds it full, he goes to the next shop. Customers randomly arrive at an average rate of $\lambda = 10$ per hour and the barber's service time is 5 minutes per customer Find P_0 and P_n.

14. A hospital emergency room can accommodate at most 5 patients. The patients arrive at a rate of 4 per hour. The single staff physician can only treat 5 patients per hour. Any patient overflow is directed to another hospital.

(i) Determine the probability of distribution for the number of patients either waiting for or receiving treatment at any given time.
(ii) Determine the mean value for the number of patients in the emergency room and number of patients waiting to see the doctor.

*15. A Telephone exchange has two long distance operators. It is observed that, during the peak load, long distance calls arrive in an Poisson fashion at an average rate of 15 per hour. The length of service on these calls is approximately exponentially distributed with mean length 5 minutes.

(i) What is the probability that a subscriber will have to wait for his long distance call during the peak hours of the day?
(ii) If the subscribers will wait and are serviced in turn, what is the expected waiting time?

[Hint: (a) $P(n \geq 2)$ (b) $W(Q) = E(Q)/\lambda$]

*16. A bank has two tellers working on savings account. The first teller handles withdrawals only. The second teller handles deposits only. It has been found that the service time distributions for both deposits and withdrawals are exponential with mean service time of 3 minutes per customer. Depositors are found to arrive in a Poisson fashion throughout the day with mean arrival rate of

16 per hour. Withdrawers also arrive in a Poisson fashion with mean arrival rate of 14 per hour. What would be the effect on the average waiting time for depositors and withdrawers if each teller handles both withdrawals and deposits?

[Hint: Withdrawals:

$\mu = 1/3$ per mt.
$= 1/3 * 60 = 20$/hr
$\lambda = 14$/hr

Use [M/M/1]: [∞/FCFS]
Find: W(Q)

Deposits:

$\mu = 1/3$ per mt.
$= 1/3 * 60 = 20$/hr
$\lambda = 16$/hr

Use [M/M/1]: [∞/FCFS]
Find: W (Q)

If both tellers consider withdrawals and deposits, then
$\lambda = 14 + 16 = 30$/hr
$\mu = 20$/hr
C = 2. Use [M/M/2]: [∞/FCFS]
Find W(Q)]

*17. A post office has 3 windows providing the same services. It receives an average of 30 customers/hr. Arrivals are Poisson distributed and service time is exponentially distributed. The post office serves 12 customers per hour.

(a) What is the probability that a customer will be served immediately?
(b) What is the probability that a customer will have to wait?
(c) What is the average number of customers in the system?
(d) What is the average total time that a customer must spend in the post office?

[Hint: (a) P_0 (b) P $(n \geq 3)$ (c) W(Q) (d) W(S)]

*18. A supermarket has two salesgirls at the counter. If the service time for each customer is exponential with an average of 3 minutes and if people arrive in a Poisson fashion at the rate of 12 per hour. Find the expected percentage of idle time for each sales girl.

*19. Given an average arrival rate of 20 per hour, is it better for a customer to get service at a single channel with mean service rate of 22 customers per hour or at one of two channels in parallel with mean service rate of 11 customers per hour for each of the two channels? Assume standard distribution for arrival and service.

[Hint: Similar to exercise problem-2. Find the waiting time W(S) of both the cases and decide the better one].

*20. An oil company is constructing a service station on a highway. Traffic analysis indicates that customer's arrivals over most of the day would approximate a Poisson distribution with a mean of 30 automobiles per hour. Previous studies show that one pump could service a mean of 10 automobiles per hour, with the service time distribution approximating the negative exponential. If 4 pumps are installed:

(i) What is the probability that an arrival would have to wait in line?
(ii) Find out the average waiting time, average time spent in the system and the average number of automobiles in the system.
(iii) For what percentage of time would a pump be idle on an average?

[Hint: (a) P $(n \geq 4)$ (b) W (Q), W (S), E (S) (c) P_0]

21. Let there be an automobile inspection station with three inspection stalls. Assume that cars wait in such a way that when a stall becomes vacant, the car at the head of the line pulls up to it. The station can accommodate at most four cars waiting (seven in the station) at a time. The arrival pattern is Poisson with mean of one car every minute during the peak hours. The service time is exponential with mean 6 minutes. Find the average number of customers in the system during peak hours, the average waiting time and the average number per hour that cannot enter the station because of full capacity.

    [Hint: (i) E(S) = 6.06 cars (ii) W(S) = 12.3 mts

    (iii) Expected number of cars that cannot enter the station = 60 * λ * P_N; N = 7 = 30.3 cars/hr]

ANSWERS TO THE EXERCISE PROBLEMS

1. (i) 33.33% (ii) 1/15 hr (iii) 0.67
2. (i) 2.25 (ii) 1/6 hr: 2/9 hr
3. (i) 1.33

   (ii) If the arrival rate is 3.6, one more counter can be opened. However it must be noted that a second counter is justified at the existing rate also.
4. (a) 13.5 mts (b) 0.133 (c) 0.75
5. λ = 1/4/mt; P ($\geq$ 12) = 0.1 => service rate is 2.48 mts
6. (i) E(S) = 5; E(Q) = 4.15

   (ii) 0.833 (iii) 16.7%
7. (a) (i) 3 (ii) 0.06 (b) (i) 4.8 (ii) 0.2
8. W(S) = 0.0625 hr; W(Q) = 0.0347 hr
9. (i) 0.067 (ii) 28 mts (iii) 13; 260 feet
10. (i) 0 0.5

    (ii)

No. of Machines (n)	0	1	2	3	4	5
$P_s(n)$	1/2	1/4	1/8	1/16	1/32	1/64
$P_s(\geq n)$	1	1/2	1/4	1/8	1/16	1/32
$P_Q(\geq n)$	1/2	1/4	1/8	1/16	1/32	1/64

    (iii) 1/8
11. (a) 20 (b) 0.67 (c) 0.65 hr
12. 0.81

13. $P_0 = 0.1925$ $P_n = 0.1925\ (5/6)^n$

(a)

n	0	1	2	3	4	5
$P_s(\geq n)$	0.271	0.217	0.173	0.139	0.111	0.0888

(b) E(S) = 1.869; E(Q) = 1.140

14. E(S) = 1.86 = 2 approximately
E(Q) = 1.13 = 1 approximately

15. 0.40 (b) 3.2 mts

16. Withdrawals W(Q) = 7 mts
Deposits W(C'I = 12 mts
Both Teller (Withdrawals & Deposits): 3.86 mts

17. (a) 0.29 (b) 0.7 (c) 6 (app) (d) 12 mts

18. 67% for each girl

19. Belter with two servers

20. (a) 0.38
(b) W(Q) = 3.05 mts

W(S) = 9.05 mts

E(S) = 4.53

(c) 24.98%

21. (i) E(S) = 6.06 cars (ii) W(S) = 12.3 mts
(iii) Expected number of cars that cannot enter the station = 60 * λ * P_N; N = 7 = 30.3 cars/hr

REVIEW QUESTIONS

Short Questions

1. Define the following.
   i) Input process
   ii) Queue discipline
   iii) Priority Queue discipline (Section 8.2.1, 8.2.2)
2. Write a short note on service mechanism (Section 8.2.3)
3. Give an example for a finite Queue. (Capacity of a bus)
4. What do you mean by system?

   A system is nothing but an integration of many subparts which represent an item.
   Example: Car can be viewed as a system.
5. Define the following terms:

   (i) Balking (ii) Reneging (Section 8.2.1)

6. Write short notes on

   (i) Queue with infinite space (ii) Queue with finite space

   (i) The queue can accommodate any number of persons.
   (ii) The queue can admit fixed number of persons.

7. Define arrival rate

   Average number of persons getting into service per unit of time. (1 hr, 1 minute, 1 day etc.,) It is denoted by λ.

8. Define service rate (μ). (Section 8.3)

   Average number of persons leaving the system after getting their service. It is denoted by μ.

9. What do you mean by the terms 'FIFO' and 'LIFO'? (Section 8.2)

   It refers the queue discipline

   FIFO: First In First Out.
   LIFO: Last In First Out.

10. Define Traffic intensity (ρ). (Section 8.3)

    It is nothing but the ratio between the arrival rate (λ) and service rate (μ). i.e., $\rho = \lambda/\mu$.

State True or False

11. M/M/1 implies that the arrival rate is Poisson, service rate is exponential and single server model. (T)
12. MIMIC implies that the arrival rate is Poisson, service rate is exponential and multi server model. (T)
13. In the [M/M/1]: [∝/FCFS] Model ρ value can be ≥ 1. (F)
14. In the [M/M/CJ: [∝/FCFS] Model ρ value can be ≥ 1. (T)
15. Using queuing theory, we can evaluate the average waiting time of the customer in the Queue. Based on this we can decide the optimum number of counters required. (T)

Fill in the Blanks

16. Queue refers to a ______ (waiting line)
17. Queuing theory originated from the work of ______. (A.K. Erlang)
18. The service mechanism may be ______ or ______. (Single, Multichannel)
19. 'FCFS' refers to ______ in Queuing theory. [First Come First Served]
20. 'LIFO' refers to ______ in Queuing theory. [Last In First Out]
21. 'SIRO' refers to ______ in Queuing theory. [Service In Random Order]

9

Dynamic Programming

9.1 INTRODUCTION

Dynamic Programming (DP) is a technique used to solve a multi-stage decision problem where decisions have to be made at successive stages. This technique is very much useful whenever if an optimization model has a large number of decision variables. It is not having any generalized formulation. That is, we have to develop a recursive equation to suit the situations. This technique was developed by Richard Bellman in 1957. This model is derived based on the Bellman's principle.

That is, 'An optimal policy has the property that whatever the initial state and the initial decisions are, the remaining decisions must constitute an optimal policy which regards to the state resulting from the first decision'.

The above stated principle forms the basis of a dynamic programming technique. When the develop a recursive equation, we have to consideir the Bellman's principle. A recursive equation expresses subsequent state conditions in terms of the processed state conditions and it is based on the fact that a policy is 'optimal' if the decision made at each stage results in overall optimality, over all the stages and not only for the current stage. It can be used to solve problems of both nature

(i) Deterministic and
(ii) Continuous.

9.1.1 Methods Used to Solve a DP

In this section we are going to discuss two methods to solve a given Dynamic Programming Problem (DPP) namely,

1. Calculus Method
2. Tabular Method

9.1.2 Characteristics of a DPP

The characteristics can be listed as follows:

1. The given problem can be subdivided into various stages with a clear policy decision to be taken at each stage.
2. Each stage comprises a number of states in it.

3. Decision taken at each stage modifies the latest stage into a state related to the next stage.
4. In each stage the state of the system is referred by the state variables.
5. If we know the current state, the optimum policy of the succeeding stages is independent of the policy taken for the processed stages.
6. Construct a recursive equation to make a decision based on the optimum policy at each stage.
7. After the construction of the recursive equation, regarding the solution, either we can move forward or backward.

9.1.3 Merits and Demerits of a DPP

Merits

(i) A complex problem can be spilitted into a series of interrelated subproblems and can be solved easily
(ii) It is highly flexible.
(iii) Built-in sensitivity analysis.
(iv) Computational savings over Computer Enumeration Procedure.

Demerits

(i) DP is conceptually powerful but high degree of expertise and insight is required for the efficient formulation of the problem.
(ii) More storage space is required in the computer.
(iii) There is no unique algorithm

9.1.4 Construction of a Recursive Equation

Additive Recursive Equation

Consider the DPP:

$$\text{Max } z = \sum_{j=1}^{n} f_j(x_j) \tag{1}$$

$$\text{s.t.} = \sum_{j=1}^{n} a_j(x_j) \le b$$

$$x_j \ge 0 \ (j = 1, 2, \ldots n)$$

Consider

$$\max_{x_1, x_2 \ldots x_{n-1}} \left[\sum_{j=1}^{n} f_j(x_j) \right] = \left[f_n(x_n) + \left[\max_{x_1\, x_2 \ldots x_{n-1}} \sum_{j=1}^{n-1} f_j(x_j) \right] \right]$$

$$\max_{x_n} \left[\max_{x_1, x_2 \ldots x_{n-1}} \left[\sum_{j=1}^{n} f_j(x_j) \right] \right] = \max_{x_n} \left[f_n(x_n) + \left[\max_{x_1\, x_2 \ldots x_{n-1}} \sum_{j=1}^{n-1} f_j(x_j) \right] \right]$$

$$\left[\max_{x_1, x_2 \ldots x_n} \left[\sum_{j=1}^{n} f_j(x_j) \right] \right] = \max_{x_n} \left[f_n(x_n) + \left[\max_{x_1\, x_2 \ldots x_{n-1}} \sum_{j=1}^{n-1} f_j(x_j) \right] \right] \tag{2}$$

Define,

$$\begin{aligned} &\max_{x_1, x_2 \ldots x_j} \sum_{j=1}^{n} f_j(x_j) = \lambda_j(c) \\ &\text{s.t.} \quad \sum_{j=1}^{n} a_j x_j \le c; \ (c \ge 0) \end{aligned} \qquad (3)$$

Using (3) in (2) we have,

$$\lambda_n(b) = \max_{\substack{x_n \\ 0 \le a_n x_n \le b}} [f_n(x_n) + \lambda^*_{n-1}(b - a_n x_n)]$$

Put n = 1, in (4),

$$\lambda_1(b) = \max_{\substack{x_1 \\ 0 \le a_1 x_1 \le b}} [f_1(x_1)]$$

Put n = 2, in (4),

$$\lambda_2(b) = \max_{\substack{x_2 \\ 0 \le a_2 x_2 \le b}} [f_2(x_2) + \lambda^*_1(b - a_2 x_2)]$$

Put n = 3, in (4),

$$\lambda_3(b) = \max_{\substack{x_3 \\ 0 \le a_3 x_3 \le b}} [f_3(x_3) + \lambda^*_2(b - a_3 x_3)]$$

Likewise we can have the recursive equation for all n stages (one stage for one variable).

(* – Optimum value)

Multiplicative Recursive Equation

Consider,

$$\text{Max } Z = f_1(x_1) * f_2(x_2) * \ldots * f_n(x_n) = \pi_{j=1}^{n} f_j(x_j)$$

$$\text{s.t.} \sum_{j=1}^{n} a_j x_j \le b; \qquad x_j \ge 0 \ (j = 1 \ldots n)$$

Similarly proceeding like the above case, we have

$$\lambda_n(b) = \max_{x_n} [f_n(x_n) * \lambda^*_{n-1}(b - a_n x_n)]$$

9.2 CALCULUS METHOD TO SOLVE A DPP

The various steps to be followed to solve a DPP using calculus method are clearly depicted in the Flowchart 9.1.

Example: 1

Divide a positive Quantity b into n parts in such a way that their product is maximum

Let $x_1, x_2 \ldots x_n$ stand for the first, second, and nth parts respectively. Then we have to

$$\text{Max } Z = x_1 \cdot x_2 \cdot x_3 \ldots x_n$$

s.t.
$$x_1 + x_2 + x_3 \ldots + x_n = b$$
$$x_j \geq 0 (j = 1 \ldots n)$$

The recursive equation is

$$\lambda_n(b) = \underset{\substack{x_n \\ 0 \leq x_n \leq b}}{\text{Max}} [x_n * \lambda^*_{n-1}[b - a_n x_n]]$$

As per this method, we consider one variable for one-stage, on the whole we are going to solve this in n-stages.

Stage 1: Consider x_1 & $\lambda_1(b)$

$$\lambda_1(b) = \underset{0 \leq x_1 \leq b}{\text{Max}} \{x_1\}$$

Since $x_1 \in [0, b]$, the maximum value attains at $x_1 = b$. That is,

$$\lambda^*_1(b) = b \tag{1}$$

Stage 2: Consider x_2 & $\lambda_2(b)$

$$\lambda_2(b) = \underset{0 \leq x_2 \leq b}{\text{Max}} \{x_2 * \lambda^*_1(b - x_2)\}$$

$$= \underset{0 \leq x_2 \leq b}{\text{Max}} \{x_2 * (b - x_2)\} \quad \text{using (1)}$$

Let $x_2 = z$ & $\phi(z) = z(b - z)$ find the optimum value of $\phi(z)$ using calculus method.

$$\phi'(z) = \frac{d}{dz}(\phi(z)) = \frac{d}{dz}[z(b - z)] = b - 2z$$

$$\phi'(z) = 0 \Rightarrow b - 2z = 0 \Rightarrow z = b/2$$

$$\phi''(z) = \frac{d}{dz}(\phi'(z)) = \frac{d}{dz}(b - 2z) = -2 < 0$$

implies at $z = b/2$ the function $\phi(z)$ is maximum.

$$\phi(z) = (b/2)(b - b/2) = (b/2)^2$$
$$\Rightarrow \lambda^*_2(b) = (b/2)^2 \tag{2}$$

---

*These questions have appeared in Indian university examination papers.

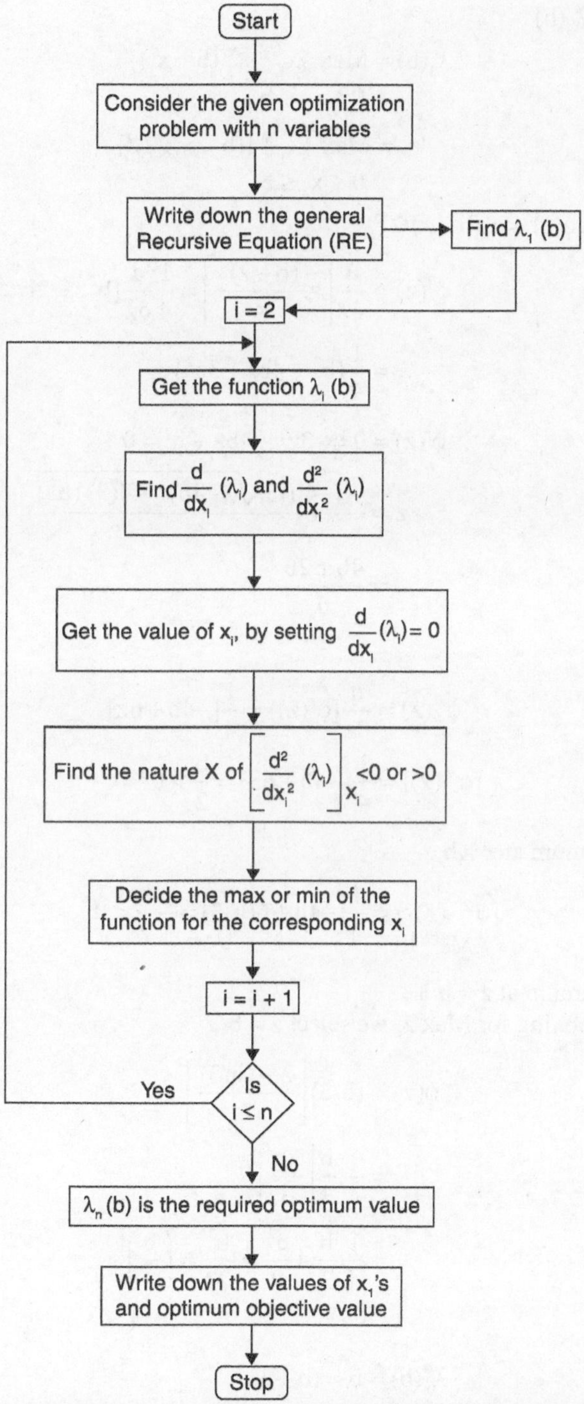

Flowchart 9.1 Calculus method to solve a DPP

Stage 3: Consider x_3 & $\lambda_3(b)$

$$\lambda_3(b) = \underset{0 \le x_3 \le b}{\text{Max}} \{x_3 * \lambda_2^*(b - x_3)\}$$

$$= \underset{0 \le x_3 \le b}{\text{Max}} \{x_3 * [(b - x_3)/2]^2\} \qquad \text{using (2)}$$

put $x_3 = z$ & let $\phi(z) = z[(b - z)/2]^2$

$$\phi'(z) = \frac{d}{dz}\left[z.\frac{(b-z)^2}{4}\right] = \frac{1}{4}\frac{d}{dz}[b^2z - 2bz^2 + z^3]$$

$$= \frac{1}{4}(b^2 - 4bz + 3z^2)$$

$$\phi'(z) = 0 \Rightarrow 3z^2 - 4bz + b^2 = 0$$

$$z = \frac{-(-4b) \pm \sqrt{(-4b)^2 - 4(3)(b^2)}}{6}$$

$$= \frac{4b \pm 2b}{6}$$

i.e. $z = b, b/3$

$$\phi''(z) = \frac{d}{dz}[\phi'(z)) = \frac{1}{4}[-4b + 6z]$$

$$[\phi''(z)\}_{z=b} = \frac{1}{4}[-4b + 6b] = \frac{b}{2} > 0$$

$\Rightarrow \phi(z)$ is minimum at $z = b$

$$[\phi''(z)\}_{z=(b/3)} = \frac{1}{4}[-4b + 6(b/3)] = -\frac{b}{2} < 0$$

$\Rightarrow \phi(z)$ is maximum at $z = b/3$
Since we are looking for Max z; we select $z = b/3$

$$\phi(z) = (b/3)\left[\frac{b - (b/3)}{2}\right]^2$$

$$= \frac{1}{4}.\frac{b}{3}\left[\frac{2b}{3}\right]^2$$

$$= \frac{1}{4}.\frac{b}{3}.4\frac{b^2}{9} = \frac{b^3}{27} = \left(\frac{b}{3}\right)^3$$

So far we have,

$$\lambda_1^*(b) = b = (b/1)^1$$
$$\lambda_2^*(b) = (b/2)^2$$
$$\lambda_3^*(b) = (b/3)^3$$

To prove the rest of the cases 4, 5 … n. We can make use of the concept of Induction. Assume this result for $\lambda^*_m(b)$ and prove it for $\lambda^*_{m+1}(b)$.

That is, $\lambda^*_m(b) = (b/m)^m$

Then prove that $\lambda^*_{m+1}(b) = (b/(m+1))^{m+1}$

We have,

$$\lambda_{m+1}(b) = \underset{0 \le x_{m+1} \le b}{\text{Max}} \{x_{m+1} * \lambda^*_m(b - x_{m+1})\}$$

$$= \underset{0 \le x_{m+1} \le b}{\text{Max}} \left[x_{m+1} * \left(\frac{b - x_{m+1}}{m} \right)^m \right]$$

$$\text{Let } x_{m+1} = z \;\&\; \phi(z) = z\left[\frac{b-z}{m}\right]^m$$

$$\phi(z) = \frac{1}{m^m}[z\,(b-z)^m]$$

$$\phi'(z) = \frac{1}{m^m}[(b-z)^m + zm\,(b-z)^{m-1}(-1)]$$

$$= \frac{1}{m^m}[(b-z)^{m-1}[b-z-zm]]$$

$$\phi'(z) = 0 \Rightarrow \frac{1}{m^m}(b-z)^{m-1}(b-(m+1)z) = 0$$

$\Rightarrow b - (m+1)\,z = 0$

$\Rightarrow z = (b/(m+1))$ be one of the roots.

$$\phi''(z) = \frac{1}{m^m}[(b-z)^{m-1}.(-(m+1)) + (b-(m+1)z)(m-1)(b-z)^{m-2}(-1)]$$

$$= \frac{(b-z)^{m-2}}{m^m}[-(b-z)(m+1) - (b-(m+1)z(m-1))]$$

$$= \frac{(b-z)^{m-2}}{m^m}[-b(m+1) + z(m+1) - b(m-1) + (m+1)z(m-1)]$$

$$= \frac{1}{m^m}(b-z)^{m-2}.(-2mb + mz\,(m+1))$$

$$= \frac{(b-z)^{m-2}}{m^m}.\,m\,[(m+1)z - 2b$$

$$= \frac{(b-z)^{m-2}}{m^{m-1}}.[(m+1)z - 2b]$$

$$\text{At } z = \frac{b}{m+1}$$

$$b - z = b - (b/(m+1)) = \frac{bm + b - b}{m+1}$$

$$= (bm/(m+1)) > 0$$

$$(m+1)\,z - 2b = (m+1)\,(b/(m+1)) - 2b = b - 2b = -b < 0$$

$$\Rightarrow (\phi''(z)) < 0$$

$$z = (b/(m+1))$$

$$\Rightarrow \text{At } z = (b/(m+1)) \text{ the function } \phi(z) \text{ is maximum}$$

$$\lambda^*_{m+1} = \left[\frac{b}{m+1}\right]\left[\frac{b-(b/(m+1))}{m}\right]^m$$

$$= \left[\frac{b}{m+1}\right]\left[\frac{mb+b-b}{(m+1)\,(m)}\right]^m$$

$$= \left[\frac{b}{m+1}\right]\left[\frac{mb}{m.(m+1)}\right]^m$$

$$= \left[\frac{b}{m+1}\right]\left[\frac{b}{(m+1)}\right]^m$$

$$= \left[\frac{b}{m+1}\right]^{m+1}$$

Hence by induction,

$$\lambda^*_n(b) = (b/n)^n$$

$$\Rightarrow x_j = b/n;\ (j = 1,\ldots n)$$

Hence the optimum solution is

$$\text{Max } Z = (b/n)^n$$

$$x_1 = x_2 = \ldots = x_n = b/n$$

Example: 2

Use DP to show that

$$Z = P_1 \log P_1 + P_2 \log P_2 + \ldots + P_n \log P_n$$

s.t. $P_1 + P_2 + \ldots + P_n = 1$ and $P_j \geq 0$ is minimum when

$$P_1 = P_2 = \ldots = P_n = 1/n$$

Consider the additive recursive equation

$$\lambda_n(1) = \text{Max } \{P_n \log P_n + \lambda_{n-1}(1 - P_{n-1})\}$$
$$0 \leq P_n \leq 1$$

Stage 1:

Consider λ_1 (1) and the variable P_1

$$\lambda_1(1) = \text{Max } \{P_1 \log P_1\}$$
$$0 \leq P_1 \leq 1$$

$$\lambda_1(1) = 1 \log 1 \quad (1)$$

Stage 2:

$$\lambda_2(1) = \underset{0 \le P_2 \le 1}{\text{Max}} \{P_2 \log P_2 + \lambda_1^*(1 - P_2)\}$$

Let $P_2 = z$ & $\phi(z) = z\log z + (1 - z) \log (1 - z)$ (2) using (1)

$$\phi'(z) = \log z + 1 - \log(1 - z) - 1 = \log z - \log (1 - z) = \log (z/(1 - z))$$

$$\phi(z) = 0 \Rightarrow \log (z/(z - 1)) = 0 \Rightarrow (z/(1 - z)) = e^0 = 1$$

$$z/(1 - z) = 1 \Rightarrow z = 1/2$$

$$\phi''(z) = \frac{d}{dz}\log\left[\frac{z}{1-z}\right] = \left(\frac{1}{z(1-z)}\right)$$

$$[\phi''(z)]_{z=1/2} = \frac{1}{(1/2)(1-(1/2))} = 4 > 0$$

$\Rightarrow$ The function is minimum,

$$\lambda_2(1) = \frac{1}{2}\left(\log\frac{1}{2}\right) + \left(1 - \frac{1}{2}\right)\log\left(1 - \frac{1}{2}\right)$$

$$= \frac{1}{2}\log(1/2) + \frac{1}{2}\log(1/2)$$

$$= 2\left[\frac{1}{2}\log(1/2)\right]$$

So, we have

$$\lambda_1^*(1) = 1\,[1 \log (1)]$$

$$\lambda_2^*(1) = 2\,[(1/2) \log (1/2)]$$

Hence, we can assume the results for

$$\lambda_m^*(1) = m\,[(1/m) . \log (1/m)]$$

and prove it for $\lambda_{m+1}^*(1) = (m + 1)[(1/(m + 1)^* \log(1/(m + 1))]$

[The proof of λ_{m+1} is left to the reader]

Hence by Mathematical induction, we have

$$\lambda_n^*(1) = (n)\,[(1/n) \log (1/n)]$$

$$= (1/n) \log (1/n) + (1/n) \log (1/n) + \ldots + n \text{ times}$$

Comparing with the objective function, we have

$$P_i = 1/n; \; [i = 1 \ldots n]$$

Hence, we have proved that the minimum Z occurs when $P_i = 1/n$; $[i = 1 \ldots n]$

***Example: 3**

$$\text{Min } Z = y_1^2 + y_2^2 + \ldots + y_n^2$$

s.t. $y_1 \cdot y_2 \cdot y_3 \ldots y_n = c,\; y_j > 0;\; (j = 1 \ldots n)$

Since, we have n-variables, we can solve this in n-stages. The recursive equation can be given as

$$\lambda_n(c) = \underset{y_n \le c}{\text{Min}} \{y_n^2 + \lambda^*_{n-1}(c/y_n)\} \tag{1}$$

Stage 1: Put n = 1 in (1)

$$\Rightarrow \lambda_1(c) = \underset{y_1 \le c}{\text{Min}} \{y_1^2\}$$

$$\lambda_1^*(c) = c^2 \tag{2}$$

Stage 2: Put n = 2 in (1)

$$\Rightarrow \lambda_2^*(c) = \underset{y_2 \le c}{\text{Min}} \{y_2^2 + \lambda_1^*(c/y_2)\}$$

$$= \underset{y_2 \le c}{\text{Min}} \{y_2^2 + (c/y_2)^2\} \quad \text{using (2)}$$

Put $y_2 = Z$ & let $\phi(z) = z^2 + (c/z)^2$

$$\phi'(z) = 2z + 2(c^2/z)(-1/z^2)$$

$$\phi'(z) = 0 \Rightarrow 2z - (2c^2/z^3) = 0$$

$$z^4 - c^2 = 0$$

$$z^4 = c^2$$

$$z = (c^{1/2})$$

$$\phi''(z) = 2 - 2c^2(-3)\frac{1}{z^4} = 2 + 6\frac{c^2}{z^4}$$

$$(\phi''(z))_{(z=c^{1/2})} = 2 + 6\frac{c^2}{c^2} = 2 + 6 = 8 > 0$$

Implies that $\phi(z)$ is minimum when $z = c^{1/2}$

$$\Rightarrow \lambda_2(c) = (c^{1/2})^2 + (c/c^{1/2})^2$$

$$\lambda_2(c) = (c^{1/2})^2 + (c^{1/2})^2 = 2c$$

$$\lambda_2^*(c) = 2c \tag{3}$$

Step 3: Put n = 3 in (1)

$$\Rightarrow \lambda_3(c) = \underset{y_3 \le c}{\text{Min}} \{y_3^2 + \lambda_2^*(c/y_3)\}$$

$$\Rightarrow \lambda_3(c) = \underset{y_3 \le c}{\text{Min}} \{y_3^2 + 2(c/y_3)\} \quad \text{using (3)}$$

Let $y_3 = z$ and $\phi(z) = z^2 + 2c/z$

$$\phi'(z) = 2z - 2c/z^2$$

$$\phi'(z) = 0$$

$$\Rightarrow 2z - 2c/z^2 = 0$$
$$\Rightarrow 2c = z^3$$
$$z = (2c)^{1/3}$$
$$\phi''(z) = 2 + (6c/z^3) > 0$$
$$\lambda_3(c) = (c^{1/3})^2 + 2.\,(c/c^{1/3})$$
$$= c^{2/3} + 2c^{2/3} = 3(c^{2/3}) \Rightarrow \lambda_3^*(c) = 3(c^{2/3}) \qquad (4)$$

Equations (2), (3) & (4) can be written as,

$$\lambda_1(c) = c^2 = 1\,(c^{2/1})$$
$$\lambda_2(c) = 2c = 2\,(c^{2/2})$$
$$\lambda_3(c) = 3c^{2/3} = 3\,(c^{2/3})$$

We can assume the value for m(m < n) and can be proved for (m + 1)

i.e., $\lambda_m(c) = m\,(c^{2/m})$

and $\lambda_{m+1}(c)$ can be proved as

$$\lambda_{m+1}(c) = (m+1)\,.\,(c^{2/m+1})$$

Hence, we have, by mathematical induction.

$$\lambda_n(c) = n\,(c^{2/n})$$

The minimum $z = n\,(c^{2/n}) = c^{2/n} + c^{2/n} + \ldots n$ times
Comparing with Min $Z = y_1^{\,2} + y_2^{\,2} + \ldots + y_n^{\,2}$

$$\Rightarrow y_j^2 = c^{2/n}\ (j = 1, 2\ldots n)$$
$$y_j^2 = (c^{1/n})^2$$
$$y_j = c^{1/n}\ (j = 1, 2\ldots n)$$

Hence, the solution is

$$\text{Min } Z = n\,(c^{2/n})$$

and $y_j = (c^{1/n});\ (j = 1, 2\ldots n)$

9.3 TABULAR METHOD TO SOLVE A DPP

Whenever if the decision variables in a given DPP are restricted to take integer values, we seek the help of this Tabular Method.

Example: 4

The owner of a chain of four grocery stores has purchased six crates of fresh strawberries. The estimated probability distribution of potential sales of the strawberries before spoilage differ among the four stores. The following table gives the estimated total expected profit at each store, when it is allocated various number of crates:

Number of Crates	Stores			
	1	2	3	4
0	0	0	0	0
1	4	2	6	2
2	6	4	8	3
3	7	6	8	4
4	7	8	8	4
5	7	9	8	4
6	7	10	8	4

For administrative reasons, the owner does not wish to split crates among stores. However, he is willing to distribute zero crates to any of his stores. Find the allocation of six crates to four stores so as to maximize the expected profit.

Step 1:

Construct the Mathematical model for the given situation.
Let x_j : The amount of crates allocated to the jth store.
Let $P_j(x_j)$: The profit of the jth store based on the allocation of x_j crates.
Then we have,

$$\text{Maximize } z = \sum_{j=1}^{4} P_j(x_j)$$

$$\text{s.t. } \sum_{j=1}^{4} x_j = 6,\ x_j \geq 0 \ \&\ \text{integers } (j = 1 \ldots 4)$$

Step 2: Introduce the state variable S_j (j = 1, 2, 3, 4) for the jth stage (j = 1, 2, 3, 4)

$\lambda_j(S_j)$ – The total profit when the system is in the S_j state of jth stage

Step 3: Construct the recursive equation

$$\lambda_n(S_n) = \underset{x_n \leq S_n}{\text{Max}} [p_n(x_n) + \lambda^*_{n-1}(S_n - x_n)]$$

When n = 1, 2, 3 & 4, we have

$$\lambda_1(S_1) = \underset{x_1 \leq S_1}{\text{Max}} [P_1(x_1)]$$

$$\lambda_2(S_2) = \underset{x_2 \leq S_2}{\text{Max}} [P_2(x_2) + \lambda^*_1(S_2 - x_2)]$$

$$\lambda_3(S_3) = \underset{x_3 \leq S_4}{\text{Max}} [P_3(x_3) + \lambda^*_2(S_3 - x_3)]$$

$$\lambda_4(S_4) = \text{Max} [P_4(x_4) + \lambda^*_3(S_4 - x_4)]$$

Step 4: Construct a table for each stage based on the decision variables.

Stage 1: Consider the variable x_1x_1 take 6 as the maximum value based on the given constraint. We know that, $x_1 \le S_1$ i.e., $S_1 \ge x_1$.

$$\lambda_1(S_1) = \underset{x_1}{\text{Max}}\{P_1(x_1)\}$$

S_j	0	1	2	3	4	5	6
x'^s_j range	0	0–1	0–2	0–3	0–4	0–5	0–6

S_1 \ x_1	0	1	2	3	4	5	6	λ_1^*	x_1^*
0	0	–	–	–	–	–	–	0	0
1	0	4	–	–	–	–	–	4	1
2	0	4	6	–	–	–	–	6	2
3	0	4	6	7	–	–	–	7	3
4	0	4	6	7	7	–	–	7	3, 4
5	0	4	6	7	7	7	–	7	3, 4, 5
6	0	4	6	7	7	7	7	7	3, 4, 5, 6

Stage 2: Consider the variable x_2 and its recursive equation x_2 can take 6 as the maximum value based on the given constraint.

	$\lambda_2(S_2) = \underset{x_2}{\text{Max}}\{P_2(x_2) + \lambda_1^*(S_2 - x_2)\}$								
S_2	0	1	2	3	4	5	6	λ_2^*	x_2^*
0	$0+\lambda_1^*(0)$ $=0+0=0$	–	–	–	–	–	–	0	0
1	$0+\lambda_1^*(1)$ $=0+4=4$	$2+\lambda_1^*(0)$ $=2+0=2$	–	–	–	–	–	4	0
2	$0+\lambda_1^*(2)$ $=0+6=6$	$2+\lambda_1^*(1)$ $=2+4=6$	$4+\lambda_1^*(0)$ $=4+0=4$	–	–	–	–	6	0,1
3	$0+\lambda_1^*(3)$ $=0+7=7$	$2+\lambda_1^*(2)$ $=2+6=8$	$4+\lambda_1^*(1)$ $=4+4=8$	$6+\lambda_1^*(0)$ $=6+0=6$	–	–	–	8	1, 2

Continued

S_2	$\lambda_2(S_2) = \text{Max}\{P_2(x_2) + \lambda_1^*(S_2 - x_2)\}$ x_2 0	1	2	3	4	5	6	λ_2^*	x_2^*
4	$0+\lambda_1^*(4) = 0+7=7$	$2+\lambda_1^*(3) = 2+7=9$	$4+\lambda_1^*(2) = 4+6=10$	$6+\lambda_1^*(1) = 6+4=10$	$8+\lambda_1^*(0) = 8+0=8$	–	–	10	2, 3
5	$0+\lambda_1^*(5) = 0+7=7$	$2+\lambda_1^*(4) = 2+7=9$	$4+\lambda_1^*(3) = 4+7=11$	$6+\lambda_1^*(2) = 6+6=12$	$8+\lambda_1^*(1) = 8+4=12$	$9+\lambda_1^*(0) = 9+0=9$	–	12	3, 4
6	$0+\lambda_1^*(6) = 0+7=7$	$2+\lambda_1^*(5) = 2+7=9$	$4+\lambda_1^*(4) = 4+7=11$	$6+\lambda_1^*(3) = 6+7=13$	$8+\lambda_1^*(2) = 8+6=14$	$9+\lambda_1^*(1) = 9+4=13$	$10+\lambda_1^*(0) = 10+0=10$	14	4

Stage 3: Consider the variable x_3 and its recursive equation. x_3 can take 6 as the maximum value.

S_3	$\lambda_3(S_3) = \text{Max}\{P_3(x_3) + \lambda_2^*(S_3 - x_3)\}$ x_3 0	1	2	3	4	5	6	λ_3^*	x_3^*
0	$0+\lambda_2^*(0) = 0+0=0$	–	–	–	–	–	–	0	0
1	$0+\lambda_2^*(1) = 0+4=4$	$6+\lambda_2^*(0) = 6+0=6$	–	–	–	–	–	6	1
2	$0+\lambda_2^*(2) = 0+6=6$	$6+\lambda_2^*(1) = 6+4=10$	$8+\lambda_2^*(0) = 8+0=8$	–	–	–	–	10	1
3	$0+\lambda_2^*(3) = 0+8=8$	$6+\lambda_2^*(2) = 6+6=12$	$8+\lambda_2^*(1) = 8+4=12$	$8+\lambda_2^*(0) = 8+0=8$	–	–	–	12	1, 2
4	$0+\lambda_2^*(4) = 0+10=10$	$6+\lambda_2^*(3) = 6+8=14$	$8+\lambda_2^*(2) = 8+6=14$	$8+\lambda_2^*(1) = 8+4=12$	$8+\lambda_2^*(0) = 8+0=8$	–	–	14	1, 2
5	$0+\lambda_2^*(5) = 0+12=12$	$6+\lambda_2^*(4) = 6+10=16$	$8+\lambda_2^*(3) = 8+8=16$	$8+\lambda_2^*(2) = 8+6=14$	$8+\lambda_2^*(1) = 8+4=12$	$8+\lambda_2^*(0) = 8+0=8$	–	16	1, 2
6	$0+\lambda_2^*(6) = 0+14=14$	$6+\lambda_2^*(5) = 6+12=18$	$8+\lambda_2^*(4) = 8+10=18$	$8+\lambda_2^*(3) = 8+8=16$	$8+\lambda_2^*(2) = 8+6=14$	$8+\lambda_2^*(1) = 8+4=12$	$8+\lambda_2^*(0) = 8+0=8$	18	1, 2

Stage 4: Consider the variable x_4 and its recursive equation. x_4 can take 6 as the maximum value. Clearly this stage is the final stage. In this there is no need to check all the possibilities of the variable S_4. That is check S_4 at the maximum, $S_4 = 6$.

$S_4 \backslash x_4$	$\lambda_4(S_4) = \underset{x_4}{\text{Max}}\{P_4(x_4) + \lambda_3^*(S_4 - x_4)\}$								
	0	1	2	3	4	5	6	λ_4^*	x_4^*
6	$0 + \lambda_3^*(6)$ $= 0 + 18 = 18$	$2 + \lambda_3^*(5)$ $= 2 + 16 = 18$	$3 + \lambda_3^*(4)$ $= 3 + 14 = 17$	$4 + \lambda_3^*(3)$ $= 4 + 12 = 16$	$4 + \lambda_3^*(2)$ $= 4 + 10 = 14$	$4 + \lambda_3^*(1)$ $= 4 + 6 = 10$	$4 + \lambda_3^*(0)$ $= 4 + 0$	18	0,1

Note: Always for the final state, it is sufficient to test the maximum value of the state variable. There is no need to check all possible sub-possibilities of the same.

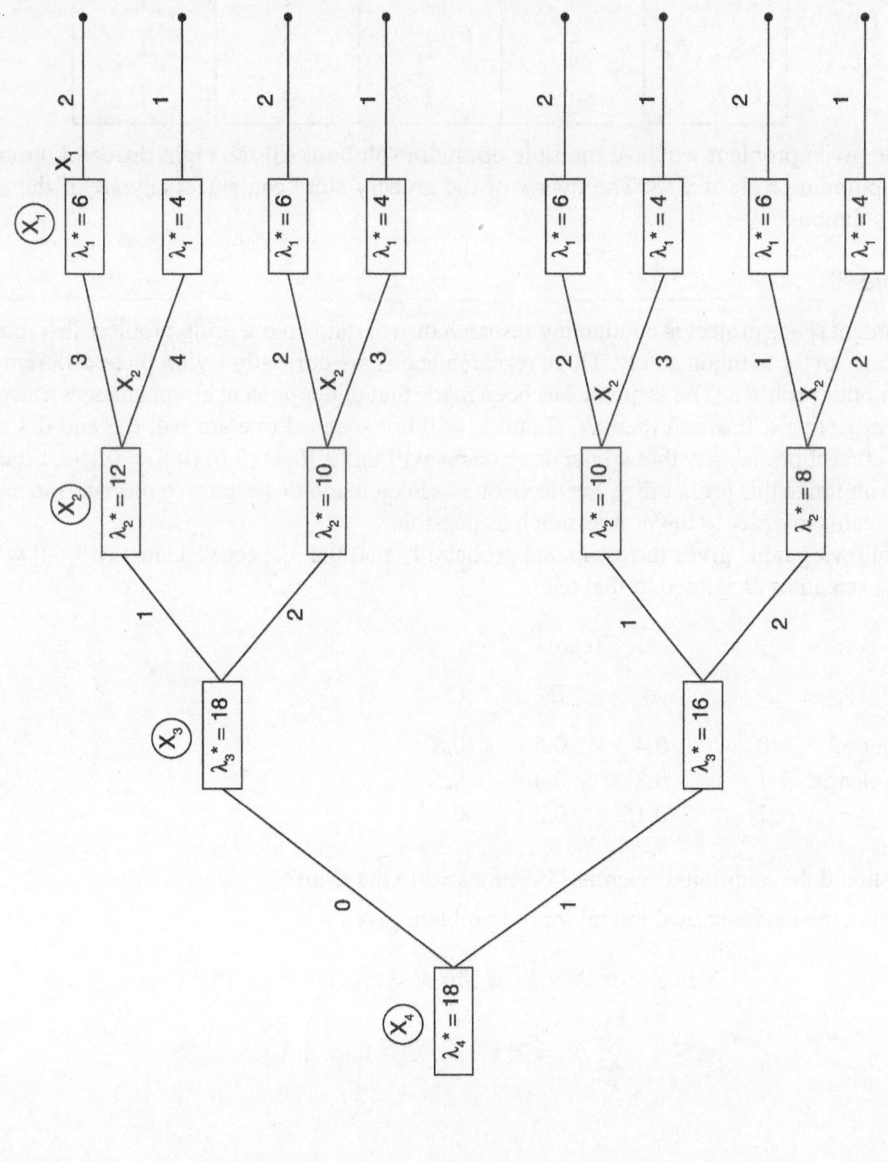

No.	x_1	x_2	x_3	x_4
1	2	3	1	0
2	1	4	1	0
3	2	2	2	0
4	1	3	2	0
5	2	2	1	1
6	1	3	1	1
7	2	1	2	1
8	1	2	2	1

For the given problem we have multiple optimum solution. All the eight different possibilities will give a maximum profit of ₹18. The owner of the grocery store can select anyone of the above stated alternate solutions.

*Example: 5

A government space project is conducting research on a certain engineering problem that must be solved before Man can fly to moon safely. Three research teams are currently trying three different approaches for solving this problem. The estimate has been made that under present circumstances, the probabilities that the respective teams-call them A, B and C will not succeed one are 0.4, 0.6 and 0.8 respectively. Thus the current probability that all the three teams will fail is (0.4) (0.6) (0.8) = 0.192. Since the objective is to minimize this probability, the decision has been made to assign two/more top scientists among the three teams in order to lower it as much as possible.

The following table gives the estimated probability that the respective teams will fail when 0.1 or 2 additional scientists are added to that team:

		Team		
		A	B	C
Number of	0	0.4	0.6	0.8
New scientists	1	0.2	0.4	0.5
	2	0.15	0.2	0.3

How should the additional scientists be allocated to the team?

Construct the mathematical model for the problem given:

$$\text{Min } Z = \underset{j=1}{\overset{3}{\pi}} P_j(x_j) \quad \text{or Min } Z = P_1(x_1) * P_2(x_2) * P_3(x_3)$$

$$\text{s.t.} \qquad \sum_{j=1}^{3} x_j = 2,\ x_1, x_2, x_3 \geq 0 \ \&\ \text{integer}$$

State (S_j): The number of scientists attached in the jth stage.
Decision variable (x_j): The number of scientists attached to the jth team.
$\lambda_j(S_j)$: The total failure probability when the system in S_j, state of the jth stage
The general recursive equation can be given as:

$$\lambda_n(S_n) = \min_{x_n \le S_n} \{P_n(x_n) * \lambda^*_{n-1}(S_n - x_n)\}$$

Stage 1: $\lambda_1(S_1) = \min_{x_1 \le S_1} \{P_1(x_1)\}$

Stage 2: $\lambda_2(S_2) = \min_{x_2 \le S_2} \{[P_2(x_2)] + [\lambda_1^*(S_2 - x_2)]\}$

Stage 3: $\lambda_3(S_3) = \min_{x_3 \le S_3} \{[P_3(x_3)] + [\lambda_2^*(S_3 - x_3)]\}$

We know that $x_j < S_j \Rightarrow S_j > x_j$ ($j = 1, 2, 3$)

When S_j	0	1	2
x_j can take	0	0–1	0–2

($j = 1, 2, 3$)

Stage 1: Team-A, consider the variable x_1 & $\lambda_1(S_1)$

S_1	$\lambda_1(S_1) = \min\{P_1(x_1)\}$, x_1 = 0	1	2	$\lambda_1^*(S_1)$	x_1^*
0	0.4	–	–	0.4	0
1	0.4	0.2	–	0.2	1
2	0.4	0.2	0.15	0.15	2

Stage 2: Team-B, consider the variable x_2 & $\lambda_2(S_2)$

S_2	$\lambda_2(S_2) = \min\{[P_2(x_2)] \times [\lambda_1^*(S_2 - x_2)]\}$, x_2 = 0	1	2	$\lambda_2^*(S_2)$	x_2^*
0	$0.6 \times [\lambda_1^*(0)] = 0.6 * 0.4 = 0.24$	–	–	0.24	0
1	$0.6 \times [\lambda_1^*(1)] = 0.6 * 0.2 = 0.12$	$0.4 \times [\lambda_1^*(0)] = 0.4 * 0.4 = 0.16$	–	0.12	0
2	$0.6 \times [\lambda_1^*(2)] = 0.6 * 0.15 = 0.09$	$0.4 \times [\lambda_1^*(1)] = 0.4 * 0.2 = 0.08$	$0.2 \times [\lambda_1^*(0)] = 0.2 * 0.4 = 0.08$	0.08	1, 2

Stage 3: Team-C, consider the variable x_3 & λ_3 (S_3)

S_3	$\lambda_3(S_3) = \text{Min}\{[P_3(x_3)] \times [\lambda_2^*(S_3 - x_3)]\}$ x_3			$\lambda_3^*(S_3)$	x_3^*
	0	1	2		
2	$0.8 \times [\lambda_2^*(2)] = 0.8 = 0.08 = 0.064$	$0.5 \times [\lambda_2^*(1)] = 0.5 * 0.12 = 0.060$	$0.3 \times [\lambda_2^*(0)] = 0.3 * 0.24 = 0.072$	0.06	1

From the table we refer that, the induction of 2 more scientists in the three teams altogether can reduce the failure probability into 0.06.

$\boxed{\lambda_3^*(S_3) = 0.06}$ —— $x_3^* = 1$ —— $\boxed{\lambda_2^*(S_2) = 0.12}$ —— $x_2^* = 0$ —— $\boxed{\lambda_1^*(S_3) = 0.2}$ —— $x_1^* = 1$

Hence the optimum solution is

$$x_1^* = 1; \quad x_2^* = 0; \quad x_3^* = 1$$

Minimum total failure probability = 0.06

That is add one scientist each in team-A and team-C. Then the total failure probability is reduced to 0.06.

*Example: 6

A truck can carry a total of 10 tons of product. Three types of product are available for shipment. Their weights and values are tabulated. Assuming that at least one of each type must be shipped, determine the loading which will maximize the total value:

Type	Value (in ₹)	Weight (in Tons)
A	20	1
B	50	2
C	60	2

Let x_1, x_2 & x_3 stand for the number of units in Type A, B & C to be shipped respectively.

Type	No. of Units	Value (in ₹)	Weight (in Tons)	Minimum No. of Units to be Shipped
A	x_1	20	1	1
B	x_2	50	2	1
C	x_3	60	2	1
		Max	5	

Maximum cargo available $= 10 - [1 \times 1 + 2 \times 1 + 2 \times 1] = 5$ tons

Then we have,

$$\text{Max } Z = 20x_1 + 50x_2 + 60x_3$$
$$\text{s.t. } x_1 + 2x_2 + 2x_3 \leq 5$$
$$x_1, x_2, x_3 \geq 0 \text{ and integers}$$

State (S_j): The number of units transported at the jth stage.
Decision variable (x_j): The number of units shipped in the jth item
λ_j (S_j): Total value when the system in S_j state of the jth stage.
The general recursive equation can be given as:

$$\lambda_n(S_n) = \underset{a_n x_n \leq S_n}{\text{Max}} \{V_n(x_n) + \lambda^*_{n-1}(S_n - a_n x_n)\}$$

Here, V_j: Value of one unit of the jth type shipped
a_j: Coefficient of the x_j[th] variable in the constraint

Stage 1: $$\lambda_1(S_1) = \underset{x_1 \leq S_1}{\text{Min}} \{20x_1\}$$

Stage 2: $$\lambda_2(S_2) = \underset{x_2 \leq S_2}{\text{Max}} \{50x_2^{th} + \lambda_1^*(S_2 - 2x_2)\}$$

Stage 3: $$\lambda_3(S_3) = \underset{x_3 \leq S_3}{\text{Max}} \{60x_3 + \lambda_2^*(S_3 - 2x_3)\}$$

$x_1 \leq S_1$ here x_1 can take, $x_1 = 0, 1 \ldots 5$

$$\Rightarrow S_1 \geq x_1$$

S_1	0	1	2	3	4	5
x_1	0	0–1	0–2	0–3	0–4	0–5

$$2x_2 \leq S_2$$
$$x_2 \leq S_2/2 \Rightarrow x_2 = 0, 1, 2$$
$$S_2 \geq 2x_2$$

S_2	0	1	2	3	4	5
x_2	0	0	0–1	0–1	0–2	0–2

$$2x_3 \leq S_3 ;\ x_3 \leq S_3/2 \Rightarrow x_3 = 0, 1, 2;\ S_3 \geq 2x_3$$

S_3	5
x_3	0–2

Stage 1: Consider the variable x_1 & λ_1 (S_1)

S_1 \ x_1	$\lambda_1(S_1) = \text{Max}(20x_1)$, $x_1 \leq S_1$							
	0	1	2	3	4	5	λ_1*	x_1
0	0	–	–	–	–	–	0	0
1	0	20	–	–	–	–	20	1
2	0	20	40	–	–	–	40	2
3	0	20	40	60	–	–	60	3
4	0	20	40	60	80	–	80	4
5	0	20	40	60	80	100	100	5

Stage 2: Consider the variable x_2 & λ_2 (S_2)

S_2 \ x_2	$\lambda_2(S_2) = \text{Max}\{50x_2 + \lambda_1{}^*(S_2 - 2x_2)\}$ $x_2 \leq S_2$				
	0	1	2	λ_2*	x_2*
0	0 + 0 = 0	–	–	0	0
1	0 + 20 = 20	–	–	20	0
2	0 + 40 = 40	50 + 0 = 50	–	50	1
3	0 + 60 = 60	50 + 20 = 70	–	70	1
4	0 + 80 = 80	50 + 40 = 90	100 + 0 = 100	100	2
5	0 + 100 = 100	50 + 60 = 110	100 + 20 = 120	120	2

Note: $S_2 - (2_t)x_2$; ∵ S_2 is interms of tonnes

Stage 3: Consider the variable x_3 & λ_3 (S_3)

S_3 \ x_3	$\lambda_3(S_3) = \text{Max}\{60x_3 + \lambda_2{}^*(S_3 - 2x_3)\}$, $x_3 \leq S_3$				
	0	1	2	λ_3*	x_3*
5	0 + 120 = 120	60 + 70 = 130	120 + 20 = 140	140	2

Note: $S_3 - (2_t)\,x_3$; ∵ S_2 is interms of tonnes

$\boxed{\lambda_3^*(S_3) = 140}$ —— $x_3^* = 2$ —— $\boxed{\lambda_2^*(S_2) = 120}$ —— $x_2^* = 0$ —— $\boxed{\lambda_1^*(S_3) = 20}$ —— $x_1^* = 1$

Hence the optimum solution for the given situation is

Product	No. of Units to be Shipped	Value
Type A	$1 + 1 = 2$	40
Type B	$1 + 0 = 1$	150
Type C	$1 + 2 = 3$	180
	Total	370

The optimum value of the product to be shipped is 370.

Example: 7

Max $Z = (x_1^2 - 2x_1) + (x_2^2 + 3x_2) + (x_3^2 + 4x_3)$

s.t. $\quad x_1^2 + 2x_2^2 + 3x_3^2 \leq 17$

$\quad x_1, x_2, x_3 \geq 0$ & integers

State (S_j): The value taken by the decision variable in the jth stage.
λ_j (S_j): Value of the objective function when the system in S_j stage of the jth stage.
x_j: The decision variable corresponds to the jth stage.
The recursive equation of various stages can be given as,

Stage 1:
$$\lambda_1(S_1) = \underset{x_1 \leq S_1}{\text{Max}} \{x_1^2 - 2x_1\}$$

Stage 2:
$$\lambda_2(S_2) = \underset{x_2 \leq S_2}{\text{Max}} \{(x_2^2 + 3x_2) + \lambda_1^*(S_2 - 2x_2^2)\}$$

Stage 3:
$$\lambda_3(S_3) = \underset{x_3 \leq S_3}{\text{Max}} \{(x_3^2 + 4x_3) + \lambda_2^*(S_3 - 3x_3^2)\}$$

We have the following conditions for x_1 & S_1 based on the constraints.

$x_1^2 \leq 17 \Rightarrow x_1 \leq 4.123 \Rightarrow x_1 = 0, 1, 2, 3, 4$

$x_1^2 \leq S_1 \Rightarrow S_1 \geq x_1^2$

x_1	0	1	2	3	4
S_1	≥0	≥1	≥4	≥9	≥16

Stage 1:

S_1 \ x_1	$\lambda_1(S_1) = \text{Max}\{x_1^2 - 2x_1\}$ x_1					λ_1*	x_1*
	0	1	2	3	4		
0	0	–	–	–	–	0	0
1	0	−1	–	–	–	0	0
2	0	−1	–	–	–	0	0
3	0	−1	–	–	–	0	0
4	0	−1	0	–	–	0	0
5	0	−1	0	–	–	0	0
6	0	−1	0	–	–	0	0
7	0	−1	0	–	–	0	0
8	0	−1	0	–	–	0	0
9	0	−1	0	3	–	3	3
10	0	−1	0	3	–	3	3
11	0	−1	0	3	–	3	3
12	0	−1	0	3	–	3	3
13	0	−1	0	3	–	3	3
14	0	−1	0	3	–	3	3
15	0	−1	0	3	–	3	3
16	0	−1	0	3	8	8	4
17	0	−1	0	3	8	8	4

Conditions for x_2 & S_2:

$$2x_2^2 \leq 17$$
$$x_2^2 \leq 8.5$$
$$x_2 \leq 2.92 \Rightarrow x_2 = 0, 1, 2$$
$$2x_2^2 \leq S_2 \Rightarrow S_2 \geq 2x_2^2$$

x_2	0	1	2
S_2	≥0	≥2	≥8

Stage 2:

S_2 \ x_2	$\lambda_2(S_2) = \text{Max }\{(x_2^2 + 3x_2)\} + \lambda_1^*(S_2 - 2x_2^2)\}$ x_2			λ_2^*	x_2^*
	0	1	2		
0	0 + 0 = 0	–	–	0	0
1	0 + 0 = 0	–	–	0	0
2	0 + 0 = 0	4 + 0 = 4	–	4	1
3	0 + 0 = 0	4 + 0 = 4	–	4	1
4	0 + 0 = 0	4 + 0 = 4	–	4	1
5	0 + 0 = 0	4 + 0 = 4	–	4	1
6	0 + 0 = 0	4 + 0 = 4	–	4	1
7	0 + 0 = 0	4 + 0 = 4	–	4	1
8	0 + 0 = 0	4 + 0 = 4	10 + 0 = 10	10	2
9	0 + 3 = 3	4 + 0 = 4	10 + 0 = 10	10	2
10	0 + 3 = 3	4 + 0 = 4	10 + 0 = 10	10	2
11	0 + 3 = 3	4 + 3 = 7	10 + 0 = 10	10	2
12	0 + 3 = 3	4 + 3 = 7	10 + 0 = 10	10	2
13	0 + 3 = 3	4 + 3 = 7	10 + 0 = 10	10	2
14	0 + 3 = 3	4 + 3 = 7	10 + 0 = 10	10	2
15	0 + 3 = 3	4 + 3 = 7	10 + 0 = 10	10	2
16	0 + 8 = 8	4 + 3 = 7	10 + 0 = 10	10	2
17	0 + 8 = 8	4 + 3 = 7	10 + 3 = 13	13	2

Conditions for x_3 & S_3.

$$3x_3^2 \le 17;\ x_3^2 \le 5.67;\ x_3 \le 2.38 \Rightarrow x_3 = 0, 1, 2;\ 3x_3^2 \le s_3;\ s_3 \ge 3x_3^2$$

x_3	0	1	2
S_3	≥0	≥3	≥12

Stage 3:

S_3 \ x_3	$\lambda_3(S_3) = \underset{x_3 \le S_3}{\text{Max}} \{(x_3^2 + 4x_3)\} + \lambda_2^*(S_3 - 3x_3^2)\}$				
	0	1	2	λ_3^*	x_3^*
17	0 + 13 = 13	5 + 10 = 15	12 + 4 = 16	16	2

$$\boxed{\lambda_3^* = 16} \xrightarrow{x_3^* = 2} \boxed{\lambda_2^* = 4} \xrightarrow{x_2^* = 2} \boxed{\lambda_1^* = 4} \xrightarrow{x_1^* = 0}$$

Hence the optimum solution is $x_1 = 0$; $x_2 = 2$; $x_3 = 2$ and Max Z = 16

9.4 DPP APPLICATION TO SOLVE AN LPP

Linear Programming Problem can also be solved by using dynamic programming technique. The various steps to be followed regarding this method are clearly explained through an example problem.

***Example:* 8**

Solve $\quad \text{Max } Z = 8x_1 + 7x_2$

s.t. $\quad 2x_1 + x_2 \le 8;$

$5x_1 + 2x_2 \le 15$

$x_1, x_2 \ge 0$

Here, number of resources = 2

Number of decision variables = 2

Then the states of the equivalent DP therefore, are

$$B_{1j} \text{ and } B_{2j} \ (j = 1, 2)$$

We can solve this using forward computational method.

Stage 1:

$$f_1(B_{11}, B_{21}) = \underset{\substack{0 \le 2x_1 \le 8 \\ 0 \le 5x_1 \le 15}}{\text{Max}} \{8x_1\}$$

$$= 8 * \underset{\substack{0 \le x_1 \le 4 \\ 0 \le x_1 \le 3}}{\text{Max}} \{x_1\}$$

$$f_1(4, 3) = 8 * \text{Min} \{4, 3\} \quad (1)$$

$$= 8 * 3 = 24$$

Stage 2:

$$f_2(B_{12}, B_{22}) = \underset{\substack{0 \le x_2 < 8 \\ 0 \le x_2 \le 15}}{\text{Max}} \{7x_2 + f_1^*((8 - x_2)/2; (15 - 2x_2)/5)\}$$

$2x_1 + x_2 = 8; \quad 2x_1 = 8 - x_2; \quad x_1 = (8 - x_2)/2$

$5x_1 + 2x_2 = 15$

$\Rightarrow 5x_1 = 15 - 2x_2$

$x_1 = (15 - 2x_2)/5$

$$= \text{Max } \{7x_2 + 8^* \min((8 - x_2)/2, (15 - 2x_2)/5)\}$$
$$0 \le x_2 \le 8$$
$$0 \le x_2 \le 7.5$$

$$= \text{Max } \{7x_2 + 8^* \min((8 - x_2)/2, (15 - 2x_2)/5)\}$$
$$0 \le x_2 \le 7.5$$

$$\text{Min}((8-x_2)/2, (15-2x_2)/5) = \begin{cases} 3 & \text{if } x_2 = 0 \\ (15-2x_2)/5 & \text{if } 0 \le x_2 \le 7.5 \end{cases}$$

$$f_2(B_{12}, B_{22}) = \begin{cases} 7x_2 + 8*3 & \text{if } x_2 = 0 \\ 7x_2 + (15-2x_2)/5 & \text{if } 0 < x_2 \le 7.5 \end{cases}$$

$$= \begin{cases} 24 & \text{if } x_2 = 0 \\ 3 + 6.6x_2 & \text{if } 0 < x_2 \le 7.5 \end{cases}$$

$$= \begin{cases} 24 & \text{if } x_2 = 0 \\ 3 + 49.5 & \text{if } x_2 = 7.5 \end{cases}$$

$$= \begin{cases} 24 & \text{if } x_2 = 0 \\ 52.5 & \text{if } x_2 = 7.5 \end{cases}$$

$f_2(B_{12}, B_{22})$ is the maximum at

$$x_2^* = 7.5$$

$$x_1^* = \text{Min}\left[\frac{8 - x_2^*}{2}, \frac{15 - 2x_2^*}{5}\right]$$

$$= \text{Min } \{1/4, 0\} = 0$$

Hence the optimum solution is,

$$x_1 = 0; \quad x_2 = 7.5; \quad \text{Max } Z = 52.5$$

*Example: 9

Max $Z = 50x_1 + 100x_2$

s.t. $10x_1 + 5x_2 \le 2500$

$4x_1 + 10x_2 \le 2000$

$x_1 + (3/2)x_2 \le 450$

$x_1, x_2 \ge 0$

Here Number of resources = 3

Number of variables = 2

Then the states of the equivalent DP therefore are

$$B_{1j}, B_{2j} \ \& \ B_{3j}; \ (j = 1, 2)$$

We can solve this using forward computational method.

Stage 1:

$$f_1(B_{11}, B_{21}, B_{31}) = \text{Max } \{50x_1\}$$
$$0 \le x_1 \le 250$$
$$0 \le x_1 \le 500$$
$$0 \le x_1 \le 450$$
$$= 50 * \text{Min } \{250, 500, 450\} \qquad (1)$$
$$= 50 * 250 = 12500$$

$$f_2(B_{12}, B_{22}, B_{32}) = \text{Max } \{100\, x_2 + f_1^* (250 - (1/2)\, x_2, 500 - (5/2)\, x_2, 450 - (3/2)\, x_2)\}$$
$$0 \le x_2 \le 500$$
$$0 \le x_2 \le 200$$
$$0 \le x_2 \le 300$$
$$= \text{Max } \{100x_2 + 50 * \text{Min } (250 - (1/2)\, x_2, 500 - (5/2)\, x_2, 50 - (3/2)\, x_2)\} \qquad (2)$$
$$0 \le x_2 \le 200 \qquad \text{using (1)}$$

When $x_2 = 0$, Min [250, 500, 450] = 250

When $x_2 = 200$, Min [150, 0, 150] = 0

Hence (2) reduced to,

$$f_2 (B_{12}, B_{22}, B_{32}) = \text{Max } \{100x_2 + 50 * \text{Min } (250 - (1/2)\, x_2, 500 - (5/2)\, x_2)\}$$
$$0 \le x_2 \le 200$$

$$\left.\begin{array}{l} 250-(1/2)x_2 \le 500-(5/2)x_2 \\ 2x_2 \le 250 \\ x_2 \le 125 \end{array}\right| \begin{array}{l} 500-(5/2)x_2 \le 250-(1/2)x_2 \\ x_2 \ge 125 \end{array}$$

$$= \text{Max} \begin{bmatrix} 100x_2 + 50[250 - x_2/2]; & 0 \le x_2 \le 125 \\ 100x_2 + 50[500 - 5x_2/2]; & 125 \le x_2 \le 200 \end{bmatrix}$$

$$= \text{Max} \begin{bmatrix} 12500 + 75x_2 & 0 \le x_2 \le 125 \\ 25000 - 25x_2 & 125 \le x_2 \le 200 \end{bmatrix}$$

$$= \begin{bmatrix} 12500 + 9375 = 21875; & \text{when } x_2 = 125 \\ 25000 - 3125 = 21875; & \text{when } x_2 = 125 \end{bmatrix}$$

$$f_2(B_{12}, B_{22}, B_{32}) = 21875 \text{ when } x_2 = 125$$
$$\Rightarrow x_2^* = 125$$
$$x_1^* = \text{Min } [250 - x_2^*/2, 500 - (5/2)\, x_2^*]$$
$$= \text{Min } [250 - 125/2, 500 - (5/2) * 125]$$
$$= \text{Min } [187.5, 187.5]$$
$$x_1^* = 187.5$$

Hence the optimum solution is,

$$x_1 = 187.5$$
$$x_2 = 125$$
$$\text{Max } Z = 21875$$

EXERCISE PROBLEMS

1. Use Bellman's principle of optimality to solve.

$$\text{Min } Z = \sum_{j=1}^{n} y_j$$

$$\text{s.t.} \quad \pi_{j=1}^{n} \; y_j = d; \qquad y_j \geq 0; \; (j = 1, 2, \dots n)$$

2. Max $Z = y_1 \, y_2 \, y_3$
   s.t. $y_1 + y_2 + y_3 = 5, \; y_j \geq 0 \; (j = 1, 2, 3)$

*3. A truck can carry 20 tons of load. Four types of products are to be loaded. Their per unit weights in tons and values in thousands of rupees are given below.

Type	Weight/Unit	Value/Unit
A	1	20
B	2	30
C	3	50
D	4	85

Atleast one unit of each type of load is to be shipped. Determine the loading which will maximize the value of the load

4. Fire station allocation planners for a city have recommended the best allocation of the fire stations to 3 districts, anywhere from 0 to 3 stations can be located in a district. The cost matrix of loss on the annual basis of the property damage due to fire is given in the following table based on the number of stations allocated to 3 districts. Differences among the districts are due to factors such as population, residential commercial mix1, socioeconomic make up and quality of construction. A budget constraint restricts the total number of allocations to 5 stations.

		Districts (₹ in Lakhs)		
		D1	D2	D3
Number of Stations	0	2	0.5	1 5
	1	0.9	0.3	1.0
	2	0.3	0.2	0.7
	3	0.2	0.1	0.3
	4	0	0	0
	5	0	0	0

The objective is that we have to allocate the five stations to three districts in such a way that the total loss should be minimum.

5. Dinesh transport company has purchased 6 additional buses that it plans to use in 3 routes. However, the company has not decided how many of the new buses to assign in each of the 3 routes. They have developed estimates of additional profit per week, for various alternatives shown in the table as follows:

Number of Buses Allocated	Additional Profits/Week (in ₹)		
	Route A	Route B	Route C
0	0	0	0
1	350	100	225
2	450	250	300
3	500	450	475
4	525	650	600
	450	700	650
6	400	750	600

a) Use DP technique to determine the optimal assignment of buses to each route.
b) Determine the optimal assignment if only 5 buses are available.
[Hint: (b) Take the maximum value of $S_3 = 5$ in stage-3 and re-do the stage-3 alone and get the modified solution]

6. Use DPP to solve:
Max $Z = x_1^2 + 2x_2^2 + 4x_3$
s.t. $x_1 + 2x_2 + x_3 \leq 8$ and $x_1 \geq 0$ & integers (j = 1, 2, 3).

*7. Cargo loading problem: A vessel is to be loaded with a stock of 3 items. Each unit of item i has a weight w_1 and value r_1. The maximum cargo weight the vessel can take is 5 and the details of the three items are as follows.

i	w_1	r_1
1	1	30
2	3	80
3	2	65

Develop the recursive equation for the above case and find the most valuable cargo load without exceeding the maximum cargo weight by using dynamic programming.
[Hint: Max $Z = 30x_1 + 80x_2 + 65x_3$ s.t. $x_1 + 3x_2 + 2x_3 \leq 5$, $x_j \geq 0$ (j = 1, 2, 3)]

*8. A number of a certain political party is making plans for an upcoming presidential election. He has received the services of six volunteer workers for district work and he wishes to assign them to three districts in such a way as to maximize their effectiveness. He feels that it would be inefficient to assign a worker to more than one district, but he is willing to assign no workers to any one of the districts if they can accomplish more in other districts. The following table gives the estimated increase in the popularity of the party candidate if it were allocated various number of workers.

Number of Workers	District		
	1	2	3
0	0	0	0
1	25	20	33
2	42	38	43
3	55	54	47
4	63	65	50
5	89	73	52
6	74	80	53

How many of the workers should be assigned to each of the three districts in order to maximize total estimated increase in the popularity of the party's candidates.

*9. Seven units of capital can be invested in four activities with a return from each activity as given in the accompanying table; Find the allocation of capital to each activity that will maximize the total return:

N	0	1	2	3	4	5	6	7
$f_1(N)$	0	2	4	6	7	8	9	9
$f_2(N)$	0	3	5	7	9	10	11	12
$f_3(N)$	0	2	3	4	5	5	5	5
$f_4(N)$	0	1	3	5	6	7	8	8

*10. An oil well servicing company has 9 experts. It has to service the oil wells in 3 zones. The revenue from each zone depends on the number of experts allocated to that zone. The expected revenue details from each zone for various levels (number) of experts allocated are summarized in the following table. Find an allocation that would maximize the revenue.

Experts Zones	Revenue in Millions of ₹									
	0	1	2	3	4	5	6	7	8	9
1	30	45	60	70	79	91	99	106	106	106
2	35	48	52	65	70	85	94	99	109	109
3	42	53	60	70	80	95	102	110	110	110

*11. Max $Z = 3x_1 + 5x_2$
s.t. $x_1 \le 4$; $x_2 \le 6$; $3x_1 + 2x_2 \le 18$; $x_1, x_2 \ge 0$

*12. Use DP technique to solve the LPP:
Max $Z = x_1 + 9x_2$
s.t. $2x_1 + x_2 \le 25$; $x_2 \le 11$; $x_1, x_2 \le 0$

*13. Solve the following linear programming problem by applying dynamic programming procedures. Explain the assumptions you make.
Max $Z = 2x_1 + 4x_2$
st. $2x_1 + 3x_2 \le 48$; $x_1 + 3x_2 \le 42$; $x_1 + x_2 \le 21$; $x_1, x_2 \ge 0$

*14. Max $Z = 2x_1 + 4x_2$
s.t. $2x_1 + x_2 \le 43$; $2x_2 \le 46$; $x_1, x_2 \ge 0$ Use DP Technique.

ANSWERS TO THE EXERCISE PROBLEMS

1. $y_1 = d^{1/n}, j = 1, 2, \ldots n$ and Min $Z = n \cdot d^{1/n}$
2. $y_1 = y_2 = y_3 = 5/3$ and Max $Z = (5/3)^3$
3. A = 1, B = 2, C = 1, D = 3 and maximum value is 385
4. Number of fire stations in
   District-1: 2
   District-2: 0
   District-3: 3 and minimum loss is ₹1.1 lakhs
5. (a) Number of buses in Route-1: 1
   Route-2: 4
   Route-3: 1
   Additional Maximum Profit: ₹1225
   (b) Number of buses in Route-1: 1
   Route-2: 3
   Route-3: 1
   Additional Maximum Profit: ₹1025
6. $x_1 = 8, x_2 = 0, x_3 = 0$ and Max $Z = 64$

7. Number of units in item-1: 1
   item-2: 0
   item-3: 2
   Maximum Revenue: 160

8. Number of persons allocated in District-1: 2
   District-2: 3
   District-3: 1

   Maximum total estimated increase in the Party's popularity: 129

9. Solution

f_1	f_2	f_3	f_4	Max	Return
1	2	4	1	0	15
2	3	3	1	0	15
3	3	4	0	0	15

10. $x_1 = 3$; $x_2 = 1$; $x_3 = 5$; Max Z = 213
11. $x_1 = 2$, $x_2 = 6$ and Max Z = 36
12. $x_1 = 7$, $x_2 = 11$ and Max Z = 106
13. $x_1 = 6$, $x_2 = 12$ and Max Z = 60
14. $x_1 = 10$, $x_2 = 23$ and Max Z = 112

REVIEW QUESTIONS

Short Questions

1. Define DPP (Section 9.1)
2. State the Bellman's principles of optimality. (Section 9.1)
3. State the methods used to solve a DPP. (Section 9.1)
4. Discuss the characteristics of a DPP. (Section 9.1.2)
5. Construct the additive recursive formula (Section 9.1.4)
6. Why do we need a DPP? (Section 9.1.3)
7. State the applications of dynamic programming
   (i) To solve optimization problem (ii) To solve an LPP
8. State the merits and demerits of a DPP. (Section 9.1.3)
9. Write short notes on the following:
   (i) decision variable (ii) state variable (iii) stage (iv) state (Section 9.1.2)

State True or False

10. We cannot solve LPP by using a DPT. (F)
11. A DPP can have a multiple optimum solution. (T)
12. The tabular method can be applied for all DPP's. (F)
13. The tabular method and the calculus method are substitute methods. (F)

Fill in the Blanks

14. Tabular method can be implemented when variables are going to take ______. (non-negative integer values)
15. Recursive formula is designed based on ______. (Bellman's principle)
16. The number of stages in a DPP is equal to ______. (the number of decision variables)
17. The final stage value refers to ______. (the optimum value)
18. We have to apply the additive recursive formula if the given problem's objective is in ______. (additive function)
19. We have to apply multiplicative recursive formula if the given problem's objective is in ______ ______ (multiplicative function)

10

Non-Linear Programming

10.1 INTRODUCTION

In the previous chapters, we have studied linear programming problems. For an LPP, our objective is to maximize or minimize a linear function subject to the linear constraints. But in many cases:

(i) The objective function may not be linear (or)
(ii) All/some of the constraints may not be linear (or)
(iii) Both the objective and the constraints may not be linear.

Such an optimization problem is called a non linear programming problem (NLPP). In this chapter, we discuss techniques to solve NLPP's.

10.2 GENERAL STRUCTURE OF AN NLPP

A general non-linear programming problem can be expressed as follows:

Objective Function

Max or Min $Z = f(x_1, x_2, \ldots x_n)$

$$\left.\begin{array}{ll} \text{s.t. } g_1(x_1, x_2, \ldots x_n) \{\leq, =, \geq\} b_1; (i = 1 \ldots m) & \\ x_j \geq 0 (j = 1, 2, \ldots n) & \ldots (1) \end{array}\right\}$$

Here, we have to evaluate the values of the decision variables x_j's ($j = 1 \ldots n$).

The above model is otherwise called a constrained NLPP. An NLPP with no constraint is called an unconstrained NLPP. The feasible region for NLPP (1) is the set of points $(x_1, x_2, \ldots x_n)$ that satisfies the m constraints of (1). A point in the feasible region is a feasible point, and a point that is not in the feasible region is an infeasible point.

Suppose the given model is a maximization one, then any point x′ in the feasible region for which $f(x') \geq f(x)$ holds for all point x in the feasible region is an optimal solution to the NLPP. [for a minimization case, x′ is the optimal solution if $f(x') \leq f(x)$ for all feasible x]

10.3 FORMULATION OF AN NLPP

***Example:* 1**

A manufacturing company produces two products, Radios and TV sets. Sales-price relationships for these two products are given below:

Product	Quantity Demanded	Unit Price
Radios	$1500\text{-}5P_1$	P_1
TV. sets	$3800\text{-}10P_2$	P_2

The total cost functions for these two products are given by $200x_1 + 0.1\ x_1^2$ and $300\ x_2 + 0.1\ x_2^2$ respectively. The production takes place on two assembly lines. Radio sets are assembled on assembly line I and TV sets are assembled on assembly line II. Because of the limitations of the assembly-line capacities, the daily production is limited to no more than 80 radio sets and 60 TV sets. The production of both types of products requires electronic components. The production of each of these sets requires five units and six units of electronic equipment respectively. The electronic components are supplied by another manufacturer, and the supply is limited to 600 units per day. The company has 160 employees, i.e., the labour supply amounts to 160 man-days. The production of one unit of radio set requires 1 man-day of labour, whereas 2 man-days of labour are required for a TV set. How many units of radio and TV sets should the company produce in order to maximize the total profit? Formulate the problem as a non-linear programming problem.

Let x_1 and x_2 stand for the number of radios and TV sets produced. Tabulate all the given information as follows.

Product	Nos. Produced	Demand	Revenue $x_1\ P_1$	Maximum Production	Electronic Component Required/u	Labour (man-day)
Radios	x_1	$x_1 = 1500 - 2P_1$ $\Rightarrow P_1 = 300 - 0.2x_1$	$x_1\ [300 - 0.2x_1]$ $= 300x_1 - 0.2x_1^{\ 2}$	80	5	1
TV sets	x_2	$x_2 = 3800 - 10P_2$ $\Rightarrow P_2 = 380 - 0.1x_2$	$x_2\ [380 - 0.1x_2]$ $= 380x_2 - 0.1x_2^{\ 2}$	60	6	2
				Max	600	160

Here, the objective is to maximize the total revenue by satisfying the available resources.
Hence the required NLPP is

$$\text{Max } Z = 300x_1 - 0.2x_1^2 + 380x_2 - 0.1x_2^2$$

s.t. Production capacity: $x_1 \leq 80;\ x_2 \leq 60$

Electronic component: $5x_1 + 6x_2 < 600$

Labour: $x_1 + 2x_2 \leq 160$

$x_1, x_2 \geq 0$

---

*These questions have appeared in Indian university examination papers.

10.4 METHODS TO SOLVE AN NLPP

Assumptions:

(i) The functions are continuous and differentiable.
(ii) The constraints can have {= or ≤ or ≥} type.
(iii) The objective may be maximization or minimization.

10.4.1 Lagrangian Method for Equality Constraints

Example: 2

Obtain the necessary and sufficient conditions for the optimum solution of the following Non-Linear programming problem:

Minimize $z = f(x_1, x_2) = 3e^{(2x_1+1)} + 2e^{(x_2+5)}$
subject to the constraints $x_1 + x_2 = 7$ and $x_1, x_2 \geq 0$

Step 1: Construct the Lagrangian function, $L(X, \lambda)$ where
X – set of all variables (x_1, x_2)
λ – The Lagrangian multiplier.

when (i) $Z = f(x_1, x_2)$ is convex and (ii) Z to be minimized. Then the necessary and sufficient conditions are:

$$\frac{\partial L}{\partial x_j} = 0;\ (j = 1, 2)$$

Here, $L(X, \lambda) = f(x_1, x_2) - \lambda[h(x_1, x_2)]$

Given:

$$g(x_1, x_2) = x_1 + x_2 = 7$$
$$h(x_1, x_2) = g(x_1, x_2) - 7 = (x_1 + x_2 - 7)$$

Hence, $L(X, \lambda) = 3e^{(2x_1+1)} + 2e^{(x_2+5)} - \lambda[x_1 + x_2 - 7]$

Step 2: Differentiate the Lagrangian function L, with respect to the variables x_1, x_2 and λ partially, we have

$$\frac{\partial L}{\partial x_1} = 6e^{(2x_1+1)} - \lambda$$

$$\frac{\partial L}{\partial x_2} = 2e^{(x_2+5)} - \lambda$$

$$\frac{\partial L}{\partial \lambda} = -[x_1 + x_2 - 7]$$

Step 3: Set $\frac{\partial L}{\partial x_1} = 0;\ j = 1, 2$ and $\frac{\partial L}{\partial \lambda} = 0$

we have,

$$6e^{(2x_1+1)} - \lambda = 0 \Rightarrow \lambda = 6e^{(2x_1+1)} \quad (1)$$
$$2e^{(x_2+5)} - \lambda = 0 \Rightarrow \lambda = 2e^{(x_2+5)} \quad (2)$$
$$-(x_1 + x_2 - 7) = 0 \Rightarrow x_1 + x_2 = 7 \quad (3)$$

Comparing (1) & (2),

$$6e^{(2x_1+1)} = 2e^{(x_2+5)}$$
$$3e^{(2x_1+1)} = e^{(x_2+5)}$$

Take log on either side,

$$\log_e(3e^{(2x_1+1)}) = \log_e(e^{(x_2+5)})$$
$$\log_e 3 + (2x_1 + 1) = (x_2 + 5)$$
$$0.094 + 2x_1 + 1 = x_2 + 5$$
$$2x_1 - x_2 = 3.906 \quad (4)$$

Adding (3) & (4)

$$\Rightarrow \quad x_1 + x_2 = 7$$
$$2x_1 - x_2 = 3.906$$
$$3x_1 = 10.906$$
$$x_1 = 3.635$$
$$(3) \Rightarrow \quad x_2 = 7 - x_1 = 7 - 3.635 = 3.365$$
$$\text{Min } Z = 3e^{[2(3.635)+1]} + 2\,e^{[3.365+5]}$$
$$= 3^*\,(3904.95) + 2^*\,(4294.11)$$
$$= 20303.07$$

Hence the solution is, $x_1 = 3.635$

$x_2 = 3.365$

Min $Z = 20303.07$

Example: 3

Obtain the set of necessary conditions for the NLPP:

Maximize: $Z = x_1^2 + 3x_2^2 + 5x_3^2$

s.t. constraints:

$$x_1 + x_2 + 3x_3 = 2$$
$$5x_1 + 2x_2 + x_3 = 5$$
$$x_1, x_2, x_3 \geq 0$$

Given,

$$\text{Max } Z = x_1^{\,2} + 3x_2^{\,2} + 5x_3^{\,2}$$
$$g_1(X) = x_1 + x_2 + 3x_3 = 2$$
$$g_2(X) = 5x_1 + 2x_2 + x_3 = 5$$
$$\Rightarrow \quad h_1(X) = g_1(X) - 2 = x_1 + x_2 + 3x_3 - 2$$
$$h_2(X) = g_2(X) - 5 = 5x_1 + 2x_2 + x_3 - 5$$

Construct the Lagrangian function $L(X, \lambda)$: Here the objective function is concave,

$$L(X, \lambda) = f(x_1\ x_2, x_3) - \lambda_1\ h_1(X) - \lambda_2\ h_2(X)$$

The necessary conditions are:

(i) $\dfrac{\partial L}{\partial x_j} = 0;\ (j = 1, 2, 3)$

(ii) $\dfrac{\partial L}{\partial \lambda_i} = 0;\ (i = 1, 2)$

$$L(X, \lambda) = x_1^2 + 3x_2^2 + 5x_3^2 - \lambda_1 [x_1 + x_2 + 3x_3 - 2] - \lambda_2 [5x_1 + 2x_2 + x_3 - 5]$$

$$\frac{\partial L}{\partial x_1} = 2x_1 - \lambda_1 - 5\lambda_2 = 0$$

$$\frac{\partial L}{\partial x_2} = 6x_2 - \lambda_1 - 2\lambda_2 = 0$$

$$\frac{\partial L}{\partial x_3} = 10x_3 - 3\lambda_1 - \lambda_2 = 0$$

$$\frac{\partial L}{\partial \lambda_1} = -[x_1 + x_2 + 3x_3 - 2] = 0$$

$$\frac{\partial L}{\partial \lambda_2} = -[5x_1 + 2x_2 + x_3 - 5] = 0$$

10.4.2 Sufficient Conditions

For an n-variable NLPP having one equality constraint, the sufficient conditions for determining the nature of resulting solution whether maximum or minimum, require us to consider the following determinant:

$$\Delta_{n+1} = \begin{vmatrix} 0 & h_{x_1} & h_{x_2} & \cdots & h_{x_n} \\ h_{x_1} & f_{x_1x_1} - \lambda h_{x_1x_1} & f_{x_1x_2} - \lambda h_{x_1x_2} & \cdots & f_{x_1x_n} - \lambda h_{x_1x_n} \\ \vdots & & & & \\ \vdots & & & & \\ \vdots & & & & \\ h_{x_n} & f_{x_nx_1} - \lambda h_{x_nx_1} & f_{x_nx_2} - \lambda h_{x_1x_1} & \cdots & f_{x_nx_n} - \lambda h_{x_nx_n} \end{vmatrix}$$

Note: $h_{x_j} = \frac{\partial h}{\partial x_j}$ and $f_{x_jx_j} - \lambda h_{x_jx_j} = \frac{\partial^2 f}{\partial x_j^2} - h\frac{\partial h}{\partial x_j^2}$; $j = 1, 2, \ldots n$

Case (i)
If Δ_3, Δ_4, Δ_5 etc are alternatively +ve and –ve, the stationary point is a local maximum.

Case (ii)
If Δ_3, Δ_4, Δ_5 etc are negative, refers to the local minimum.

Example: 4

Determine the optimal solution for the following NLPP and verify whether it maximizes or minimizes the objective function.

Optimize $Z = x_1^2 - 10x_1 + x_2^2 - 6x_2 + x_3^2 - 4x_3$
s.t. $x_1 + x_2 + x_3 = 7$, $x_j \geq 0$ $(j = 1, 2, 3)$

Step 1: Construct the Lagrangian function

$$L(X, \lambda) = x_1^2 - 10x_1 + x_2^2 - 6x_2 + x_3^2 - 4x_3 - \lambda [x_1 + x_2 + x_3 - 7]$$

The necessary conditions for Z optimum are

$$\frac{\partial L}{\partial x_1} = 2x_1 - 10 - \lambda = 0$$

$$\frac{\partial L}{\partial x_2} = 2x_2 - 6 - \lambda = 0$$

$$\frac{\partial L}{\partial x_3} = 2x_3 - 4 - \lambda = 0$$

$$\frac{\partial L}{\partial \lambda} = -[x_1 + x_2 + x_3 - 7] = 0$$

This implies that,

$$x_1 = 4; \quad x_2 = 2; \quad x_3 = 1 \text{ and } \lambda = -2$$

The next step is to know whether it refers to the maximum or minimum point. For this we have to test the Δ_3 and Δ_4.

$$\Delta_3 = \begin{bmatrix} 0 & h_{x_1} & h_{x_2} \\ h_{x_1} & f_{x_1x_1} - \lambda h_{x_1x_1} & f_{x_1x_2} - \lambda h_{x_1x_2} \\ h_{x_2} & f_{x_2x_1} - \lambda h_{x_2x_1} & f_{x_2x_2} - \lambda h_{x_2x_2} \end{bmatrix}$$

We have,

$$f = x_1^2 - 10x_1 + x_2^2 - 6x_2 + x_3^2 - 4x_3$$
$$h = x_1 + x_2 + x_3 - 7$$

$$f_{x_1} = \frac{\partial f}{\partial x_1} = 2x_1 - 10 \qquad h_{x_1} = \frac{\partial h}{\partial x_1} = 1$$

$$f_{x_2} = 2x_2 - 6 \qquad h_{x_2} = 1$$

$$f_{x_3} = 2x_3 - 4 \qquad h_{x_3} = 1$$

$$f_{x_1x_1} = \frac{\partial}{\partial x_1}(f_{x_1}) = 2 \qquad h_{x_1x_1} = \frac{\partial}{\partial x_1}(hf_{x_1}) = 0$$

$$f_{x_1x_2} = \frac{\partial}{\partial x_1}(f_{x_2}) = 0 \qquad h_{x_1x_2} = 0$$

$$f_{x_2x_1} = 0; \; f_{x_2x_2} = 2 \qquad h_{x_2x_1} = 0; \; h_{x_2x_2} = 0$$

$$\Delta_3 = \begin{vmatrix} 0 & 1 & 1 \\ 1 & 2 & 0 \\ 1 & 0 & 2 \end{vmatrix}$$

$$\Delta_3 = \begin{vmatrix} 0 & 1 & 1 \\ 0 & 2 & -2 \\ 1 & 0 & 2 \end{vmatrix} R_2 -> R_2 - R_3$$

$$\Delta_3 = 1^*\begin{vmatrix} 1 & 1 \\ 2 & -2 \end{vmatrix} = (-2-2) = -4$$

$$\Delta_4 = \begin{vmatrix} 0 & h_{x_1} & h_{x_2} & h_{x_3} \\ h_{x_1} & f_{x_1x_1} - \lambda h_{x_1x_1} & f_{x_1x_2} - \lambda h_{x_1x_2} & f_{x_1x_3} - \lambda h_{x_1x_3} \\ h_{x_2} & f_{x_2x_1} - \lambda h_{x_2x_1} & f_{x_2x_2} - \lambda h_{x_2x_2} & f_{x_2x_3} - \lambda h_{x_2x_3} \\ h_{x_3} & f_{x_3x_1} - \lambda h_{x_3x_1} & f_{x_3x_2} - \lambda h_{x_3x_2} & f_{x_3x_3} - \lambda h_{x_3x_3} \end{vmatrix}$$

$$\Delta_4 = \begin{vmatrix} 0 & 1 & 1 & 1 \\ 1 & 2 & 0 & 0 \\ 1 & 0 & 2 & 0 \\ 1 & 0 & 0 & 2 \end{vmatrix}$$

$$\Delta_4 = \begin{vmatrix} 0 & 1 & 1 & 1 \\ 0 & 2 & -2 & 0 \\ 0 & 0 & 2 & -2 \\ 1 & 0 & 0 & 2 \end{vmatrix} \begin{matrix} \\ R_2 -> R_2 - R_3 \\ R_3 -> R_3 - R_4 \\ \\ \end{matrix}$$

$$\Delta_4 = \begin{vmatrix} 0 & 1 & 1 & 1 \\ 0 & 0 & -4 & -2 \\ 0 & 0 & 2 & -2 \\ 1 & 0 & 0 & 0 \end{vmatrix} R_2 -> R_2 - 2R_1$$

$$= -1^*\begin{vmatrix} 1 & 1 & 1 \\ 0 & -4 & -2 \\ 0 & 2 & -2 \end{vmatrix}$$

$$\Delta_4 = (-1)\,(1)\begin{vmatrix} -4 & -2 \\ 2 & -2 \end{vmatrix} = -[8-2(-2)] = -12$$

Note: Using row transformation technique, we reduce the determinant such a fashion only one element is 1 and all others are zero; which in turn simplifies the determinant evaluation.

Since, both the principle minors Δ_3 & Δ_4 are negative, implies that the point refers to the minimum. Hence the solution is,

$$x_1 = 4, \quad x_2 = 2, \quad x_3 = 1$$

and

$$\begin{aligned} Z &= (4)^2 - 10(4) + (2)^2 - 6(2) + (1)^2 - 4(1) - (-2)\,(4 + 2 + 1 - 7) \\ &= 16 - 40 + 4 - 12 + 1 - 4 \\ &= 21 - 56 \\ &= -35 \end{aligned}$$

$$\text{Min } Z = -35$$

10.4.3 Constrained Optimization with Two or More Equality Constraints

Consider the NLPP.

Max/Min $z = f(x)$,

s.t. $g_1(X) = b_i$ $(i = 1, 2 \ldots m)$

$$X \geq 0; \quad X \in R^n$$

wkt, $$h_i(X) = g_i(X) - b_i = 0; \quad (i = 1, 2, \ldots m)$$

Then,

$$L(X, \lambda) = f(X) - \sum_{i=1}^{m} \lambda_i h_i(X)$$

where $\lambda_i (i = 1, 2, \ldots m)$ are the Lagrangian Multipliers.

The necessary condition for optimality can be given as:

$$\frac{\partial L}{\partial x_j} = 0; \qquad (j = 1 \ldots n) \; \& \; \frac{\partial L}{\partial \lambda_i} = 0; \quad (i = 1 \ldots m)$$

The sufficiency condition for optimality can be given as:

$$H_8 = \left[\begin{array}{c|c} O & K \\ \hline K^T & M \end{array}\right]$$

where,

O – is an $m \times m$ null matrix

K – Coefficient matrix of the variables presented in the constraints

$$M - \begin{vmatrix} L_{x_1x_1} & L_{x_1x_2} & \ldots & L_{x_1x_n} \\ L_{x_2x_1} & L_{x_2x_2} & \ldots & L_{x_2x_n} \\ : & & & \\ : & & & \\ L_{x_nx_1} & L_{x_nx_2} & \ldots & L_{x_nx_n} \end{vmatrix}$$

H_B^* is the corresponding boarded Hession matrix based on the stationary point (X^*, λ^*). The maxima and minima is determined by the signs of the last (n – m) principle minors of H_B^*, starting with the principal minor of the order (2m + 1). If the signs are alternate then X^* maximizes f(X), starting with $(-1)^{m+n}$ on the other hand, X* minimizes the function if all the signs are the same and of the type $(-1)^m$.

***Example: 5**

Solve the non-linear programming problem:

Optimize $Z = 4x_1^2 + 2x_2^2 + x_3^2 - 4x_1x_2$

s.t. $x_1 + x_2 + x_3 = 15$

$2x_1 - x_2 + 2x_3 = 20; \; x_1, x_2 \geq 0$

Construct the Lagrangian function L(X, λ)

$$L(X, \lambda) = 4x_1^2 + 2x_2^2 + x_3^2 - 4x_1x_2 - \lambda_1(x_1 + x_2 + x_3 - 15) - \lambda_2(2x_1 - x_2 + 2x_3 - 20)$$

$$Lx_1 = 8x_1 - 4x_2 - \lambda_1 - 2\lambda_2 \qquad (1)$$

$$L_{x_1x_1} = \frac{\partial}{\partial x_1}(L_{x_1}) = 8$$

$$L_{x_1x_2} = \frac{\partial}{\partial x_1}(L_{x_2}) = -4$$

$$L_{x_1x_3} = 0$$

$$L_{x_2} = 4x_2 - 4x_1 - \lambda_1 + \lambda_2 \quad (2)$$

$$L_{x_2x_2} = 4; \quad L_{x_2x_1} = -4; \quad L_{x_2x_3} = 0$$

$$L_{x_3} = 2x_3 - \lambda_1 - 2\lambda_2 \quad (3)$$

$$L_{x_3x_3} = 2; \quad L_{x_3x_1} = 0; \quad L_{x_3x_2} = 0$$

$$L_{\lambda_1} = -[x_1 + x_2 + x_3 - 15] \quad (4)$$

$$L_{\lambda_2} = -[2x_1 - x_2 + 2x_3 - 20] \quad (5)$$

Here $m = 2$ & $n = 3$

$$O = (\text{null matrix})_{2\times 2} = \begin{bmatrix} 0 & 0 \\ 0 & 0 \end{bmatrix}$$

$$K = \begin{bmatrix} 1 & 1 & 1 \\ 2 & -1 & 2 \end{bmatrix}$$

$$K^T = \begin{bmatrix} 1 & 2 \\ 1 & -1 \\ 1 & 2 \end{bmatrix}$$

$$M = \begin{bmatrix} 8 & -4 & 0 \\ -4 & 4 & 0 \\ 0 & 0 & 2 \end{bmatrix}$$

Then the Hessian matrix H_B^* can be given as:

$$H_B^* = \left[\begin{array}{c|c} O & K \\ \hline K^T & M \end{array}\right] = \begin{bmatrix} 0 & 0 & 1 & 1 & 1 \\ 0 & 0 & 2 & -1 & 2 \\ 1 & 2 & 8 & -4 & 0 \\ 1 & -1 & -4 & 4 & 0 \\ 1 & 2 & 0 & 0 & 2 \end{bmatrix}$$

Equating the equations, (1), (2), (3), (4) and (5) to zero and solving for x_1, x_2, x_3, λ_1 & λ_2, we have,

$$x_1 = 3.67; \quad x_2 = 3.33; \quad x_3 = 8; \quad \lambda_1 = 4.44; \quad \lambda_2 = 5.77$$

Find $$H_B^* = \begin{vmatrix} 0 & 0 & 1 & 1 & 1 \\ 0 & 0 & 2 & -1 & 2 \\ 1 & 2 & 8 & -4 & 0 \\ 1 & -1 & -4 & 4 & 0 \\ 1 & 2 & 0 & 0 & 2 \end{vmatrix}$$

$$= \begin{vmatrix} 0 & 0 & 1 & 1 & 1 \\ 0 & 0 & 2 & -1 & 2 \\ 0 & +3 & +12 & -8 & 0 \\ 0 & -3 & -4 & 4 & -2 \\ 1 & 2 & 0 & 0 & 2 \end{vmatrix} \begin{matrix} R_3 -> R_3 - R_4 \\ R_4 -> R_4 - R_5 \end{matrix}$$

$$= \begin{vmatrix} 0 & 0 & 1 & 1 & 1 \\ 0 & 0 & 2 & -1 & 2 \\ 0 & 0 & 8 & -4 & -2 \\ 0 & -3 & -4 & 4 & -2 \\ 1 & 2 & 0 & 0 & 2 \end{vmatrix} \quad R_3 -> R_3 - R_4$$

$$= 1*(-1)(3)* \begin{vmatrix} 1 & 1 & 1 \\ 2 & -1 & 2 \\ 8 & -4 & -2 \end{vmatrix}$$

$$= 3*2* \begin{vmatrix} 1 & 1 & 1 \\ 2 & -1 & 2 \\ 4 & -2 & -1 \end{vmatrix} = 6* \begin{vmatrix} 1 & 1 & 1 \\ 0 & -3 & 0 \\ 0 & -6 & -5 \end{vmatrix} \begin{matrix} R_2 -> R_2 - 2R_1 \\ R_3 -> R_3 - 4R_1 \end{matrix}$$

$$= 90 > 0$$

Since, $n = 3$, $m = 2$, $n - m = 1$ and $2m + 1 = 5$, we have to solve only one principle minor of the order 5.
$\therefore$ $|H_B^*| > 0$ implies that, the solution X^*, minimizes the objective function.

The optimum solution is, $x_1 = 3.67$; $x_2 = 3.33$; $x_3 = 8$
Min $Z = 4(3.67)^2 + 2(3.33)^2 + (8)^2 - 4(3.67)(3.33) = 91.169$

10.5 CONSTRAINED OPTIMIZATION WITH INEQUALITY CONSTRAINTS

10.5.1 Kuhn-Tucker Conditions

Maximize $Z = f(X)$
s.t. $g(X) \leq b$
$X \geq 0$; $X = (x_1\, x_2, \ldots x_m) \in R^m$
Let $h(X) = g(X) - b \leq 0$ (1)

Convert (1) into an equation by introducing a slack variable δ. Since we are not very sure of the nature of δ ie) $\delta > 0 = 0$ or < 0 we introduce δ^2.

Then (1) reduced to

$h(x) + \delta^2 = 0$ construct the Lagrangian function

$L(X, \lambda) = f(X) - \lambda[h(X) + \delta^2]$; where λ is the Lagrangian Multiplier. Then the necessary conditions for stationary points can be given as,

(i) $\dfrac{\partial L}{\partial x_j} = 0;\ (j-1 \ldots n) \Rightarrow \quad \dfrac{\partial f}{\partial x_j} - \lambda \dfrac{\partial h}{\partial x_j} = 0; \qquad (j = 1, 2, \ldots n)$ (2)

(ii) $\dfrac{\partial L}{\partial \lambda} = 0 \Rightarrow \quad -[h(X) + \delta^2] = 0$ (3)

(iii) $\dfrac{\partial L}{\partial \delta} = 0 \Rightarrow \quad -2\delta\lambda = 0$ (4)

Equation (4) $\Rightarrow \quad -2\delta\lambda = 0$

$\Rightarrow \quad \delta\lambda = 0$

$\Rightarrow$ either $\quad \delta = 0$ or $\lambda = 0$

if $\delta = 0; \Rightarrow \lambda \neq 0$

Equation (3) $\Rightarrow -h(X) = 0$

$\Rightarrow \lambda h(X) = 0$

if $\delta \neq 0; \quad \Rightarrow \lambda \neq 0$

Equation (3) $\Rightarrow h(X) \leq 0$

$\Rightarrow$ When $h(X) < 0 \Rightarrow \lambda = 0$ when $\lambda > 0;\ h(X) = 0$

The necessary condition for the point X to be a point of maximum can be given as,

$$\left.\begin{aligned} &\frac{\partial f}{\partial x_j} - \lambda \frac{\partial h}{\partial x_j} = 0; \qquad (j = 1, \ldots n) \\ &\lambda[h(x)] = 0 \\ &h(x) \leq 0 \\ &\lambda \geq 0 \end{aligned}\right] \qquad \text{(I)}$$

All the four conditions of the system (I) are known as Kuhn-Tucker conditions. (KT-conditions)

Note: If the NLPP is of minimization structure, that is,

Minimize $Z = f(X)$

s.t. $\quad g(X) \geq b$

$X \geq 0; \quad X \in R^n$

We have the conditions as:

$$\begin{aligned} &\frac{\partial f}{\partial x_j} - \lambda \frac{\partial h}{\partial x_j} = 0 \\ &\lambda h(X) = 0 \\ &h(X) \geq 0 \\ &\lambda \geq 0 \end{aligned} \qquad \text{(II)}$$

All the four conditions of the system (II) are known as KT-conditions.

Example: 6

Use Kuhn-Tucker conditions to solve the following NLPP

Max $Z = 2x_1^2 + 12x_1x_2 - 7x_2^2$

s.t. $\quad 2x_1 + 5x_2 \leq 98,\ x_1, x_2 \geq 0$

Let, $h(X) = 2x_1 + 5x_2 - 98 \le 0$
Construct the Lagrangian function:

$$L(X, \lambda) = 2x_1^2 + 12x_1x_2 - 7x_2^2 - \lambda\,[2x_1 + 5x_2 - 98]$$

The K-T conditions are

$$\frac{\partial L}{\partial x_1} = 0 \;\Rightarrow\; 4x_1 + 12x_2 - 2\lambda = 0$$

$$\frac{\partial L}{\partial x_2} = 0 \;\Rightarrow\; 12x_1 + 14x_2 - 5\lambda = 0 \quad (1)$$

$$\lambda h(X) = 0 \;\Rightarrow\; \lambda[2x_1 + 5x_2 - 98] = 0 \quad (2)$$

$$h(X) \le 0 \;\Rightarrow\; 2x_1 + 5x_2 - 98 \le 0 \quad (3)$$

$$\lambda \ge 0 \quad (4)$$

$\therefore$ λ can be either 0 or >0, we have two cases:

Case (i): $\lambda = 0$

(1) reduces to, $4x_1 + 12x_2 = 0$ and $12x_1 - 14x_2 = 0$. This implies that $x_1 = 0$ and $x_2 = 0$ since the values [0, 0] are not satisfying the condition (3), $-98 \le 0$.

We discard this case.

Case(ii): $\lambda \ne 0$

$$(1) \Rightarrow \quad 4x_1 + 12x_2 - 2\lambda = 0 \Rightarrow 2x_1 + 6x_2 - \lambda = 0 \quad (5)$$

$$12x_1 - 14x_2 - 5\lambda = 0 \quad (6)$$

$$(2) \Rightarrow \quad 2x_1 + 5x_2 - 98 = 0 \quad (7)$$

Solve for x_1, x_2 & λ

$$(5)*5 \;\Rightarrow\; 10x_1 + 30x_2 - 5\lambda = 0$$

$$(-) \quad (+) \quad (+)$$

$$\Rightarrow\; 12x_1 - 14x_2 - 5\lambda = 0$$

Subtract: $\quad -2x_1 + 44x_2 = 0$

$$x_1 = 22x_2 \quad (8)$$

Using (8) in (7) $\Rightarrow 2(22x_2) + 5x_2 - 98 = 0$

$\Rightarrow 49x_2 = 98 \quad \Rightarrow x_2 = 98/49 = 2 \quad (9)$

Using (9) in (8) $\Rightarrow x_1 = 22(2) = 44$

Using the values of x_1 & x_2 in (5) $\Rightarrow 2(44) + 6(2) - \lambda = 0$

$\Rightarrow \lambda = 88 + 12 = 100$

Max $Z = 2(44)^2 + 12(44)(2) - 7(2)^2 = 4900$

Hence the solution is.

Max $Z = 4900$, $x_1 = 44$ & $x_2 = 2$

Example: 7

Find the optimum solution using K-T conditions

Maximize $Z = 2x_1 + 3x_2 - x_1^2 - 2x_2^2$

s.t. $\quad x_1 + 3x_2 \le 6$

$\quad 5x_1 + 2x_2 \le 10,\; x_1, x_2 \ge 0$

Here, $$h_1(X) = x_1 + 3x_2 - 6 \le 0$$
$$h_2(X) = 5x_1 + 2x_2 - 10 \le 0$$

Construct the Lagrangian function,

$$L(X, \lambda) = 2x_1 + 3x_2 - x_1^2 - 2x_2^2 - \lambda_1 [x_1 + x_2 - 6] - \lambda_2 [5x_1 + 2x_2 - 10]$$

Where λ_1 & λ_2 are the Lagrange Multipliers. The necessary conditions are,

$$\left.\begin{aligned} \frac{\partial L}{\partial x_1} = 0 \ &\Rightarrow 2 - 2x_1 - \lambda_1 - 5\lambda_2 = 0 \\ \frac{\partial L}{\partial x_2} = 0 \ &\Rightarrow 3 - 4x_2 - 3\lambda_1 - 2\lambda_2 = 0 \end{aligned}\right\} \quad (1)$$

$$\left.\begin{aligned} \lambda_1[x_1 + 3x_2 - 6] = 0 \\ \lambda_2[5x_1 + 2x_2 - 10] = 0 \end{aligned}\right\} \quad (2)$$

$$\left.\begin{aligned} x_1 + 3x_2 - 6 \le 0 \\ 5x_1 + 2x_2 - 10 \le 0 \end{aligned}\right\} \quad (3)$$

$$\left.\begin{aligned} \lambda_1 \ge 0 \\ \lambda_2 \ge 0 \end{aligned}\right\} \quad (4)$$

Since, we have two values of λ, they lead to the following four cases.

Case (i): $\lambda_1 = 0$; $\lambda_2 = 0$

System (1) is reduced to,
$$2 - 2x_1 = 0 \ \Rightarrow \ x_1 = 1$$
$$3 - 4x_2 = 0 \ \Rightarrow \ x_2 = 3/4$$

Verify the system (3):

$$1 + 3(3/4) - 6 = \frac{4 + 9 - 24}{4} = \frac{-11}{4} \le 0$$

$$5 + 2(3/4) - 10 = \frac{20 + 6 - 40}{4} = -14/4 \le 0$$

Clearly system (3) satisfies

$$Z = 2x_1 + 3x_2 - x_1^2 - 2x_2^2 = 2 + 3(3/4) - 1 - 2(3/4)^3$$
$$Z = 2.125$$

Case (ii): $\lambda_1 \ne 0$, $\lambda_2 = 0$

System (1) is reduced to,

$$2 - 2x_1 - \lambda_1 = 0 \quad (5)$$
$$3 - 4x_2 - 3\lambda_1 = 0 \quad (6)$$

System (2) $\Rightarrow$ $$x_1 + 3x_2 - 6 = 0 \quad (7)$$

Solve for x_1, x_2 & λ_1

3^* (5) $\Rightarrow$ $6 - 6x_1 - 3\lambda_1 = 0$

(–) (+) (+)

(6) $\Rightarrow$ $3 - 4x_2 - 3\lambda_1 = 0$

subtract ------------------------

$3 - 6x_1 + 4x_2 = 0$ (8)

6^* (7) $\Rightarrow$ $6x_1 + 18x_2 - 36 = 0$

(8) $\Rightarrow$ $-6x_1 + 4x_2 + 3 = 0$

Add ------------------------

$22x_2 - 33 = 0$

$x_2 = 3/2$

Using the value of x_2 in (7)

$\Rightarrow$ $x_1 + 3(3/2) - 6 = 0$

$x_1 = 6 - (9/2) = 3/2$

Verify the system (3):

$$5x_1 + 2x_2 - 10 = 5(3/2) + 2(3/2) - 10$$

$$= \frac{15+6-20}{2} = 1/2 \le 0$$

$\therefore$ System (3) does not satisfy, discard this case.

Case (iii): $\lambda_1 = 0$ & $\lambda_2 \neq 0$

System (1) is reduced to,

$2 - 2x_1 - 5\lambda_2 = 0$ (9)

$3 - 4x_2 - 2\lambda_2 = 0$ (10)

System (2) is reduced to, $5x_1 + 2x_2 - 10 = 0$ (11)

2^* (9) $\Rightarrow$ $4 - 4x_1 - 10\lambda_2 = 0$

(–) (+) (+)

5^* (10) $\Rightarrow$ $15 - 20x_2 - 10\lambda_2 = 0$

subtract: ------------------------

$-11 + 20x_2 - 4x_1 = 0$ (12)

10^* (11) $\Rightarrow$ $50x_1 + 20x_2 - 100 = 0$

(+) (–) (+)

(12) $\Rightarrow$ $-4x_1 + 20x_2 - 11 = 0$

subtract ------------------------

$54x_1 - 89 = 0$

$x_1 = 89/54 = 1.648$

Using x_1 value in (11),

(11) $\Rightarrow$ $2x_2 = 10 - 5x_1 = 10 - 5\,(1.648)$

$2x_2 = 1.7593 \Rightarrow x_2 = 0.88$

Verify the system (3):

$$x_1 + 3x_2 - 6 = 1.648 + 3(0.88) - 6$$

$$= -1.712 \le 0$$

System (3) satisfies.
Verify the system (4):
Using the value of x_2 in (10)

$$\Rightarrow \quad 3 - 4(0.88) - 2\lambda_2 = 0$$
$$2\lambda_2 = -0.52$$
$$\lambda_2 = -0.26 < 0$$

Clearly system (4) does not satisfy. So discard this case.

Case (iv): $\lambda_1 \neq 0$ & $\lambda_2 \neq 0$
System (1) is reduced to,

$$2 - 2x_1 - \lambda_1 - 5\lambda_2 = 0 \quad (13)$$
$$3 - 4x_2 - 3\lambda_1 - 2\lambda_2 = 0 \quad (14)$$

System (2) is reduced to,

$$x_1 + 3x_2 - 6 = 0 \quad (15)$$
$$5x_1 + 2x_2 - 10 = 0 \quad (16)$$

5* (15) $\Rightarrow$ $5x_1 + 15x_2 - 30 = 0$

(–) (–) (+)

(16) $\Rightarrow$ $3x_1 + 2x_2 - 10 = 0$

Subtract

$$13x_2 - 20 = 0$$
$$x_2 = 20/13 = 1.539$$

(15) $\Rightarrow$ $x_1 = 6 - 3x_2 = 6 - 3(1.539) = 1.385$

Using the values of x_1 & x_2 in (13) & (14), we have

(13) $\Rightarrow$ $\lambda_1 + 5\lambda_2 = 2 - 2(1.385) = -0.77$

(14) $\Rightarrow$ $3\lambda_1 + 2\lambda_2 = 3 - 4x_2$

$$= 3 - 4(1.539) = -3.156$$
$$\lambda_1 + 5\lambda_2 = -0.77 \quad (17)$$
$$3\lambda_1 + 2\lambda_2 = -3.156 \quad (18)$$

3* (17) $\Rightarrow$ $3\lambda_1 + 15\lambda_2 = -2.31$

(–) (–) (+)

(18) $\Rightarrow$ $3\lambda_1 + 2\lambda_2 = -3.156$

----

subtract: $13\lambda_2 = 0.846$

$$\lambda_2 = 0.065$$

(17) $\Rightarrow$ $\lambda_1 = -0.77 - 5^* 0.065 = -1.095$

Clearly system (4) is violated. Hence discard this case.
Out of all the four cases three cases, that is, case (ii), case (iii) & case (iv) are discarded.
Hence the required optimum solution is.

$$x_1 = 1; \quad x_2 = 3/4; \quad \lambda_1 = 0; \quad \lambda_1 = 0 \text{ and Max } Z = 2.125$$

Note: If more than one case exists, select the best solution.

10.6 QUADRATIC PROGRAMMING PROBLEM (QPP)

Consider an NLPP with non-linear objective function and with one/more linear constraints.
That is, MAX/MIN $Z = f(x)$

s.t. $AX \leq b$ and $x \geq 0$

where $f(X)$ = Non-linear function

A = Real valued matrix of order $(m \times n)$

$C, X^T \in R^n$ and $b^T \in R^m$

The above NLPP can be called a QPP, if it is possible for us to rewrite the objective function in the following fashion.

$$\text{MAX/MIN } Z = f(X) = CX + (1/2)\, X^T QX$$

where

Q-is a real valued matrix of order $(n \times n)$

Hence the general structure of a QPP is,

$$\left.\begin{array}{ll} \text{MAX/MIN } Z = CX + (1/2)\, X^T QX & \\ \text{s.t} \quad AX \leq b & \\ \quad X \geq 0 & \end{array}\right\} \qquad (1)$$

(I) is said to be a QPP in standard form.

***Example:* 8**

Convert the following NLPP into QPP form:

$\text{Max } Z = 4x_1 + 6x_2 - 2x_1^2 - 2x_1x_2 - 2x_2^2$

st. $x_1 + 2x_2 \leq 2;\quad x_1, x_2 \geq 0$

$C^T = [4 \quad 6]$ Generally

$$X^T = [x_1 \quad x_2] \qquad Q = \begin{bmatrix} \text{Co-efficient of } x_1^2 & (1/2) \text{ co-efficient } (x_1\ x_2) \\ (1/2) \text{ co-efficient of } (x_1\ x_2) & \text{Co-efficient } x_2^2 \end{bmatrix}$$

$b^T = [2]$

$A = [1 \quad 2]$

$$\text{Max } z = [4 \quad 6]\begin{bmatrix} x_1 \\ x_2 \end{bmatrix} + [x_1 \quad x_2]\begin{bmatrix} -2 & -1 \\ -1 & -2 \end{bmatrix}\begin{bmatrix} x_1 \\ x_2 \end{bmatrix}$$

$$[1 \quad 2]\begin{bmatrix} x_1 \\ x_2 \end{bmatrix} \leq [2]$$

$x_1, x_2 \geq 0$

10.7 WOLFE'S METHOD TO SOLVE A QPP

The various steps to be followed while solving a QPP using Wolfe's method are clearly depicted in the Flowchart 10.1.

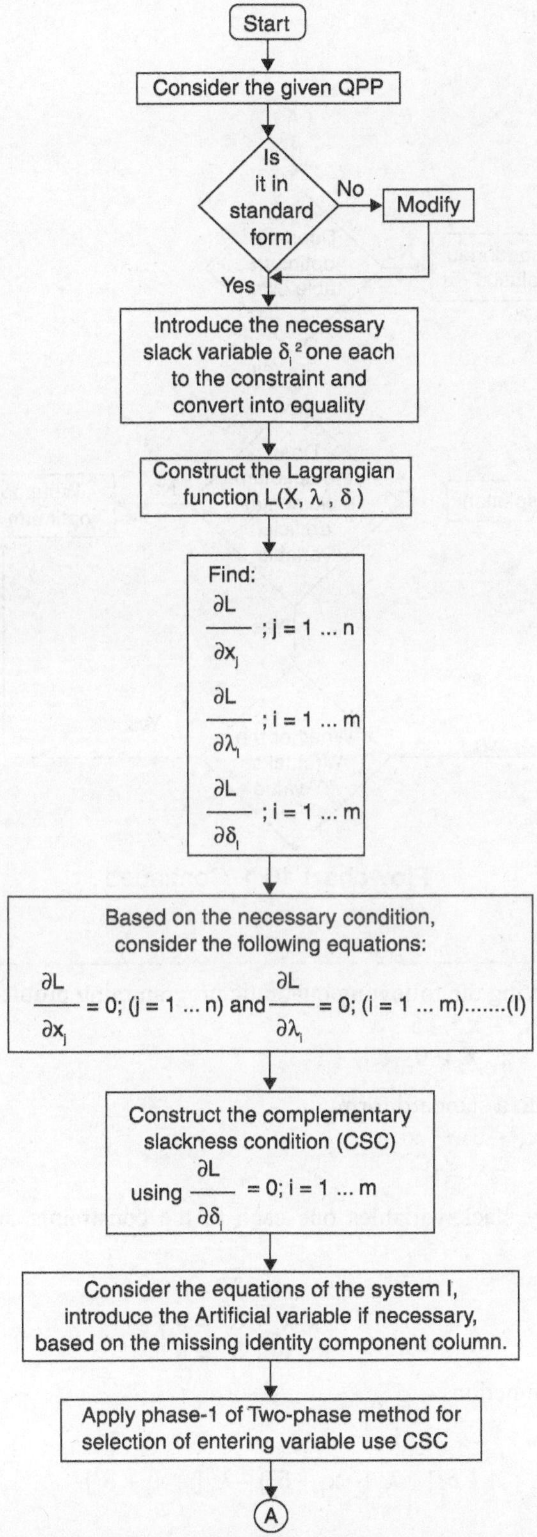

Flowchart 10.1 Wolfe's method to solve a QPP

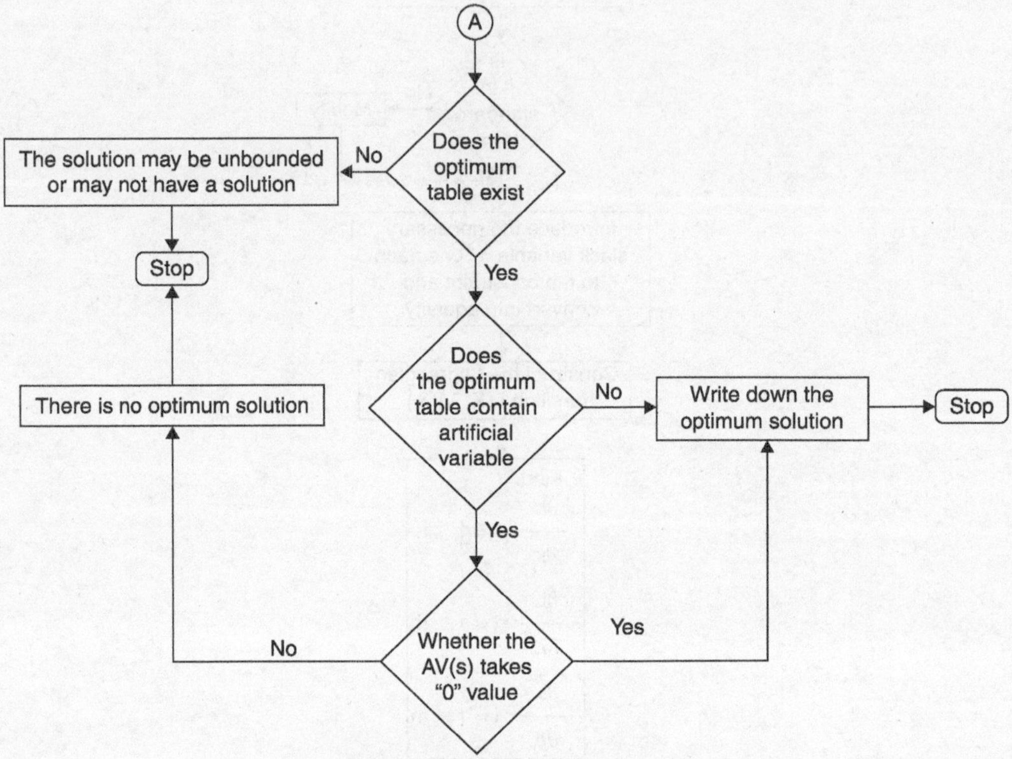

Flowchart 10.1 Continued

***Example:* 9**

Use Wolfe's method in solving the following quadratic programming problem:
Maximize $Z = 4x_1 + 2x_2 - x_1{}^2 - x_2{}^2 - 5$
s.t. constraints $x_1 + x_2 \le 4$; $x_1, x_2 \ge 0$

Consider the given QPP into a standard form:
$\text{Max } Z = 4x_1 + 2x_2 - x_1{}^2 - x_2{}^2 - 5$
s.t. $x_1 + x_2 \le 4$; $-x_1 \le 0$; $-x_2 \le 0$

Introduce the necessary slack variables one each to the constraints, including the non-negative constraints.

$$x_1 + x_2 - 4 + \delta_1{}^2 = 0$$
$$-x_1 + \delta_2{}^2 = 0$$
$$-x_2 + \delta_2{}^2 = 0$$

Construct the Lagrangian function,

$$L(X, \lambda, \delta) = (4x_1 + 2x_2 - x_1^2 - x_2^2 - 5)$$
$$- \lambda_1[x_1 + x_2 - 4 + \delta_1^2] - \lambda_2[-x_1 + \delta_2^2] - \lambda_3[-x_2 + \delta_3^2]$$

Find the necessary conditions:

$$\frac{\partial L}{\partial x_j} = 0; \qquad j = 1,2$$

$$\frac{\partial L}{\partial \lambda_i} = 0; \qquad i = 1,2,3$$

$$\frac{\partial L}{\partial \delta_i} = 0; \qquad i = 1,2,3$$

$$\left.\begin{aligned} \frac{\partial L}{\partial x_1} &= 4 - 2x_1 - \lambda_1 + \lambda_2 = 0 \\ \frac{\partial L}{\partial x_2} &= 2 - 2x_2 - \lambda_1 + \lambda_3 = 0 \\ \frac{\partial L}{\partial \lambda_1} &= -[x_1 + x_2 - 4 + \delta_1^2] = 0 \end{aligned}\right] \qquad \text{(I)}$$

$$\left.\begin{aligned} \frac{\partial L}{\partial \lambda_2} &= [-x_1 + \delta_2^2] = 0 \\ \frac{\partial L}{\partial \lambda_3} &= [-x_2 + \delta_3^2] = 0 \end{aligned}\right] \qquad \text{(II)}$$

$$\left.\begin{aligned} \frac{\partial L}{\partial \delta_1} &= [-2\lambda_1 \delta_1] = 0 \\ \frac{\partial L}{\partial \delta_2} &= [-2\lambda_2 \delta_2] = 0 \\ \frac{\partial L}{\partial \delta_3} &= [-2\lambda_3 \delta_3] = 0 \end{aligned}\right] \qquad \text{(III)}$$

Let us take $\delta_1^{\,2} = x_3 \quad (x_3 \geq 0)$ (IV)

System (1) is reduced to,

$$\Rightarrow \qquad \left.\begin{aligned} 2x_1 + \lambda_1 - \lambda_2 &= 4 \\ 2x_2 + \lambda_1 - \lambda_3 &= 2 \\ x_1 + x_2 + x_3 &= 4 \end{aligned}\right] \qquad \text{(V)}$$

System (II) is reduced to,

$$\Rightarrow \qquad \delta_2^{\,2} = x_1; \ \delta_3^{\,2} = x_2 \qquad \text{(VI)}$$

Consider the system (III) & modify:

$-2\lambda_1\,\delta_1 = 0$; Multiply by $(\delta_1/(-2))$; $\Rightarrow \lambda_1\,\delta_1^{\,2} = 0 \Rightarrow \lambda_1 x_3 = 0$ (Using IV)

$-2\lambda_2\,\delta_2 = 0$; $\Rightarrow \lambda_2\,\delta_2^{\,2} = 0 \Rightarrow \lambda_2 x_1 = 0$ (Using VI)

$-2\lambda_2\,\delta_3 = 0$; $\Rightarrow \lambda_3\,\delta_3^{\,2} = 0 \Rightarrow \lambda_3 x_2 = 0$ (Using VI)

Merging all the three we have,

$$\lambda_1 x_3 + \lambda_2 x_1 + \lambda_3 x_2 = 0 \quad \text{(VII)}$$

The system (VII) is known as the required Complementary Slackness Conditions (CSC). Consider the system V. Introduce the necessary artificial variable and construct the dummy objective function based on the artificial variables. (Similar to Phase-1 of Two-Phase method).

Introduce the artificial variables x_4 & x_5 one each in the first and second equation of the system (V).

Max $Z^* = -x_4 - x_5$

s.t. $2x_1 + \quad x_4 + \lambda_1 - \lambda_2 = 4$

$2x_2 + \quad x_5 + \lambda_1 - \lambda_3 = 2$

$x_1 + x_2 + x_3 \quad = 4$

$x_j \geq 0$ (j = 1, 2, … 5) $\lambda_i \geq 0$ (i = 1, 2, 3) and the CSC is, $\lambda_1 x_3 + \lambda_2 x_1 + \lambda_3 x_2 = 0$

Enter into a simplex table and apply the simplex process.

Since some of the $(Z_j - C_j)$ entries are < 0 in table-1, implies optimality is not attained. That is, we have to modify, the existing solution. The most (–ve) $\{(Z_j - C_j)\}$ corresponds to x_1, x_2 & λ_1. There is a tie in the selection of entering variable. As it is, we can select any one variable. But here we have to make the decision using the complementary slackness condition.

Most (–ve) Value of $(Z_j - C_j)$	Variable or Variables	Complementary Slackness Condition (CSC)	Variables Presented in B at +ve Level	Qualified or Not to Enter	Selected Variable
– 2	x_1	$\lambda_2 x_1 = 0$	–	Yes	x_1
	x_2	$\lambda_3 x_2 = 0$	–	Yes	
	λ_1	$\lambda_1 x_3 = 0$	x_3	No	

*Regarding λ_1; $\lambda_1 x_3 = 0$ implies that $\lambda_1 = 0$ or $x_3 = 0$; Since x_3 is in the table already at +ve level ($x_3 \neq 0$) $\Rightarrow \lambda_1 = 0$, ie., λ_1 cannot enter into the basis. Clearly x_1 & x_2 are qualified to enter. We select x_1 to enter into the basis.

In Table: 2

Most (–ve) $(Z_j - C_j)$	Variable	(CSC)	Variable Presented in B	Qualified or Not to Enter the Basis
– 2	x_2	$\lambda_3 x_1 = 0$	–	Yes

Clearly, x_2 enters into the basis.

Table Number	C_j	0	0	0	−1	−1	0	0	0				
	x_1, λ_1	x_1	x_2	x_3	x_4	x_5	λ_1	λ_2	λ_3	X_B	B	C_B	θ
1		(2) PE	0	0	1	0	1	−1	0	4	x_4	−1	2 ⇐ PR
		0	2	0	0	1	1	0	−1	2	x_5	−1	–
		1	1	1	0	0	0	0	0	4	x_3	0	4
	z_j	−2	−2	0	−1	−1	−2	1	1				
	$Z_j - C_j$	−2↑ PC	−2	0	0	0	−2	1	1				
2	PR	1	0	0	1/2	0	1/2	−1/2	0	2	x_1	0	–
	R_2	0	(2) PE	0	0	1	1	0	−1	2	x_5	−1	1 ⇐ PR
	R_3 − PR	0	1	1	−1/2	0	−1/2	1/2	0	2	x_3	0	2
	z_j	0	−2	0	0	−1	−1	0	1				
	$Z_j - C_j$	0	−2↑ PC	0	1	0	−1	0	1				
3	R_1	1	0	0	1/2	0	1/2	−1/2	0	2	x_1	0	
	PR	0	1	0	0	1/2	1/2	0	−1/2	1	x_2	0	
	R_3 − PR	0	0	1	−1/2	−1/2	−1	1/2	1/2	1	x_3	0	
	z_j	0	0	0	0	0	0	0	0				
	$Z_j - C_j$	0	0	0	1	1	0	0	0				

All $Z_j - C_j \geq 0$ implies that the optimum table is attained. Hence the optimum solution is,

$x_1 = 2, x_2 = 1$

Max $Z = 4\,(2) + 2\,(1) - (2)^2 - (1)^2 - 5 = 8 + 2 - 4 - 1 - 5 = 0$

Note:

NLPP's Constraints Having	The Nature of Corresponding Lagrangian Multiplier	Necessary Steps to be Taken Before Applying Phase-l of Two-phase Method
$\leq$	$\lambda \geq 0$	Keep as such
$\geq$	$\lambda \leq 0$	$\lambda = -\beta;\ \beta \geq 0$
$=$	λ unrestricted	$\lambda = \alpha - \beta;\ \alpha, \beta \geq 0$

***Example:* 10**

Solve the following NLPP using Wolfe's Method:

Max $Z = 2x_1 - x_1^{\,2} + x_2$

s.t. $2x_1 + 3x_2 \leq 6;\ 2x_1 + x_2 \leq 4;\ x_1, x_2 \geq 0$

Convert the NLPP into a standard form

Max $Z = 2x_1 - x_1^2 + x_2$

s.t. $2x_1 + 3x_2 - 6 \leq 0;\ 2x_1 + x_2 - 4 \leq 0;\ -x_1 \leq 0;\ -x_2 \leq 0$

Introduce the necessary slack variable δ_1^2 (i = 1, 2, 3, 4) one each to the constraints, including the non-negative constraints.

$$2x_1 + 3x_2 - 6 + \delta_1^2 = 0$$
$$2x_1 + x_2 - 4 + \delta_2^2 = 0$$
$$-x_1 + \delta_3^2 = 0$$
$$-x_2 + \delta_4^2 = 0$$

Construct the Lagrangian function:

$$\begin{aligned} L(X, \lambda, \delta) = 2x_1 - x_1^2 + x_2 &- \lambda_1[2x_1 + 3x_2 - 6 + \delta_1^2] \\ &- \lambda_2[2x_1 + x_2 - 4 + \delta_2^2] \\ &- \lambda_3[-x_1 + \delta_3^2] \\ &- \lambda_4[-x_2 + \delta_4^2] \end{aligned}$$

Find the necessary conditions:

$$\left.\begin{aligned} \frac{\partial L}{\partial x_1} &= 0 \Rightarrow 2 - 2x_1 - 2\lambda_1 - 2\lambda_2 + \lambda_3 = 0 \\ \frac{\partial L}{\partial x_2} &= 0 \Rightarrow 1 - 3\lambda_1 - \lambda_2 + \lambda_4 = 0 \\ \frac{\partial L}{\partial \lambda_1} &= 0 \Rightarrow -[2x_1 + 3x_2 - 6 + \delta_4^2] = 0 \\ \frac{\partial L}{\partial \lambda_2} &= 0 \Rightarrow -[2x_1 + x_2 - 4 + \delta_2^2 = 0 \end{aligned}\right] \quad \text{(I)}$$

$$\left.\begin{aligned} \frac{\partial L}{\partial \lambda_3} &= 0 \Rightarrow [-x_1 + \delta_3^2] = 0 \\ \frac{\partial L}{\partial \lambda_4} &= 0 \Rightarrow -[-x_2 + \delta_4^2] = 0 \end{aligned}\right] \quad \text{(II)}$$

$$\left.\begin{aligned} \frac{\partial L}{\partial \delta_1} &= -2\lambda_1\delta_1 \\ \frac{\partial L}{\partial \delta_2} &= -2\lambda_2\delta_2 \\ \frac{\partial L}{\partial \delta_3} &= -2\lambda_3\delta_3 \\ \frac{\partial L}{\partial \delta_4} &= -2\lambda_4\delta_4 \end{aligned}\right] \quad \text{(III)}$$

Let us take

$$\delta_1^2 = x_3 \text{ and } \delta_2^2 = x_4; \ (x_3, x_4 \geq 0) \quad \text{(IV)}$$

The system (I) is reduced to,

$$\left.\begin{aligned} 2x_1 + 2\lambda_1 + 2\lambda_2 - \lambda_3 &= 2 \\ 3\lambda_1 + \lambda_2 - \lambda_4 &= 1 \\ 2x_1 + 3x_2 + x_3 &= 6 \\ 2x_1 + x_2 + x_4 &= 4 \end{aligned}\right] \quad \text{(V) using (IV)}$$

The system (II) is reduced to,

$$\left.\begin{array}{l}\delta_3^2 = x_1 \\ \delta_4^2 = x_2\end{array}\right] \qquad \text{(VI)}$$

Consider the system (III) and modify multiply by $(\delta_1/(-2))$

$-2\lambda_1\,\delta_1 = 0 \quad \Rightarrow \quad \lambda_1\,\delta_1^{\,2} = 0 \quad \Rightarrow \quad \lambda_1\,x_3 = 0$

Similarly we have, $\lambda_2\,\delta_2^{\,2} = 0 \quad \Rightarrow \quad \lambda_2\,x_4 = 0$

$\lambda_3\,\delta_3^{\,2} = 0 \quad \Rightarrow \quad \lambda_3\,x_1 = 0$

$\lambda_4\,\delta_4^{\,2} = 0 \quad \Rightarrow \quad \lambda_4\,x_2 = 0$

Merging all the four equations we have,

$$\lambda_1\,x_3 + \lambda_2\,x_4 + \lambda_3\,x_1 + \lambda_4\,x_2 = 0 \qquad \text{(VII)}$$

Then system (VII) is known as the required complementary slackness condition. Consider the system (V), introduce the necessary artificial variable and construct the dummy objective function based on the artificial variables.

Introduce the artificial variables x_5 & x_6 one each in the first and the second equation of the system (V).

Max $Z^* = -x_5 - x_6$

s.t. $2x_1 + x_5 + 2\lambda_1 + 2\lambda_2 - \lambda_3 = 2$

$x_6 + 3\lambda_1 + \lambda_2 - \lambda_4 = 1$

$2x_1 + 3x_2 + x_3 = 6$

$2x_1 + x_2 + x_4 = 4$

$x_j \geq 0$ (j = 1 … 6), $\lambda_i \geq 0$ (i = 1, 2, 3, 4) and the CSC is,

$\lambda_1\,x_3 + \lambda_2\,x_4 + \lambda_3\,x_1 + \lambda_4\,x_2 = 0$

Enter into a simplex table and apply the simplex process. Some of the $(Z_j - C_j)$ entries are < 0 in table-1, implies that the optimality is not attained. That is, modify the existing solution.

Most (–ve) Value of $(Z_j - C_j)$	Variable or Variables	Complimentary Slackness Condition (CSC)	Variables Presented in B at the +ve Level	Qualified or Not to Enter	Variable Selected to Enter
–3	λ_2	$\lambda_2\,x_4 = 0$	x_4	No	–
next most (–ve) –2	x_1	$\lambda_3\,x_1 = 0$	–	Yes	x_1
	λ_1	$\lambda_1\,x_3 = 0$	x_3	–	

Some of the $(Z_j - C_j)$ entries are <0 in table –2, implies that the optimality is not attained. That is, modify the existing solution.

Most (–ve) Value of $(Z_j - C_j)$	Variable	(CSC)	Variables Presented in B at +ve Level	Qualified or Not to Enter	Variable Selected to Enter
– 3	λ_1	$\lambda_1 x_3 = 0$	x_3	No	–
next most (–ve) – 1	λ_2	$\lambda_2 x_4 = 0$	x_4	No	–
next most (–ve) 0*	x_2	$\lambda_4 x_2 = 0$	–	Yes	x_2

*Since there are no strict negative elements, select any zero entry of the non-basic variables for further processing.

Some of the $(Z_j - C_j)$ entries are <0 in table-3, implies that the modification is essential.

Most (–ve) Value of $(Z_j - C_j)$	Variable	(CSC)	Variables Presented in B at +ve Level	Qualified or Not to Enter	Variable Selected to Enter
– 3	λ_1	$\lambda_1 x_3 = 0$	–	Yes	λ_1

Table Number	C_j	0	0	0	0	–1	–1	0	0	0	0				
	x_j & λ_1	x_1	x_2	x_3	x_4	x_5	x_6	λ_1	λ_2	λ_3	λ_4	X_B	B	C_B	θ
1		(2) PE	0	0	0	1	0	2	2	–1	0	2	x_5	–1	1 ⇐ PR
		0	0	0	0	0	1	3	1	0	–1	1	x_6	–1	–
		2	3	1	0	0	0	0	0	0	0	6	x_3	0	3
		2	1	0	1	0	0	0	0	0	0	4	x_4	0	2
	z_j	–2	0	0	0	–1	–1	–2	–3	1	1				
	$z_j - c_j$	–2↑PC	0	0	0	0	0	–2	–3	1	1				
2	PR	1	0	0	0	1/2	0	0	1	–1/2	0	1	x_1	0	–
	R_2	0	0	0	0	0	1	1	3	0	–1	1	x_6	– 1	–
	R_3 – 2 PR	0	(3) PE	1	0	–1	–1	0	– 2	1	0	4	x_3	0	4/3 ⇐ PR
	R_4 – 2PR	0	1	0	1	–1	–1	0	– 2	1	0	2	x_4	0	1
	z_j	0	0	0	0	0	–1	–3	–1	0	1				
	$z_j - c_j$	0	0↑PC	0	0	1	0	–3	–1	0	1				
3	R_1	1	0	0	0	1/2	0	1	1	–1/2	0	1	x_1	0	1
	R_2	0	0	0	0	0	1	(3) PE	1	0	–1	1	x_6	–1	1/3 ⇐ PR
	PR	0	1	1/3	0	–1/3	0	– 2/3	–2/3	1/3	0	4/3	x_2	0	
	R_4 – PR	0	0	–1/3	1	–2/3	0	– 4/3	–4/3	2/3	0	2/3	x_4	0	
	z_j	0	0	0	0	0	–1	–3	–1	0	1				
	$z_j - c_j$	0	0	0	0	1	0	– 3↑PC	–1	0	1				

Table Number	C_j	0	0	0	0	−1	−1	0	0	0	0				
	x_j & λ_i	x_1	x_2	x_3	x_4	x_5	x_6	λ_1	λ_2	λ_3	λ_4	X_B	B	C_B	θ
4	R_1 − PR	1	0	0	0	1/2	−1/3	0	2/3	−1/2	1/3	2/3	x_1	0	
	PR	0	0	0	0	0	1/3	1	1/3	0	−1/3	1/3	λ_1	0	
	R_3 + (2/3) PR	0	1	1/3	0	−1/3	2/9	0	−4/9	1/3	−2/9	14/9	x_2	0	
	(R_4 + (4/3) PR	0	0	−1/3	1	−2/3	4/9	0	−8/9	2/3	−4/9	10/9	x_4	0	
	z_j	0	0	0	0	0	0	0	0	0	0				
	$z_j - c_j$	0	0	0	0	1	1	0	0	0	0				

All the $(Z_j - C_j)$ entries are ≥ 0, implies that the optimality is reached. Hence the optimum solution is, $x_1 = 2/3$; $x_2 = 14/9$; $x_4 = 10/9$; $\lambda_1 = 1/3$

Max $Z = 2(2/3) - (2/3)^2 + (14/9)$

$= 4/3 - 4/9 + 14/9 = (12 - 4 + 14)/9 = 22/9$

10.8 BEALS METHOD TO SOLVE A QPP

The various steps to be followed while solving a QPP using Beals Method are clearly depicted in the following Flowchart 10.2.

Notations Used

Max – Maximum
Min – Minimum
m – Number of constraints
n – Number of variables

***Example:* 11**

Solve using Beals Method:
Max $f(X) = 2x_1 + 3x_2 - x_1^2$
s.t. $x_1 + 2x_2 \leq 4$, x_1 $x_2 \geq 0$

The objective of the given problem is maximization. Convert the constraint into an equality by introducing the slack variable x_3. We have,

$$x_1 + 2x_2 + x_3 = 4$$

Here m = 1 & n = 3. This system can have only one basic variable (Since m = 1). We can select any one variable as basic. Let us take x_3 as basic and x_1 & x_2 as non-basic ones.

$$\Rightarrow x_1 = 0 \ \& \ x_2 = 0$$

Express the basic variable x_3 in terms of the non-basic variables x_1 & x_2. We have,

$$x_3 = 4 - x_1 - 2x_2 \quad (1)$$

expresses the objective function in terms of the non-basic variables x_1 & x_2. We have,

$$f(X) = 2x_1 + 3x_2 - x_1^2 \text{ (As it is)}$$

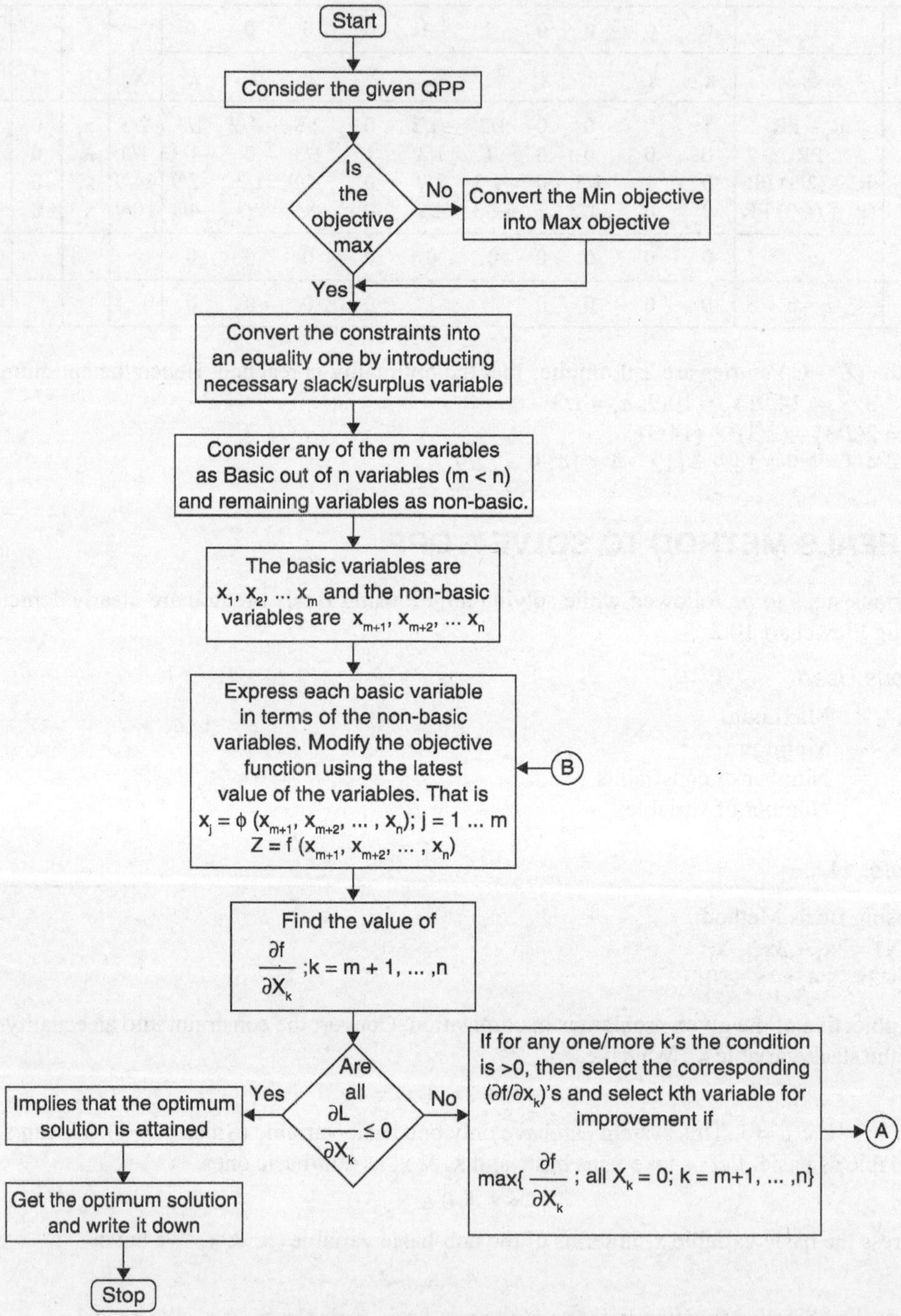

Flowchart 10.2 Beal's method to solve a QPP

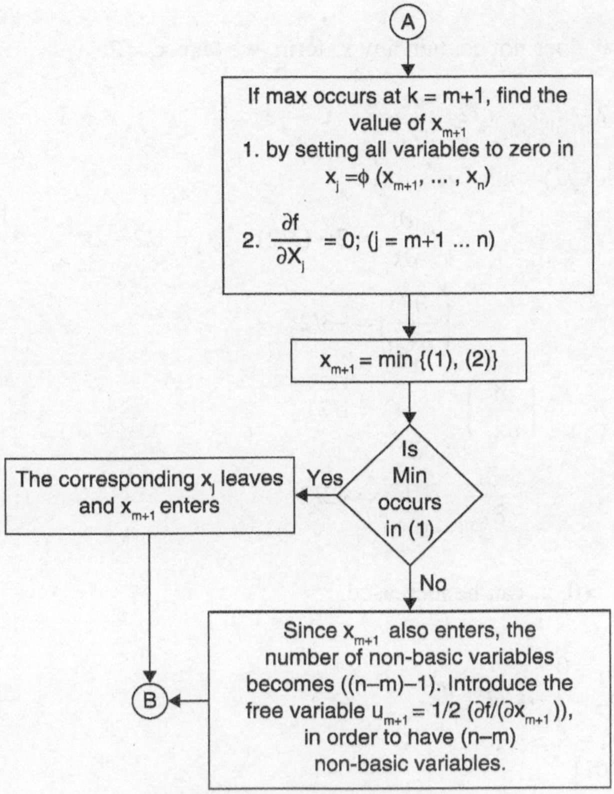

Flowchart 10.2 Continued

find
$$\frac{\partial f}{\partial x_1} \ \& \ \frac{\partial f}{\partial x_2} \text{ at } x_1 = 0 \ \& \ x_2 = 0$$

$$\frac{\partial f}{\partial x_1} = 2 - 2x_1; \left(\frac{\partial f}{\partial x_1}\right)_{(x_1 = 0 = x_2)} = 2 > 0$$

$$\frac{\partial f}{\partial x_2} = 3; \ \frac{\partial f}{\partial x_2}_{(x_1 = 0 = x_2)} = 3 > 0$$

Both the elements are +ve (>0). implies that the solution is not optimal. We have to improve it. At the level of $x_1 = 0 = x_2$, select

$$\max\left\{\frac{\partial f}{\partial x_1}, \frac{\partial f}{\partial x_2}\right\} = \text{Max } \{2, 3\} = 3$$

⇒ The corresponding variable x_2 can be improved. Put $x_1 = 0$ in (1),

$$(1) \Rightarrow x_3 = 4 - 2x_2$$

⇒ x_2 can take the maximum value as 2 ($x_3 \geq 0$; min $x_3 = 0$).

Since at $\left(\frac{\partial f}{\partial x_2}\right)_{(x_1=0)} = 0$ does not contain any x_2 term, we take $x_2 = 2$.

Now (1) becomes $x_3 = 2 - x_1/2 - x_3/2$ (2)

f(x) becomes,

$f(x) = 2x_1 + 3(2 - x_1/2 - x_3/2) - x_1^2$

$$\frac{\partial f}{\partial x_1} = 2 - (3/2) - 2x_1 = 1/2 - 2x_1$$

$$\left(\frac{\partial f}{\partial x_3}\right) = -3/2$$

$$\left(\frac{\partial f}{\partial x_1}\right)_{(x_1=0=x_3)} = 1/2$$

$$\frac{\partial f}{\partial x_3}{}_{(x_1=0=x_3)} = -3/2$$

Since $\left(\frac{\partial f}{\partial x_1}\right)_{(x_1=0=x_3)} > 0$, x_1 can be increased.

Equation (2) ⇒ when $x_3 = 0$

$x_2 = 2 - x_1/2$

⇒ $x_1 = 4$ (3)

[Since $x_2 \geq 0$; min $x_2 = 01$]

Also, $\frac{\partial f}{\partial x_1} = 0 \Rightarrow 1/2 - 2x_1 = 0 \Rightarrow x_1 = 1/4$ (4)

$x_1 = \min\{4, 1/4\} = 1/4$

Since the minimum value lies in the partial derivatives x_2 can not leave. But x_1 enters. Since we have two non-basic variables for this system, we introduce a non-basic free variable.

$$U_1 = 1/2\left(\frac{\partial f}{\partial x_1}\right)$$

$$U_1 = 1/2\,(1/2 - 2x_1) = 1/4 - x_1$$

$$x_1 = 1/4 - U_1 \quad (4)$$

$$x_2 = 2 - (1/4 - U_1)(1/2) - x_3/2 \quad (5)$$

$$= 2 - 1/8 + 1/2U_1 - x_3/2$$

$$x_2 = 15/8 + 1/2U_1 - x_3/2$$

As we express f(X) using x_3 & U_1, we have,

$$f(X) = 2(1/4 - U_1) + 3(15/8 + U_1/2 - x_3/2) - (1/4 - U_1)^2$$

$$\frac{\partial f}{\partial U_1} = -2 + 3/2 - 2(1/4 - (U_1))\,(-1) = -2y_1$$

$$\frac{\partial f}{\partial x_1} = -3/2$$

At $x_3 = 0$ & $U_1 = 0$; both the derivatives take negative values implies that the optimality is attained. To get the optimum solution put $x_3 = 0$ & $U_1 = 0$ in equating (3) & (4)

$$\Rightarrow \quad x_1 = 1/4$$
$$x_2 = 15/8$$
$$Z = 2x_1 + 3x_2 - x_1^2$$
$$= 2(1/4) + 3(15/8) - (1/4)^2$$
$$= (1/2) + (45/8) - (1/16)$$
$$= (8 + 90 - 1)/16 = 97/16$$

Max $Z = 97/16$

Hence the optimum solution is, $x_1 = 1/4$, $x_2 = 15/8$ and Max $Z = 97/16$

Example: 12

Max $Z = 10x_1 + 25x_2 - 10x_1^2 - x_2^2 - 4x_1x_2$

s.t. $x_1 + 2x_2 \le 10$; $x_1 + x_2 \le 9$; $x_1, x_2 \ge 0$

Here the objective is Maximization. Introduce the slack variables x_3 & x_4 one each in the first and second constraint. We have,

$$x_1 + 2x_2 + x_3 = 10$$
$$x_1 + x_2 + x_4 = 9$$

Here m = 2 & n = 4. Number of basic variable = 4 – 2 = 2. Let us select x_3 & x_4 as basic variables. This implies that

$$x_3 = 10 - x_1 - 2x_2 \tag{1}$$
$$x_4 = 9 - x_1 - x_2 \tag{2}$$

Express Z in terms of the non-basic variables x_1 & x_2.

$$Z = 10x_1 + 25x_2 - 10x_1^2 - x_2^2 - 4x_1 x_2$$

$$\left(\frac{\partial z}{\partial x_1}\right) = 10 - 2x_1 - 4x_2$$

$$\left(\frac{\partial z}{\partial x_2}\right) = 25 - 2x_2 - 4x_1$$

$$\left(\frac{\partial z}{\partial x_1}\right)_{(x_1 = 0 = x_2)} = 10$$

$$\left(\frac{\partial z}{\partial x_2}\right)_{(x_1 = 0 = x_2)} = 25$$

Since both the values are >0, we have to improve the solution. Select the Max {10, 25} = 25, which corresponds to x_2. Improve x_2.

equation (1) $\Rightarrow$ $x_2 = 5$ ($x_3 = 0$ level and $x_1 = 0$)

equation (2) $\Rightarrow$ $x_2 = 9$ ($x_4 = 0$ level and $x_1 = 0$)

$$\frac{\partial f}{\partial x_2} = 0 \quad \Rightarrow \quad x_2 = 12.5 \quad (x_1 = 0)$$

$$x_2 = \min\{5, 9, 12.5\}$$

$\Rightarrow$ $x_2 = 5$,
$\Rightarrow$ x_3 becomes non-basic & x_2 enters
Equation (1) $x_2 = 5 - (1/2)\,x_1 - (1/2)\,x_3$ (3)
Equation (2) is reduced to $x_4 = 9 - (5 - (1/2)\,x_1) - ((1/2)\,x_1) - x_1$ (4)
Express in terms of x_1 & x_3

$$Z = 10x_1 + 25(25 - x_1/2 - x_3/2) - 10x_1^2 - (5 - x_1/2 - x_3/2)^2 - 4x_1(5 - x_1/2 - x_3/2)$$

$$\left(\frac{\partial z}{\partial x_1}\right) = 10 - 25/2 - 20x_1 - 2(5 - x_1/2 - x_3/2)(-1/2) - 4(5 - x_1/2 - x_3/2) - 4x_1(-1/2)$$

$$\left(\frac{\partial z}{\partial x_1}\right)_{(x_1 = 0 = x_3)} = 10 - 25/2 + 5 - 20 = -35/2$$

$$\left(\frac{\partial z}{\partial x_3}\right) = 25(-1/2) - 2(5 - x_1/2 - x_3/2)(-1/2) - 4x_1(-1/2)$$

$$\left(\frac{\partial z}{\partial x_3}\right)_{(x_1 = 0 = x_3)} = -25/2 - 10 = -45/2$$

Both the derivatives are negative, implies that the optimally is attained.
put $x_1 = 0$ & $x_3 = 0$ in (3) & (4)
Equation (3) $\Rightarrow x_2 = 5$
Equation (4) $\Rightarrow x_4 = 9$
Max $Z = 25\,(5) - (5)^2 = 100$
Hence the optimum solution is,
Max $Z = 100$; $x_2 = 5$; $x_1 = 0$

EXERCISE PROBLEMS

*1. Use the method of Lagrangian multipliers to solve the following non-linear programming problem.
Minimize $z = 2x_1^2 + x_2^2 + 3x_3^2 + 10x_1 + 8x_2 + 6x_3 - 100$
s.t. the constraints $x_1 + x_2 + x_3 = 20$
$x_1, x_2, x_3 \geq 0$

*2. Min $f(x_1, x_2) = 3x_1^2 + x_2^2 + 2x_1x_2 + 6x_1 + 2x_2$
s.t the constraints: $2x_1 - x_2 = 4$

3. Solve the NLPP,
Optimize $Z = 4x_1 + 9x_2 - x_1^2 - x_2^2$
s.t. $4x_1 + 3x_2 = 15$
$3x_1 + 5x_2 = 14$; $x_1, x_2 \geq 0$

4. Min $Z = x_1^2 + x_2^2 + x_3^2$
   s.t. $x_1 + x_2 + 3x_3 = 2$
   $5x_1 + 2x_2 + x_3 = 5,\ x_j \geq 0 \quad (j = 1, 2, 3)$
*5. Use the Kuhn-Tucker conditions to solve the following NLPP: Max $Z = 2x_1 - x_2^2 + x_2$
   s.t. $2x_1 + 3x_2 \leq 6;\ 2x_1 + x_2 \leq 4;\ x_1, x_2 \geq 0$
*6. Maximize $Z = -x_1^2 - x_2^2 - x_3^2 + 4x_1 + 6x_2$
   s.t. $x_1 + x_2 \leq 2;\ 2x_1 + 3x_2 \leq 12;\ x_1, x_2 \geq 0$
*7. Maximize $Z = 8x_1^2 + 2x_2^2$
   s.t. $x_1^2 + x_2 \leq 9;\ x_1 \leq 2;\ x_1, x_2 \geq 0$
*8. Maximize $Z = 7x_1^2 - 6x_1 + 5x_2^2$
   s.t. $x_1 + 2x_2 \leq 10;\ x_1 - 3x_2 \leq 9;\ x_1, x_2 \geq 0$
*9. Consider the problem:
   Maximize $Z = 4x_1 + 6x_2 - 2x_1^2 - 2x_1x_2 - 2x_2^2$
   s.t $x_1 + 2x_2 \leq 2$
   $x_1, x_2 \geq 0$
   Solve this problem by the quadratic programming technique.
*10. Use Wolfe's method to solve the QPP
   Maximize $Z = 2x_1 + 3x_2 - 2x_1^2$
   s.t. $x_1 + 4x_2 \leq 4,\ x_1 + x_2 \leq 2$
   $x_1, x_2 \geq 0$
*11. Use Wolfe's method in solving the following QPP.
   Min $Z = 6 - 6x_1 + 2x_1^2 - 2x_1x_2 + 2x_2^2$
   s.t. $x_1 + x_2 \leq 2,\ x_1, x_2 \geq 0$
12. Solve using Wolfe's method:
   Max $Z = 2x_1 + 3x_2 - 2x_1^2$
   s.t. $x_1 + 4x_2 \leq 4,\ x_1 + x_2 \leq 2$
   $x_1, x_2 \geq 0$
*13. A manufacturing concern produces a product consisting of two raw materials, say A_1 and A_2. The production function is estimated as
   $Z = f(x_1, x_2) = 3.6x_1 - 0.4x_1^2 + 1.6x_2 - 0.2x_2^2$
   where z represents the quantity (in tones) of the product produced, and x_1 and x_2 designate the input amount of raw materials A_1 and A_2. The company has ₹50,000 to spend on these two raw materials. The unit price of A_1 is ₹10,000 and that of A_2 is ₹5,000. Determine how much input amounts of A_1 and A_2 are to be decided so as to maximize the production output.
   [Hint: Construct the constraint based on the given information
   $10000x_1 + 5000x_2 \leq 50000$
   $2x_1 + x_2 \leq 10;\ x_1, x_2 \geq 0$ Solve it using QPP]
*14. Use Beal's method to solve Quadratic Programming Problem:
   Max $f(x) = 1/4\,(2x_3 - x_1) - 1/2(x_1^2 + x_2^2 + x_3^2)$
   s.t the constraints $x_1 - x_2 + x_3 = 1, \qquad x_1, x_2, x_3 \geq 0$
*15. Max $f(x) = 2x_1 + 3x_2 - 2x_2^2$
   s.t. $x_1 + 4x_2 \leq 4; \qquad x_1 + x_2 \leq 2; \qquad x_1, x_2 \geq 0$
*16. Use Beal's method to solve the following NLPP
   Min $Z = 6 - 6x_1 + 2x_1^2 - 2x_1x_2 + 2x_2^2$
   s.t. $x_1 + x_2 \leq 2; \qquad x_1, x_2 \geq 0$

ANSWERS TO THE EXERCISE PROBLEMS

1. $x_1 = 5, x_2 = 11, x_3 = 4$ and Min $Z = 281$
2. $x_1 = 1, x_2 = -2$ and Min $Z = 5$
3. $x_1 = 3, x_2 = 1$ and Min $Z = 11$
4. $x_1 = 0.81, x_2 = 0.35, x_3 = 0.28$ and Min $Z = 0\ 857$
5. $x_1 = 2/3, x_2 = 14/9$ and Max $Z = 22/9$
6. $x_1 = 1/2, x_2 = 3/2, x_3 = 0$ and Max $Z = 17/2$
7. $x_1 = 2, x_2 = 2.236$ (or) $\sqrt{5}$ and Max $Z = 42$
8. $x_1 = 48/5, x_2 = 1/5$ and Max $Z = 587.72$
9. $x_1 = 1/3, x_2 = 5/6$ and Max $Z = 25/6$
10. $x_1 = 5/16, x_2 = 59/64$ and Max $Z = 3.19$
11. $x_1 = 3/2, x_2 = 1/2$ and Min $Z = 1/2$
12. $x_1 = 5/16, x_2 = 59/64$ and Max $Z = 3.19$
13. $x_1 = 3.5, x_2 = 3$ and Max $Z = 10.7$
14. $x_1 = 1/8, x_2 = 0, x_3 = 7/8$ and Max $Z = 1/64$
15. $x_1 = 0, x_2 = 3/4$ and Max $Z = 9/8$
16. $x_1 = 3/2, x_2 = 1/2$ and Max $Z = 1/2$

REVIEW QUESTIONS

Short Questions

1. Define the term NLPP. (Section 10.1)
2. Write down the general structure of an NLPP. (Section 10.2)
3. State the methods used to solve an NLPP. (Section 10.4)
4. Write the general structure of the H_8. (Section 10.4.3)
5. Write the general structure of the constrained optimization with inequality constraints. (Section 10.4.3)
6. State the KT conditions. (Section 10.5.1)
7. Define the term QPP. (Section 10.6)

State True or False

8. Wolfe's and Beals methods are substitutes. (T)
9. The maximum and the minimum for a given model is decided by H_B^* value. (T)

10. In an NLPP, the objective must be a non-linear function. (F)
11. In an NLPP, there is no chance for a constraint to be non-linear. (F)
12. If an NLPP contains equality constraint, we have to make use of the Lagrangian Method. (T)

Fill in the Blanks

13. If an NLPP contains inequality constraint, we can solve it by using ______. (KT-conditions).
14. Wolfe's method is used to solve a ______. (QPP)
15. Beal's method is a substitute for ______. (Wolfe's method)
16. The nature of λ in the Lagrangian function is based on the ______. [$\leq$/$\geq$/= symbol in the constraints)
17. If the constraint of an NLPP has the λ symbol, then the nature of λ is ______. (< 0)

Appendix A

(Area Under the Standard Normal Curve from 0 to z)

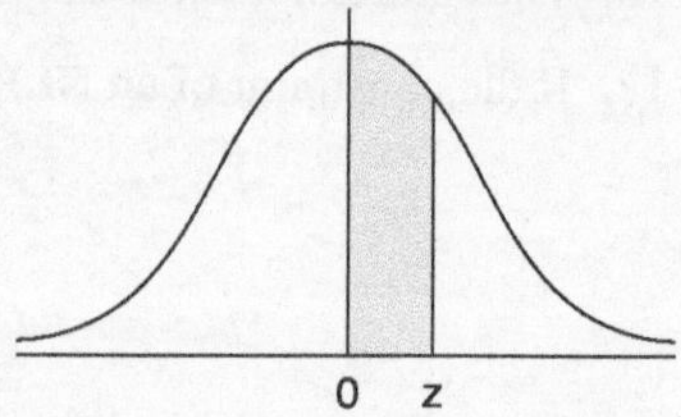

z	0	1	2	3	4	5	6	7	8	9
0.0	.0000	.0040	.0080	.0120	.0160	.0199	.0239	.0279	.0319	.0359
0.1	.0398	.0438	.0478	.0517	.0557	.0596	.0636	.0675	.0714	.0754
0.2	.0793	.0832	.0871	.0910	.0948	.0987	.1026	.1064	.1103	.1141
0.3	.1179	.1217	.1255	.1293	.1331	.1368	.1406	.1443	.1480	.1517
0.4	.1554	.1591	.1628	.1664	.1700	.1736	.1772	.1808	.1844	.1879
0.5	.1915	.1950	.1985	.2019	.2054	.2088	.2123	.2157	.2190	.2224
0.6	.2258	.2291	.2324	.2357	.2389	.2422	.2454	.2486	.2518	.2549
0.7	.2580	.2612	.2642	.2673	.2704	.2734	.2764	.2794	.2823	.2852
0.8	.2881	.2910	.2939	.2967	.2996	.3023	.3051	.3078	.3106	.3133
0.9	.3159	.3186	.3212	.3238	.3264	.3289	.3315	.3340	.3365	.3389
1.0	.3413	.3438	.3461	.3485	.3508	.3531	.3554	.3577	.3599	.3621
1.1	.3643	.3665	.3686	.3708	.3729	.3749	.3770	.3790	.3810	.3830
1.2	.3849	.3869	.3888	.3907	.3925	.3944	.3962	.3980	.3997	.4015
1.3	.4032	.4049	.4066	.4082	.4099	.4115	.4131	.4147	.4162	.4177
1.4	.4192	.4207	.4222	.4236	.4251	.4265	.4279	.4292	.4306	.4319
1.5	.4332	.4345	.4357	.4370	.4382	.4394	.4406	.4418	.4429	.4441
1.6	.4452	.4463	.4474	.4484	.4495	.4505	.4515	.4525	.4535	.4545
1.7	.4554	.4564	.4573	.4582	.4591	.4599	.4608	.4616	.4625	.4633
1.8	.4641	.4649	.4656	.4664	.4671	.4678	.4686	.4693	.4699	.4706
1.9	.4713	.4719	.4726	.4732	.4738	.4744	.4750	.4756	.4761	.4767
2.0	.4772	.4778	.4783	.4788	.4793	.4798	.4803	.4808	.4812	.4817
2.1	.4821	.4826	.4830	.4834	.4838	.4842	.4846	.4850	.4854	.4857
2.2	.4861	.4864	.4868	.4871	.4875	.4878	.4881	.4884	.4887	.4890
2.3	.4893	.4896	.4898	.4901	.4904	.4906	.4909	.4911	.4913	.4916
2.4	.4918	.4920	.4922	.4925	.4927	.4929	.4931	.4932	.4934	.4936

z	0	1	2	3	4	5	6	7	8	9
2.5	.4938	.4940	.4941	.4943	.4945	.4946	.4948	.4949	.4951	.4952
2.6	.4953	.4955	.4956	.4957	.4959	.4960	.4961	.4962	.4963	.4964
2.7	.4965	.4966	.4967	.4968	.4969	.4970	.4971	.4972	.4973	.4974
2.8	.4974	.4975	.4976	.4977	.4977	.4978	.4979	.4979	.4980	.4981
2.9	.4981	.4982	.4982	.4983	.4984	.4984	.4985	.4985	.4986	.4986
3.0	.4987	.4987	.4987	.4988	.4988	.4989	.4989	.4989	.4990	.4990

Random Numbers

51772	74640	42331	29044	46621	62898	93582	04186	19640	87056
24033	23491	83587	06568	21960	21387	76105	10863	97453	90581
45939	60173	52078	25424	11645	55870	56974	37428	93507	94271
30586	02133	75797	45406	31041	86707	12973	17169	88116	42187
03585	79353	81938	82322	96799	85659	36081	50884	14070	74950
64937	03355	95863	20790	65304	55189	00745	65253	11822	15804
15630	64759	51135	98527	62586	41889	25439	88036	24034	67283
09448	56301	57683	30277	94623	85418	68829	06652	41982	49159
21631	91157	77331	60710	52290	16835	48653	71590	16159	14676
91097	17480	29414	06829	87843	28195	27279	47152	35683	47280
50532	25496	95652	42457	73547	76552	50020	24819	52984	76168
07136	40876	79971	54195	25708	51817	36732	72484	94923	75936
27989	64728	10744	08396	56242	90985	28868	99431	50995	20507
85184	73949	36601	46253	00477	25234	09908	36574	72139	70185
54398	21154	97810	36764	32869	11785	55261	59009	38714	38723
65544	34371	09591	07839	58892	92843	72828	91341	84821	63886
08263	65952	85762	64236	39238	18776	84303	99247	46149	03229
39817	67906	48236	16057	81812	15815	63700	85915	19219	45943
62257	04077	79443	95203	02479	30763	92486	54083	23631	05825
53298	90276	62545	21944	16530	03878	07516	95715	02526	33537

B

Appendix

Index